D1591397

Monarchs Ministers and Maps

THE KENNETH NEBENZAHL, JR., LECTURES IN THE HISTORY OF CARTOGRAPHY
PUBLISHED FOR THE HERMON DUNLAP SMITH CENTER FOR THE HISTORY OF CARTOGRAPHY
THE NEWBERRY LIBRARY

PREVIOUSLY PUBLISHED
Series Editor, David Woodward

Maps: A Historical Survey of Their Study and Collecting
by R. A. Skelton (1972)

British Maps of Colonial America
by William P. Cumming (1974)

Five Centuries of Map Printing
edited by David Woodward (1975)

Mapping the American Revolutionary War
by J. B. Harley, Barbara Bartz Petchenik, and Lawrence W. Towner (1978)

Series Editor, David Buisseret

Art and Cartography
edited by David Woodward (1987)

Monarchs Ministers and Maps

Edited by
David Buisseret

The Emergence of Cartography as a Tool of Government in Early Modern Europe

The University of Chicago Press
Chicago & London

DAVID BUISSERET is director of the Hermon Dunlap Smith Center for the History of Cartography at the Newberry Library. He is the author of *Historic Illinois from the Air* and editor of *From Sea Charts to Satellite Images: Interpreting North American History through Maps,* both published by the University of Chicago Press.

The University of Chicago Press, Chicago 60637
The University of Chicago Press, Ltd., London
01 00 99 98 97 96 95 94 93 92 5 4 3 2 1
ISBN (cloth) 0-226-07987-2

Library of Congress Cataloging-in-Publication Data

Monarchs, ministers, and maps : the emergence of cartography as a tool of government in early modern Europe / edited by David Buisseret.
p. cm. — (The Kenneth Nebenzahl, Jr., lectures in the history of cartography)
Includes bibliographical references and index.
1. Cartography—Europe—History. I. Buisseret, David. II. Series.
GA781.M76 1992
526′.094—dc20 91-36088
CIP

This book is printed on acid-free paper.

Contents

Illustrations

Editor's Note

The essays published here derive from the invited lectures that made up the eighth series of Kenneth Nebenzahl, Jr., Lectures in the History of Cartography, given at the Newberry Library in November 1985. There were originally seven lectures, but the seventh, given by William Goetzmann of the University of Texas at Austin, and concerning map use by the "fathers of the republic," was regretfully omitted in order to preserve the volume's coherence as an account of mapmaking in Europe.

Even within this area, our work cannot pretend to completeness; it would have been good to have some coverage of the United Netherlands, for instance, and of Prussia and Sweden. However, a lecture series is in its essence an invitation to reflection and research rather than an attempt at encyclopedic coverage of a theme, and we hope that these papers will indeed encourage other scholars to tackle adjacent subjects.

Most of the authors were able to alter and strengthen their contributions substantially in the light of criticisms from a variety of readers. This was not possible in the case of James Vann, who tragically died not long after giving the lecture for us; his paper has therefore been retained much as he gave it.

The lecture series and this publication are made possible through the contributions of Mr. and Mrs. Kenneth Nebenzahl to the Kenneth Nebenzahl, Jr., Fund, and it is as always both a duty and a great pleasure to acknowledge the Nebenzahls' generosity.

Introduction

David Buisseret

One of the great problems in the history of cartography—and, indeed, in the intellectual history of early modern Europe—is this: how did it come about that whereas in 1400 few people in Europe used maps, except for the Mediterranean navigators with their portolan charts, by 1600 maps were essential to a wide variety of professions?

This change, which amounted to a revolution in the European way of "seeing" the world, no doubt emerged from a variety of causes. Some scholars would emphasize the role of the Renaissance and of the fashionable admiration for antiquity, exemplified in cartography by the work of Ptolemy. His *Geography,* which began circulating in manuscript at the beginning of the fifteenth century and was printed in many editions after the first one of 1477, no doubt accustomed many Europeans to a new view of the world, and to the idea that it might be described under a system of mathematical coordinates.

Other scholars have dwelt on the whole movement of thought known as the Scientific Revolution, with its emphasis on quantification and measurement. In this sense, the advent of maps was more or less inevitable, once people began thinking in terms of pinning down locations in terms of figures of latitude and longitude. Just as the human body came to be delineated with a new accuracy by Vesalius, so the geographical features of the natural world were brought under a newly precise and intensive scrutiny.

Another strand leading to the development of a new map consciousness can be followed back into the artistic developments of the fifteenth century, particularly in France and the Low Countries. Here artists like the Van Eycks and the Limbourg brothers had begun to delineate rural and urban scenes with hitherto unparalleled "realism," so encouraging the later emergence of utilitarian topographical views and, eventually, of a profusion of maps of country and town.

Cities were of course easier to delineate recognizably, in a gross way, than were countrysides. Here the great expansion in mapping activity seems to go back to the later sixteenth century, when, particularly in England and the Low Countries, landowners began commissioning "estate plans" to help them manage their holdings. The emergence of the estate plan, which was the theme of the Nebenzahl Lectures of 1988, seems to have been intimately linked with economic developments, for such plans thrived in precisely those regions where the first signs of the Agricultural Revolution were beginning to appear.

In this book we examine yet another element in the emergence of map consciousness, and that is the role of political units. Of course, by 1500 the process of "national" consolidation had been going on for centuries in many countries of western Europe, and it had occasionally been accompanied by appreciation of the possible role of maps; Saint Louis of France, for instance, is said to have used them in his Mediterranean campaigns. However, by and large it would be very difficult to show many examples of such use in 1450. The great change begins to be perceptible toward the end of the fifteenth century, and our authors have done their best to identify the stages of its advent.

They have tried to answer one central question: just when did monarchs and ministers in various countries begin to perceive that maps could be useful in government? Other questions then suggest themselves: For what purposes were these maps commissioned? How accurate and useful did they actually prove to be? How did the new cartographic knowledge strengthen the hand of central governments in dealing with provincial autonomies? These are the types of questions that our six authors have tried to answer, for different times and places. There is a further problem, of great interest to the history of cartography, but that we have not tried to tackle here, and that has to do with the relative importance of "state" activity in the general progress of mapmaking.

We begin in Italy, because it was long assumed that this was where all European map development began. However, John Marino's chapter warns us not to make this assumption too easily. Only in Venice, it seems, were maps used for governmental purposes as early as the fifteenth century, and even there the examples are rare. According to Marino, it was not until the third quarter of the sixteenth century that governments commissioned substantial numbers of maps in Milan, Florence, or the Papal States. Marino relates this development partly to the great increase in numbers of printed maps at that time, the result of the explosion in the number of presses, and partly to a marked quickening of economic activity. His chapter ends with an analysis of the way in which a revival of the economy in Naples during the later seventeenth and eighteenth centuries led to a renewed burst of mapmaking there.

The next two chapters concern England from medieval times to 1625. Peter Barber has cast his net more widely than some of the other contributors, including medieval *mappaemundi*, globes, paintings of royal palaces, and court symbols. However, even without this broader treatment England would have needed more than one chapter. That country's archival cartographic resources seem to be remarkably rich, and the secondary literature is equally profuse. For none of the other countries under consideration is there such a quantity and quality of recent work, published by scholars like John Andrews, H. M. Colvin, Peter Eden, Sir John Hale, P. D. Harvey, R. A. Skelton, Sarah Tyacke, and Helen Wallis. Barber makes full use of this primary and secondary material in chapters that suggest that the mapping impulse was extraordinarily active in sixteenth-century England—and Ireland.

In chapter 4, we move to France. Here the early sources are not nearly so rich, but from what remains of them it looks as if developments roughly paralleled those in England. François I (1515–47) was the equivalent of England's Henry VIII (1509–47) as the first king with a strong mapping sense, and they commissioned maps for the same kinds of purpose: fortifying the frontiers, planning campaigns, mounting overseas expeditions, and so forth. Because of the civil wars between 1560 and the 1590s, there was no French equivalent of Christopher Saxton, but once peace had been restored the military services of Henry IV began to make up for lost time in internal mapping. Henry also brought new support for French overseas ventures, largely neglected during the civil wars, and one of these resulted in the remarkable cartographic achievements of Samuel de Champlain. Richelieu commissioned further maps of ill-known parts of France, so that by the middle of the seventeenth century that country was poised for the extraordinary cartographic developments of the reign of Louis XIV.

In the lands of the Spanish Habsburgs, covered by Geoffrey Parker in chapter 5, the story was somewhat different. Parker is able to identify some remarkable cartographic ventures fostered by Philip II, such as the Escorial atlas, the Wyngaerde series of city views, and the maps drawn by Spain's military and naval commanders. But in spite of these brilliant initiatives, Parker's view is that there was no steady, long-term royal plan for mapping the peninsula. It may well be that Spain's cartographic resources were heavily diverted to meeting the needs of its overseas empire, not addressed in this chapter.

Cartography in the Austrian Habsburg lands, according to James Vann in chapter 6, faithfully reflected political conditions there. Thus although Maximilian I (1493–1519) was one of the earliest European rulers to see the usefulness of maps in war and administration, neither he nor his immediate successors commissioned maps of the empire as a whole, preferring to delineate only such separate constituents as the Tyrol or Lower Austria. All this eventually changed, according to Vann, and the imperial power began to call for imperial maps. But that was not until the eighteenth century.

In the final chapter, Michael Mikoś details a rather similar relationship between cartographic activity and political reality in Poland. There, monarchs like Stefan Batory in the sixteenth century and Stanislas Augustus in the eighteenth did their best to produce not only local and military maps but also maps of their lands as a whole. But in this as in other governmental activities they encountered many difficulties, so that it is to their rivals, the great magnates, that we have to look for such achievements as Mikołaj Krzysztof Radziwiłł's great map of Lithuania in the sixteenth century and Józef Aleksander Jabłonowski's maps of Poland in the eighteenth century.

The chapters as a whole offer varied insights into both technological and political conditions in the countries they cover. Some regions of great interest, such as the Sweden of Gustav Adolf, the Prussia of the Great Elector, and the United Provinces of Maurice of Nassau, had to be left out, with the hope that others will take up these themes. Some good leads, too, could not be followed up in the space available. For instance, just how did governmental demand affect cartographic techniques? Was there conscious "image making" in all royal cartography? How did extensive royal cartographic publications influence public perceptions of the monarchy?

For all these questions, and many others, there was no room. But the contributions seem, nevertheless, to have thrown some light on the theme of monarchs and maps, a subject so neglected until now that there has been no general treatment of it. Our authors enable us to see with new eyes some developments in both the history of cartography and general European history. In the history of cartography, we can now be sure that governmental activity was one of the main ways in which Europeans became habituated to the use of maps, and so to the use of a new way of both "seeing" the world and changing it.

In general or political history, we now know much better than before what kinds of maps were available to early modern rulers. On the whole, they were much better provided cartographically than we might have thought possible; Philip II and Henry IV of France, for instance, had access to maps that showed even small villages in the whole of their lands. However, this abundance of cartographic information did not automatically and immediately translate into the power to influence events. It was one thing to "know" the territory and another to exercise tight control over it, particularly when communications were quite slow. The restraints on absolute rule were both political and technical: it might take many weeks to send a letter or an emissary to some provincial body with a long tradition of independent action. But we can hardly doubt that the existence of accurate cartographic information eventually proved a powerful tool in the imposition of central authority on the recalcitrant periphery.

ONE

Administrative Mapping in the Italian States

John Marino

If the precocity of Italian cartography in portolan charts and nautical maps from the fourteenth century[1] and the monumental magnificence of political-strategic maps like Egnazio Danti's designs in Florence's Palazzo Vecchio or the Vatican's Hall of Maps[2] led us to believe that mapping was commonplace in Renaissance Italy, we would be greatly mistaken. Not a long interest in exploration, nor a long tradition of state rationalization and bureaucratization, nor innovation in the arts and sciences, nor a propensity to depict the world "naturalistically" with linear perspective, nor even the drawn-out military maneuvers during the French invasions of the first third of the sixteenth century left traces in everyday maps in the Italian states. Maps appear only sporadically before 1560 in the state archives of Venice, Florence, and Naples and in published collections from Milan and the Vatican. Mapping as a normal administrative way of looking at the world dates only from the third quarter of the sixteenth century, at the same time that technical changes in mapmaking and map publishing transformed the market for cartographic products.

From our modern point of view, maps could have aided states in the visualization of spatial relationships for a host of reasons, but mapping was not yet a common conceptual tool in the early Renaissance. Two ideal subjects not mapped, "the complexities of the ownership of arable strips in open-field farming" and "directions for a journey," for example, continued to be expressed through verbal details of boundaries and topological lists by travel days.[3] Though map consciousness, nevertheless, distinguished Italy from the rest of medieval Europe, the meaning of such a vision can be generally interpreted as an idealized and moralized geography, a kind of "integrated cosmography" of spiritual and geographical knowledge.[4] The purposes of these early maps (that is, why they were made) still remain conjectural—celebratory in the Humanist quest for fame, didactic in a program for good government of the city, and sometimes military-strategic.

From the earliest maps of the twelfth and thirteenth centuries, cartographers of the northern Italian plain led the way in focusing on three subjects: district maps, maps and plans of particular localities and plots of land, and plans of bird's-eye views of towns.[5] According to P. D. A. Harvey's account, only two extant district maps predate the fifteenth century: a 1291 manuscript map of an area southeast of Turin, around Alba and Asti, and a map of Lake Garda (c. 1380), which clearly marks lakeshore settlements and fortresses.[6] Fifteenth-century district maps from the areas around Brescia,

Verona, Padua, and Parma, and general views of Lombardy all emphasize military and strategic details. During the same period, artists drew mural maps in the Palazzo Comunale at Siena in 1413–14 and in the Doge's Palace in Venice in 1474, but with a slightly different emphasis on city walls and pictures of outstanding monuments in a kind of humanist archaeology.[7] Such decorative maps, like bird's-eye views of towns, derived from another tradition that includes the ancient city ideogram, the picture-map, and plans of towns in the Holy Land.[8]

The earliest state-sponsored maps appear to have been those commissioned by the Venetial Council of Ten in 1460.[9] Only three surviving maps may date from the decree, showing the territories around Padua (1465), Brescia (1469–70), and Verona (1479–83). All three owe much to the earlier district maps. Similarly, an interesting map of the whole Venetian Terraferma (c. 1496–99) could be described as a military map because of its detailing of the fortified walls of cities and layout of defenses.[10] Venice appears to have been the only state in fifteenth-century Europe to employ maps for administrative purposes. The Venetians relied upon a century-old tradition of the northern Italian plain to aid in military planning, defense, and government of their newly acquired Terraferma empire.[11] In Venice, maps eventually were executed to define borders, to aid in water and lagoon management, to illustrate and clarify ambassadorial dispatches, for defense and fortress designs, and to resolve disputes in court cases.[12] These state functions were not newly created in the late fifteenth century; rather, only then was the need or possibility to use maps for practical geographical information perceived and initiated, and only in the late sixteenth century was the technical expertise of surveyors and cartographers joined to replace artist mapmakers.[13]

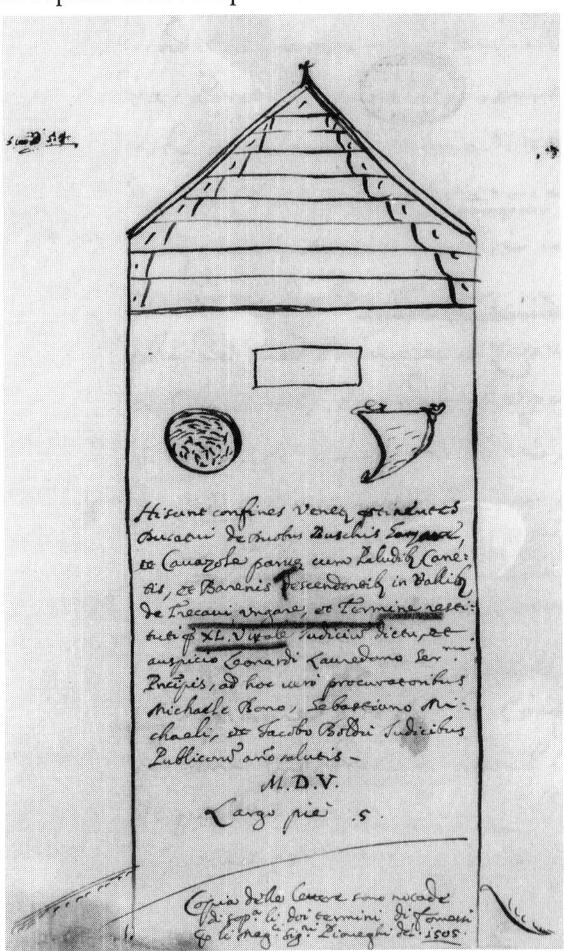

FIG. 1.1 Design from a border dispute over two woods. *Rason Vecchie,* 350, dis. 1112 (1508). No. 23/1992, Sezione di fotoriproduzione, Archivio di Stato, Venezia (ASV)

Mapping in Venice

Before hazarding an explanation for this tardy transition to the modern way of "seeing," let us turn to the maps themselves in Venice, the leading Italian state not only in map consciousness but also in power and wealth, and examine their appearance and purpose. I have reviewed four out of thirty volumes (a sample of about 13 percent) of the roughly ten thousand maps preserved in the Venetian state archives. Out of more than one thousand maps, then, I have found only fifteen, fewer than 1.5 percent, dated earlier than 1565. And of these fifteen, seven come from the 1560s, roughly the beginning of more common map use.

The earliest map in my sample, really a design, dates from 1505 (figure 1.1).[14] The original measures 198 × 278 mm and concerns a border dispute over the rental of two woods, Harzere and Cavazole(?), in the Venetian territory near Caorle. I read it as a design of a building, possibly a church because of the cross mounted on the roof. But I find no direct relevance for the adjudication of the dispute in question.

A more typical map comes from the Office of the Border Commissioners in 1538 (figure 1.2).[15] It depicts the land between Strasoldo, Cervignano, Aquileia, and the Aussa River, with clear demarcation of proprietorship between Venice and Germany (*de' Tedeschi*). In addition to the towns, the Aquileia road joining them, the river and its tributaries, the one bridge crossing it, and the boundaries, note the specificity of jurisdictions. Some are named after cities like Strasoldo and Malisana, others after lords like the monks of Aquileia or the Savorgnani. Even generic, descriptive names like "lo pizot," for the finger of land between the Chiastra and Aqua di Mezzo fork, are given. Territorial information distinguishes such maps.

These local maps could get quite large. A November 1564 map of part of the valleys of San Michele and Tripona at the headwaters of the Brenta River near Angarona (Vicenza) measures 1621 × 850 mm, plus a 118 × 391-

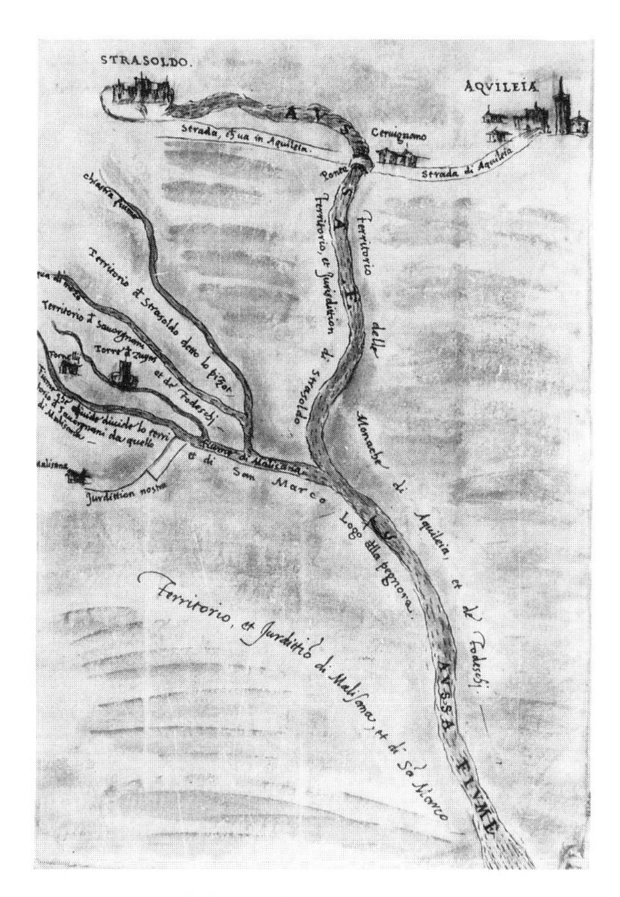

Fig. 1.2 Map of the Border Commissioners tracing the Aussa River. *Proveditori dei Confini*, B. 132, C. 127r (1538). ASV

mm section (figure 1.3).[16] Rivers, mountains, passes, houses, mills, and jurisdictions are all clearly labeled. Cross sections of these large landscapes could provide even more information. In March 1564, Pompeio Caneparo, the same cartographer, had drawn a 1038 × 624-mm map of the torrent Silano near Angarano.[17] It gives an enlargement of the upper right half of the larger map and obviously suggests how territorial maps were put together as composites from studies of smaller areas.

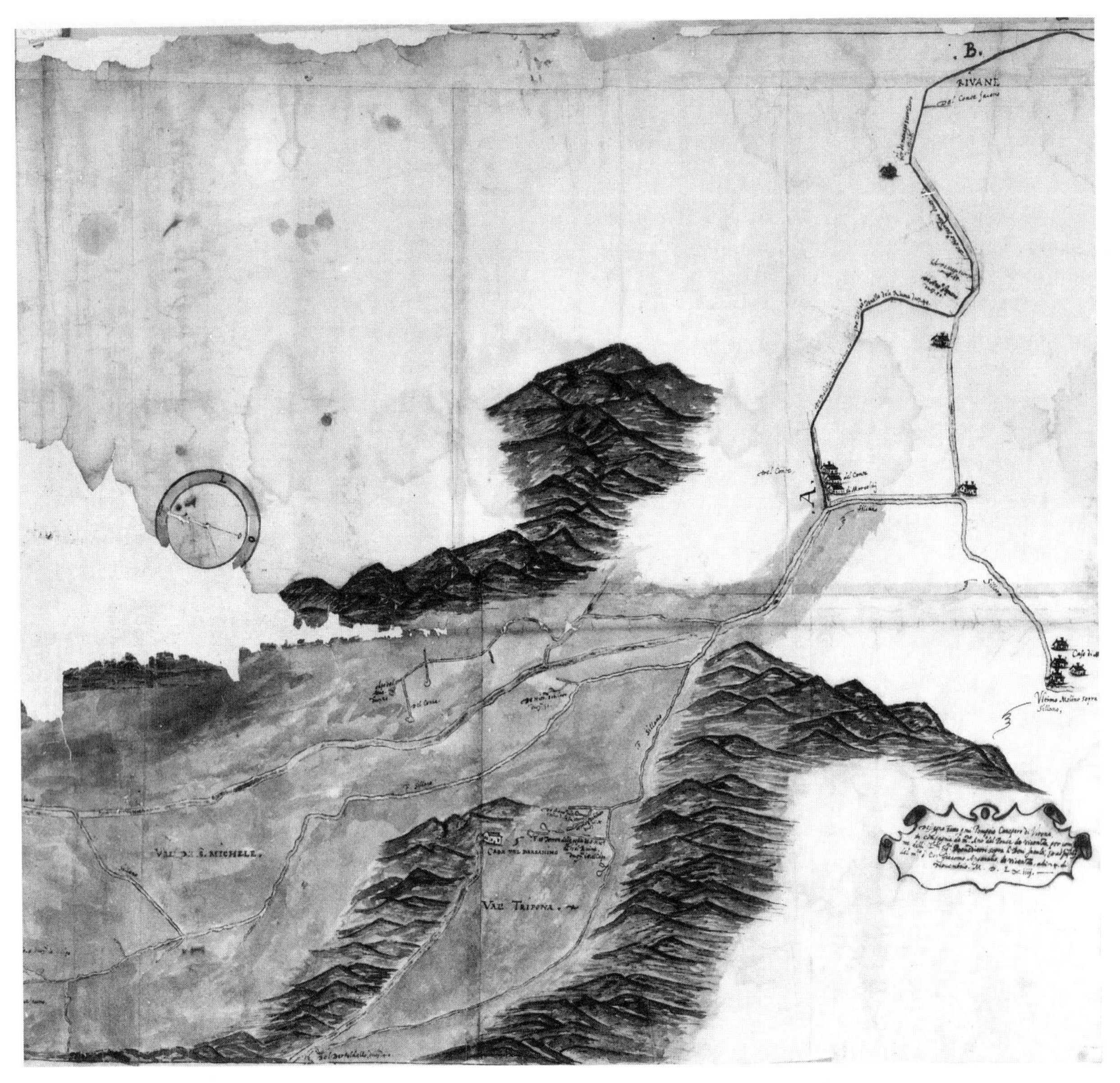

FIG. 1.3 Detail from a Rural Land Office map of the Brenta River headwaters. *Beni inculti, Vicenza,* not. 234, maz. 49, dis. 1, C, 15v (1564). ASV

Much earlier, in 1515, the second earliest map in my sample gives a sense of the detail provided. A huge 4,660 × 1,120-mm map of the Brenta from Angarano to the north of Nove follows both river and roads.[18] Every stray building along the road and some off it appear to be recorded. A 1558 map of the territory along the Adige River between Valle Serego and Beccacivetta also employs a large format, 2,640 × 840 mm.[19] Here the countryside, even to the differentiation of fields as we might see in an aerial photograph, is emphasized. Another large map (1340 × 797 mm) from the same source, the Office of Rural Lands, gives more detail of an unspecified section of the Adige in 1564 (Figure 1.4).[20] The focal point is not what is along the river but what lies between it and the canal of Santa Caterina. A sense of the irrigation system and landlords is given. All three of these wide-angle views find their resolution in maps of property held by individual landlords. A miscellaneous map from 1563 of the territory of the Barbarigo family is also exemplary, with its detailed numerical and figurative reproduction of the plot in the lower left corner.[21] Here the particulars of landhc ing are defined precisely in a kind of real est title map.

Two other maps from the Water Man ment Board for the Lagoon are most illumi ing as we try to define the starting date for administrative cartography. The first dates from 1542 and depicts a southwest section of the lagoon and Terraferma from Chioggia to Malamoco.[22] The second (figure 1.5) dates from 1556 and focuses on a small section of the canals coming off the lagoon near the Sile River.[23] From 1224, the Venetian government had established a magistracy in charge of canals.[24] Water management to prevent river silting and human incursions had distinguished Venice from earlier capitals in the northern Adriatic like Ravenna and Aquileia. Why did administrative mapping of ongoing legislative and judicial decisions delay for more than three hundred years? Something had happened by the mid-sixteenth century to create a map culture and a map market.

FIG. 1.4 Detail from a Rural Land Office map of the Adige River. *Beni inculti, Verona,* not. 50, maz. 45/B, dis. 1 (1564). ASV

A final design from Venice, a 1555 ambassadorial drawing of the organization of treaty negotiations at Calais, vividly demonstrates this new visual mentality (figure 1.6).[25] The diagram emphasizes the equality between the parties—English, French, imperial, and papal legates—and the neutrality of their common meeting place. Verbal description of these arrangements might take one sentence. Why fill an entire page of the dispatch with such a figure? Some kind of clarity must be assumed in a visualization that sacrifices economy of space

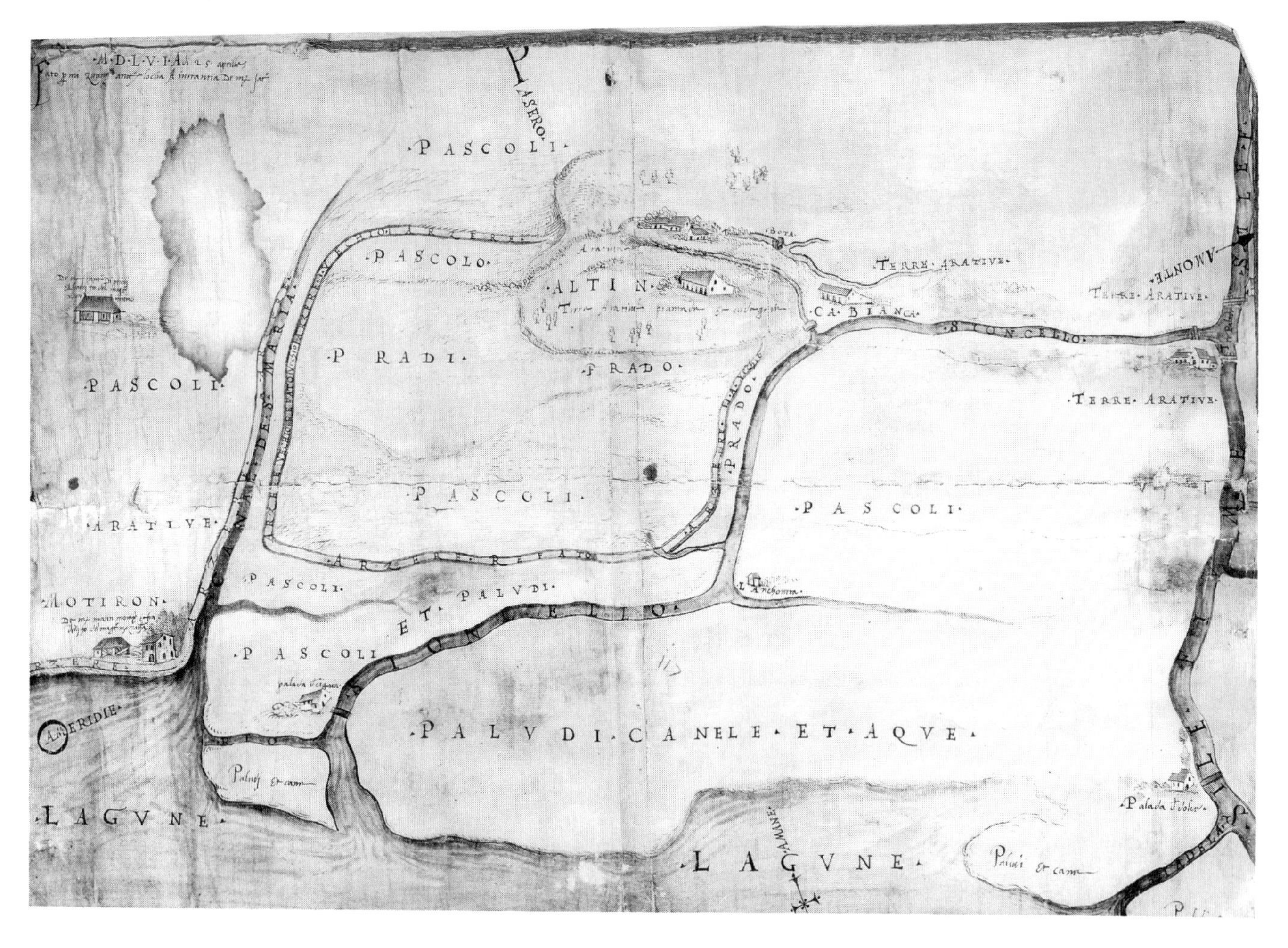

Fig. 1.5 Water Management Board, lagoon map of canals near the Sile River. *Savi Esecutori Acqua, Laguna*, dis. 13 (1556). ASV

in a document meant to be transported over such a long distance.

Mapping in Florence, Milan and the Papal States

The state archives in Florence reinforce the mapping portrait and chronology in Venice. Ten inventories list 9,856 maps.[26] Excluding some half-dozen early nautical maps, I have found only six inventoried maps dated before 1565.[27] Despite Tuscany's advanced bureaucracy,[28] the bulk of archival maps in Florence date from the seventeenth century.

In a 1911 survey of maps in the communal archives of Milan, only 17 out of 175 cataloged items predated 1560.[29] The earliest bird's-eye view of the city of Milan is Antonio Lafreri's 1560 map published in Rome. The earliest general view of Lombardy comes from another Roman map, that of Vincenzo Lucchini

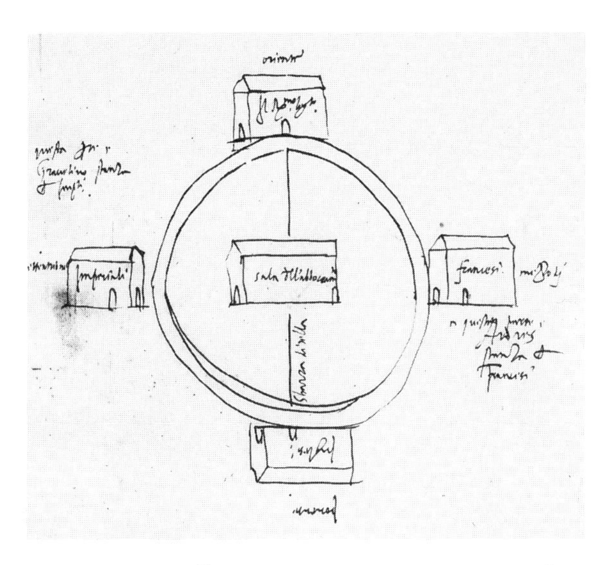

Fig. 1.6 Design illustrating treaty negotiations at Calais. *Senato dispacci, Inghilterra,* f. 1, c. 38 (1555). ASV

in 1558. While this collection may focus on printed materials, manuscript sources are considered. But earlier maps tended to be schematic—those of the city sketched major monuments only, while those of the countryside remained incomplete. The Milanese state archives are much vaster, with some 76,291 maps.[30] But offices typically commissioning maps have the same sixteenth-century trajectory as in Venice: *Acque* (sixteenth century–1801), *Acque e strade* (1574–1801), *Confini* (1518–1802). Nevertheless, although we would expect to find earlier maps in the collection, the important point for the present inquiry is the sixteenth-century link between diffuse mapping and state administration.

For the Papal States, the profile is similar. A small number of general maps of central Italy and a few regional designs of Lazio, the March of Ancona and Umbria date from the fifteenth century. Enough materials to construct what we might call a modern atlas, however, did not exist before the last quarter of the sixteenth century.[31]

The indefatigable Robert Almagià, Italy's leading twentieth-century cartographic scholar, pinpointed 1565–70 as the beginning of such published collections, with Venice and Rome as their seats.[32] The conjuncture is substantiated by the dates of the leading Venetian atlases—Gastaldi (1548), Rucelli (1561), Magini (1596); by the *Theatrum Orbis Terrarum* of Ortelius (1570), which signaled the birth of the modern atlas; and by the Ptolemaic atlas of Mercator (1578), which marked "the end of the reign of Greek geography."[33] In Rome, the Burgundian mapmaker Antoine Lafrère (c. 1512–77) became Antonio Lafreri when he set up shop there about 1540. In fact, one modern collation of thirty-five collections from 1553 to 1580 counted some 614 different maps.[34] There is no doubt that printed maps and the mapmaking business took off in the third quarter of the sixteenth century.

The Example of Naples

To test the simultaneity of administrative interest in mapmaking and the birth of the map trade, let us turn to the Italian South where the stereotypically backward Kingdom of Naples fell under Spanish rule after 1503. Conforming even more to the mid-sixteenth-century time frame than materials in the state archives of Venice, Florence, Milan, and Rome, the Neapolitan state archive's 1,130 extant items include nothing predating 1563.[35] Thus, the relative lateness of mapping for jurisdictional plots and travel itineraries still holds in Naples through midcentury. In the Neapolitan Mesta, the sheep customhouse at Foggia, cadastral surveys conducted in 1533 and 1548–53 and surveys of the sheep walks in 1574 and 1601–11

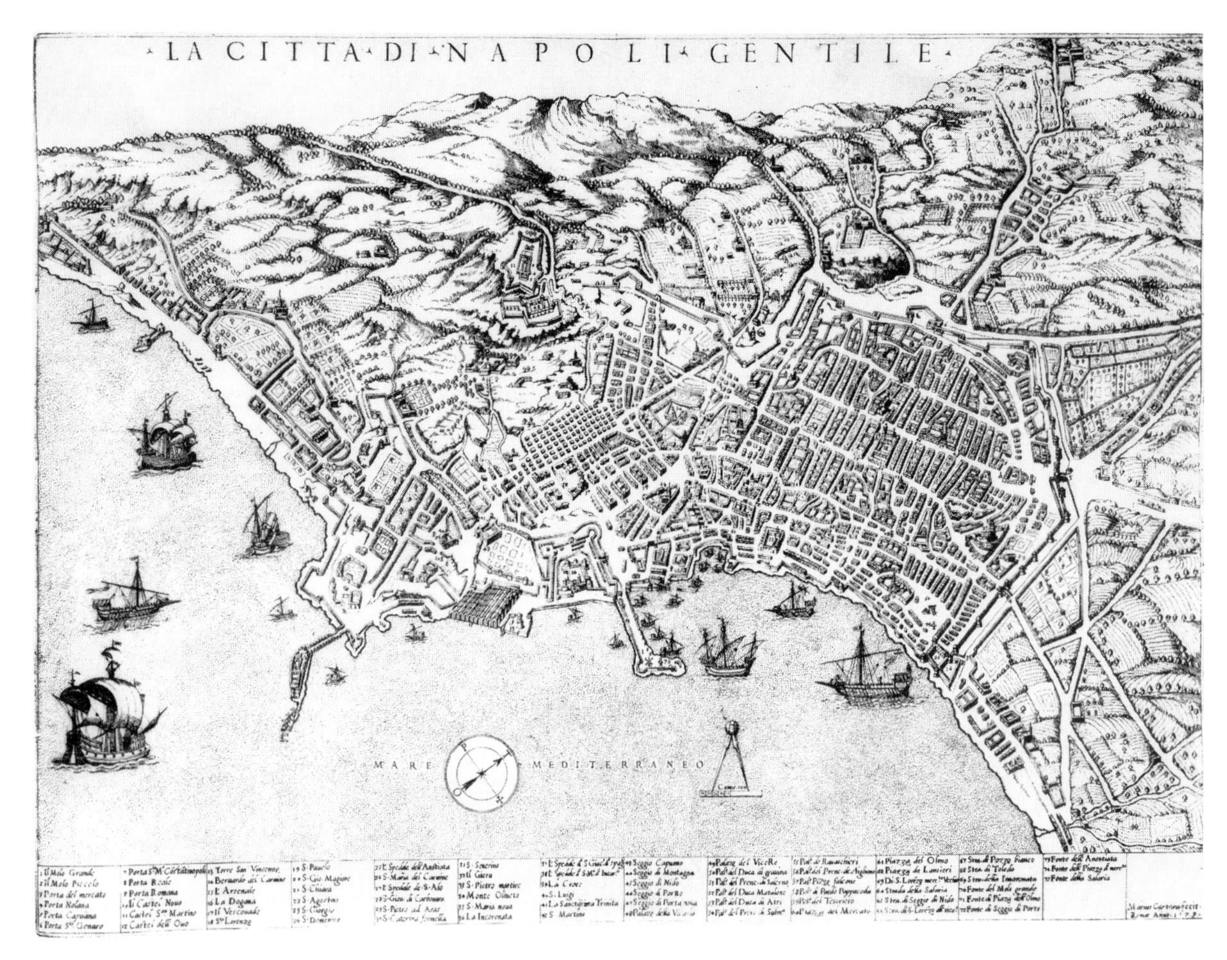

FIG. 1.7 Mario Cartaro, Map of the City of Naples. Biblioteca Angelica Rome, *Bancone stampe*, n.s. 56 (1579)

included no maps.[36] An itinerary of the Kingdom of Naples compiled in 1559 continued the practice of listing distances and places without maps.[37] Nevertheless, a late fifteenth-century map of the northern border of the kingdom is drawn to scale in Roman miles—probably the earliest itinerary map based on a measured survey.[38] Naples shared the political and cultural world of the rest of the Italian states through the Renaissance, as it magnified the factional struggles and contradictions found everywhere in the peninsula.

Mario Cartaro (1540–1614), the official cartographer of the Spanish viceroyalty of Naples from 1583 to 1594, is our best witness to the spread of surveyor cartography for civil-administrative purposes in the Italian South.[39] Cartaro's 1579 map of "La Città di Napoli Gentile" (figure 1.7) was made at Rome and clearly copies Lafreri's 1566 bird's-eye view.[40] The angle of view, the area mapped, the design of buildings, roads, and boats in the bay, the seventy-five places itemized below the frame, and even the compass and scale are similar. Cartaro replaced Lafreri's inserted legend with another boat and corrected the error in the numbering of items, but these are minimal changes. The work of Lafreri held its place in the mainstream of cartographic reproduction and imitation into the seventeenth century.

Cartaro associated himself with another Neapolitan scientist, Nicolà Antonio Stigliola, who was a keen observer of the well-known, pathbreaking cartographers of his times.[41] Giacomo Gastaldi's 1561 map of the Kingdom of Naples was the first to shift its north-south axis and place the peninsula diagonally with the boot in kicking position, rather than horizontal as it was placed in earlier views.[42] Stigliola and Cartaro followed the Gastaldi example in their map of the kingdom (c. 1590).[43] Similarly, the only surviving model of a provincial map for Stigliola and Cartaro was a 1567 Gastaldi map of the province of Terra d'Otranto.[44] The Stigliola-Cartaro map, like the Gastaldi exemplar, is noteworthy for its use of non-Ptolemaic elements, its town locations, and the realistic horizontal placement of the heel (figure 1.8).

The most extraordinary administrative use of the Stigliola-Cartaro maps can be seen in the present reproductions. The maps remained unpublished, and soon after their production, numbers were penned onto eight out of twelve of them, near the name of each town. These numbers correspond to the number of hearths counted in each town for a hearth tax, which constituted the kingdom's largest single source of income.[45]

The 1594 completion of Stigliola and Cartaro's base maps coincided with the new hearth count completed in 1595. Although hearth counts were supposed to be conducted every fifteen years since their establishment in 1447, Philip II's 1575 bankruptcy spawned a unique proposal to suspend that year's count and subsequent annual payments in exchange for two immediate, discounted prepayments.[46] From 1561 to 1595, then, there was no accurate record of hearth numbers. Mapping the number of hearths by the kingdom's towns would, therefore, have provided an easy means to visualize the changes and expectations of revenues for the royal patrimony after the extraordinary late sixteenth-century lacuna.

Cartaro, in turn, became a model to copy. A map of "Napoli Gentile" by Claudio Duchetti, Lafreri's heir and successor, was dated 1585, Rome, and copies not only the title but every detail of Cartaro's city map, down to each of the boats in the bay.[47] Duchetti's only emendation added two more items to the bottom place list. Similarly, Cartaro's original maps of the kingdom's twelve provinces, drawn between 1590 and 1594, in 1602–4 became the models for Giovanni Antonio Magini's Neapolitan maps, posthumously published in his 1620 atlas, *Italia*.[48] The map of Capitanata provides a good comparison (figure 1.9). Cartaro's map measured 36 × 51 cm and Magini's 37 × 42.5 cm. Rivers, mountains, and town identification are the distinctive features of both. Magini's error in omitting the name of the Fortore River and mislabeling the Biferno River as "Biferna or Fortore" continued to be repeated into the mid-eighteenth century.

The province of Capitanata was also the site of the kingdom's largest source of indirect taxation, the sheep customhouse at Foggia. This Mesta-like institution administered winter pasture in the plain around Foggia, main-

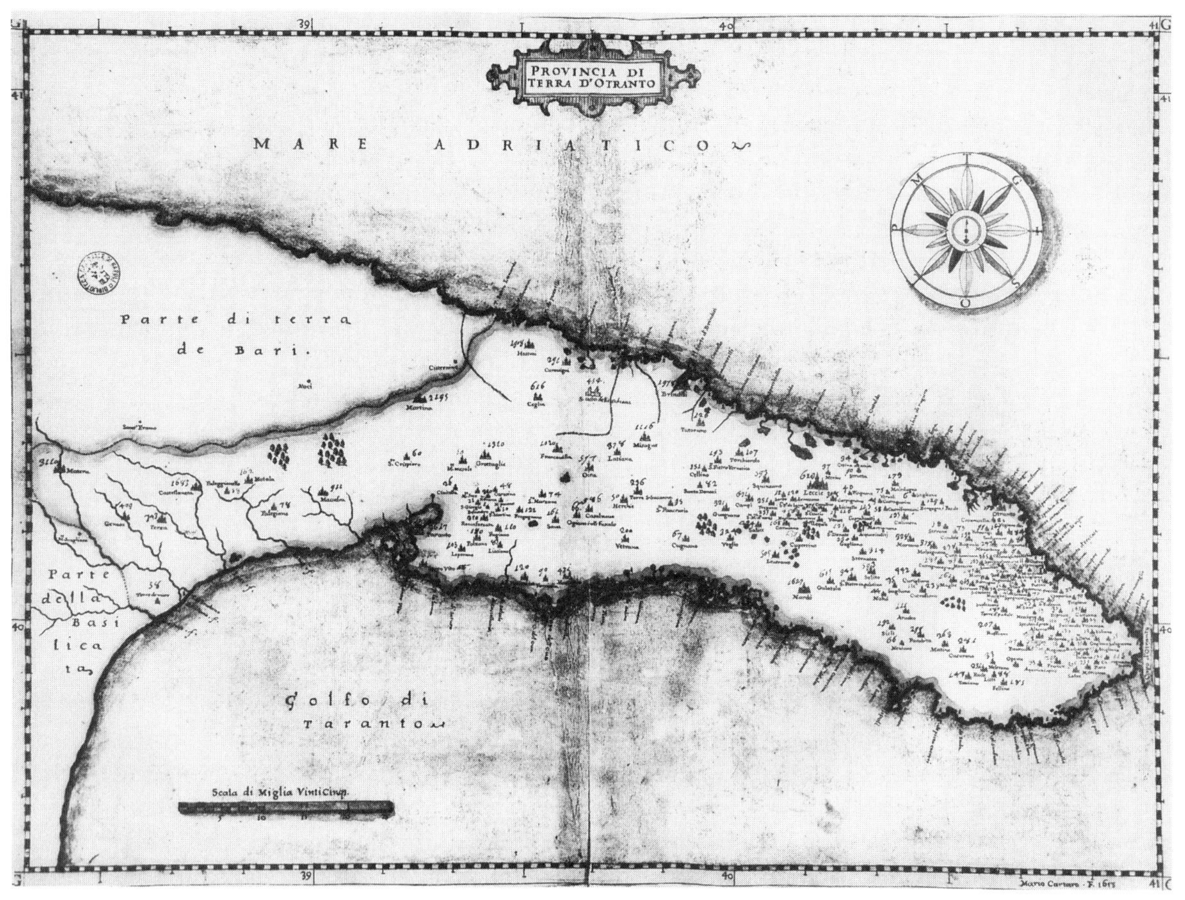

FIG. 1.8 Mario Cartaro, Map of Terra d'Otranto. From Mazzetti, *Cartografia*, 2 vols. (Naples, 1972), ii, tav. xxi

tained sheep walks between these winter grazing locations and summer pastures in the Abruzzi, rented agricultural lands on short-term contracts, adjudicated disputes in civil and criminal cases, and oversaw commercial transactions at the annual spring fair.[49] Although its origins stretched back to the twelfth or thirteenth century, the modern foundation charter dated from 1447.

Not surprisingly, as I have already mentioned in keeping with our earlier mapping examples, no regular mapping appears before the seventeenth century, even though surveying and land management were important institutional functions from the beginning.[50] Typically, the earliest institutional maps sketched the reclamation of sheep walks and boundaries of winter grazing locations. Yet no composite map of the whole administrative system dates before that of the customhouse's royal surveyor, Agatangelo della Croce, around 1765.[51] It would seem that only the parts, not the whole, were mapped earlier. Because the state administration centered on one-to-one obliga-

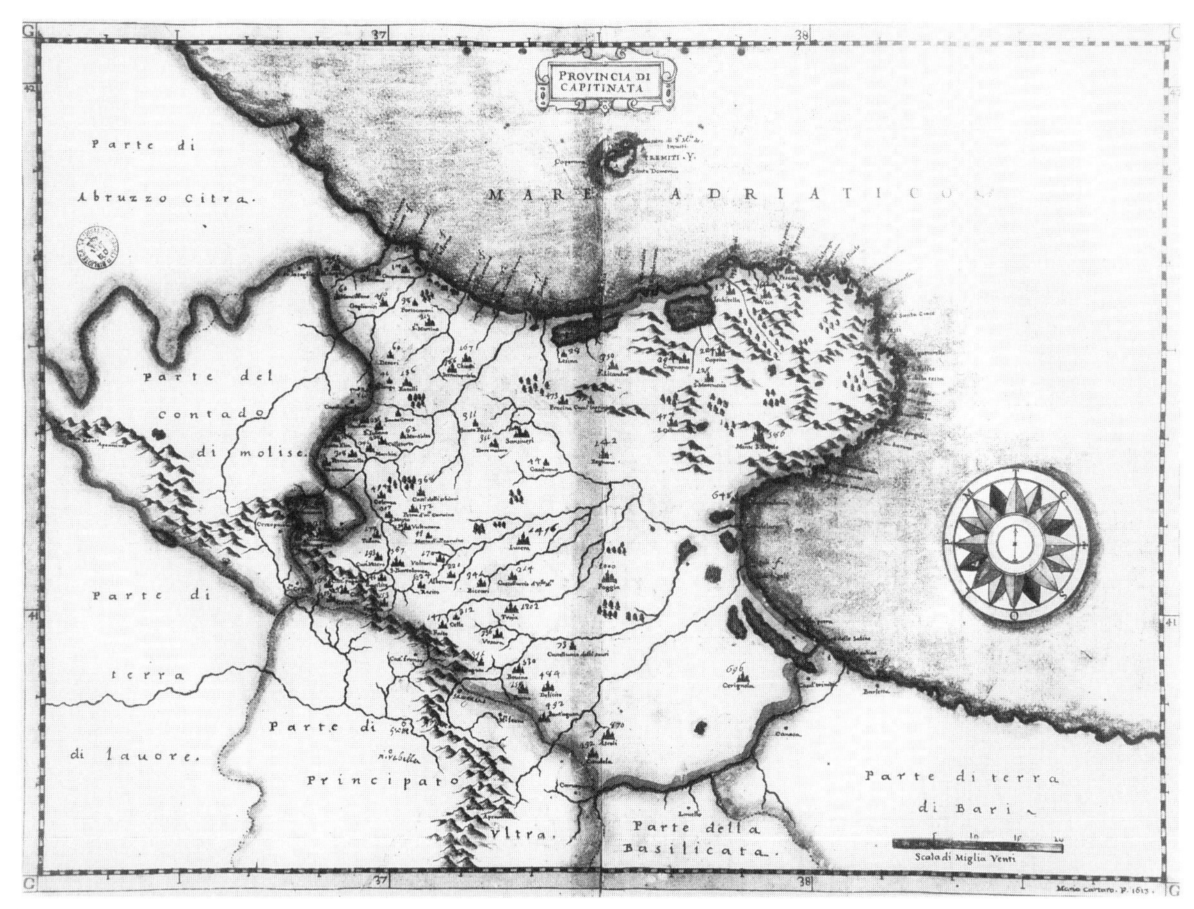

FIG. 1.9 Mario Cartaro, Map of Capitanata. From Mazzetti, *Cartografia,* ii, tav. xviii

tions and duties of individual sheepowners, the administration saw no need for a view of the whole.

The Ettore Capecelatro reintegration of sheep walks in 1647–51 led to the first complete visual rendering of the customhouse roadways.[52] A 542-folio manuscript with twenty-seven composite maps divided into numerous double-page sections along the sheep walk, however, still relies predominantly upon verbal description. Surveyor-mapmakers traveled along the sheep walks and kept journals filled with their drawings. The trailhead of the Celano-Foggia sheep walk, lying between the walled town and castle of Celano and the northern shore of Lake Fucino, gives some sense of the detail of administrative mapping in the double-page format (figure 1.10). The 111.11-meter sheep walk spread out around the old Roman road, while individual plots, forests, streams, settlements, and buildings like churches, way stations, and mills along the way are marked. Numbers on the map correspond to numbers in the text to aid in identifying spe-

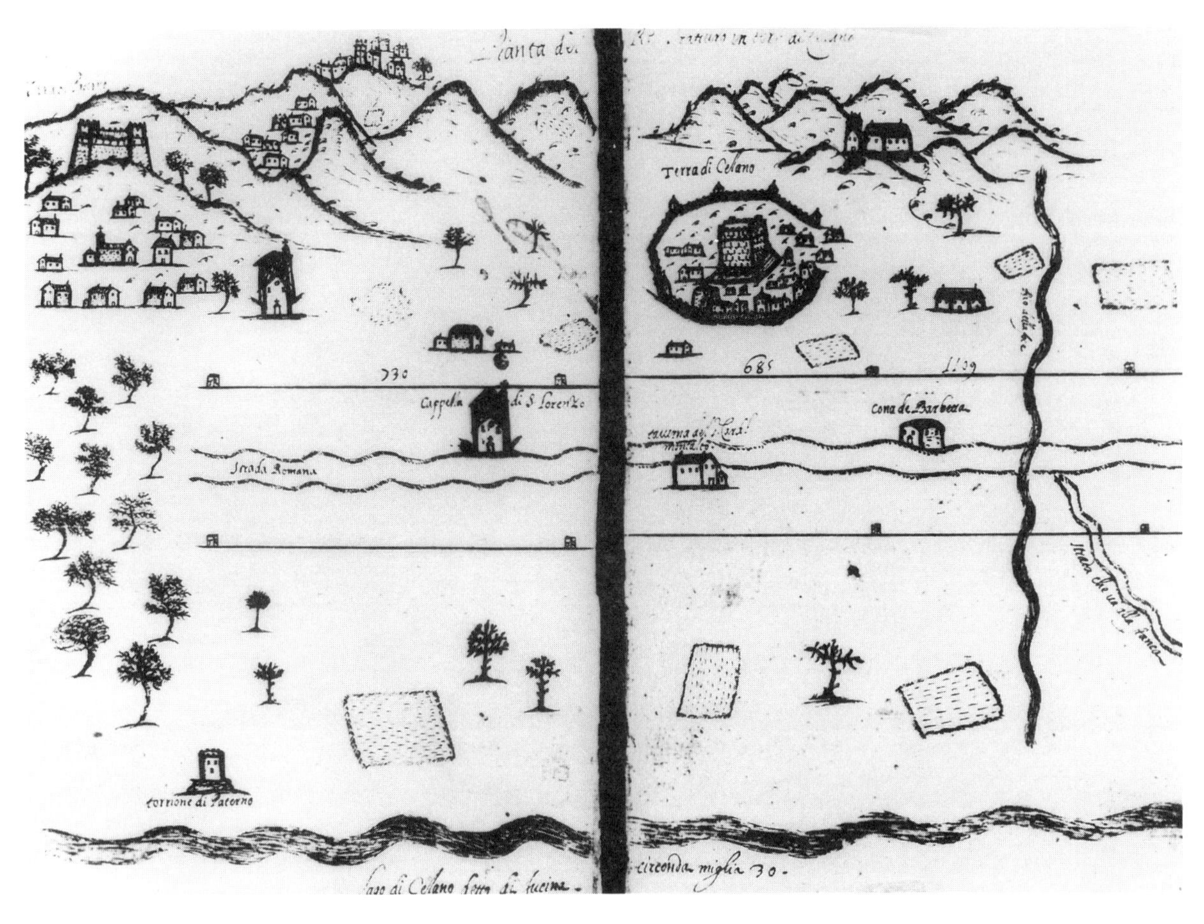

Fig. 1.10 Celano-Foggia sheep-walk trailhead. Archivio di Stato, Foggia (ASFg), *Dogana delle pecore,* serie I, fas. 18, 325v–328r (1648–51)

cific boundary and encroachment questions. Following maps trace the sheep walk to its end outside Foggia.

The Antonio de Centellas reintegration of 1668–86 provides another map of the end of the trail.[53] Here the customhouse port, Manfredonia, can be seen in the right corner of a descending mountain landscape. The difference in land use between pastoral and agricultural farms is evident. A more detailed view of an individual plot from 1736 gives a greater sense of the diversity of land use even within uniform holdings (figure 1.11).[54] In this map of the fief of Visciglito near Lucera, draft animal pasture (*mezzane*) is distinguished from arable land (*portata*).

Sheep transhumance was a capital-intensive enterprise, a variant of capitalist agriculture adapted to exploit the marginal zone of sparse rainfall. It could exist only in conjunction with agricultural production. In order to moderate competition and foster collaboration, the state customhouse continually had to monitor winter grazing assignments in the

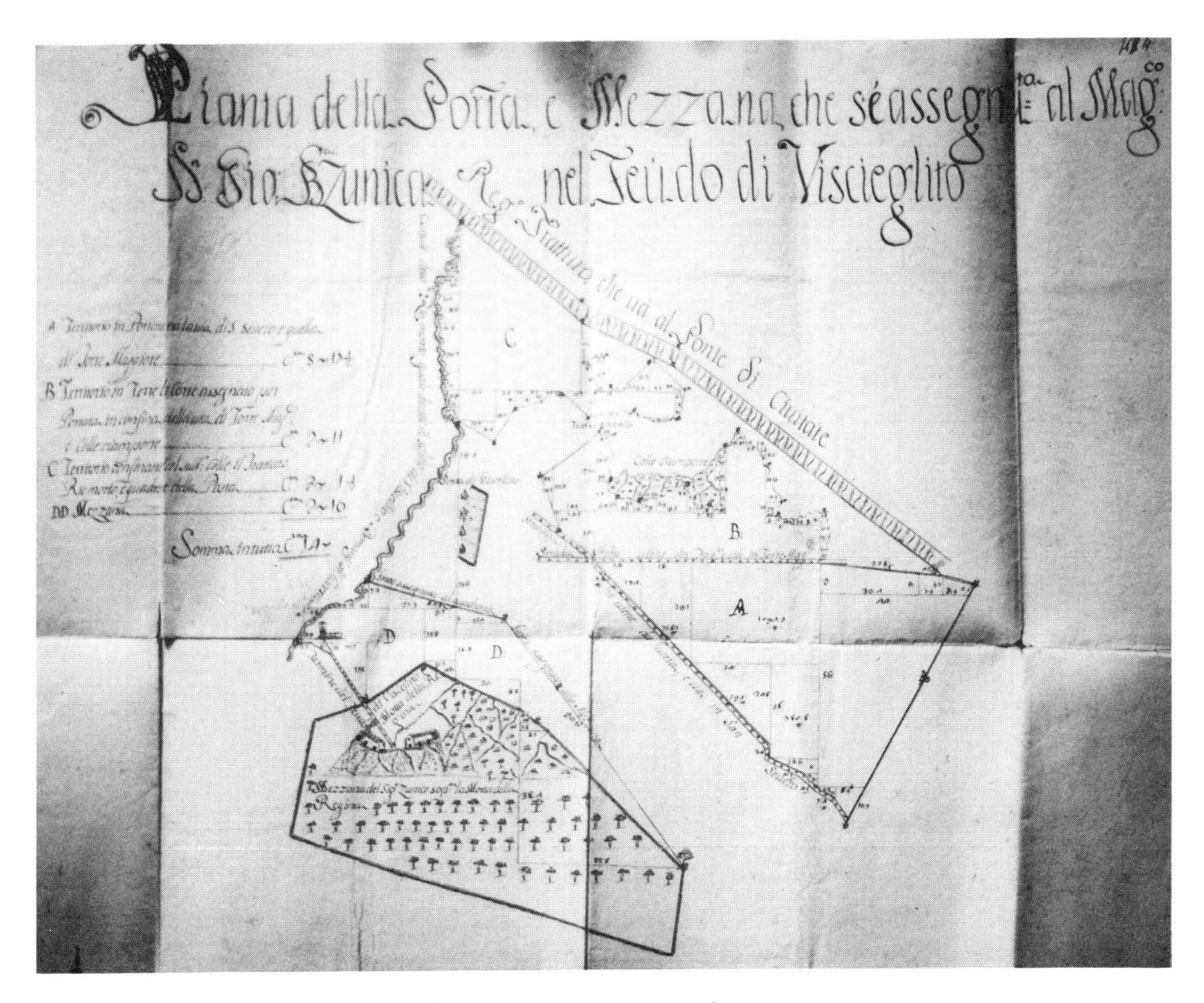

FIG. 1.11 Farmland and pasture at Visciglitto. ASFg, *Dogana,* serie I, fas. 18 (1738)

twenty-three ordinary locations in its 4,300-square-kilometer jurisdiction spread around Foggia. Again—though surprisingly—the first maps of these winter locations date only from 1686, with the twenty-eight maps (75 × 50 cm) of Antonio and Nunzio Michele.[55]

The 1686 maps corresponded to a significant governmental success: the restarting of the pastoral economy after the seventeenth-century crisis. The viceroy, marchese del Carpio, should receive special mention for his military efforts in combating endemic rural banditry in the mountains of the Abruzzi.[56] The maxim inserted as a legend on the map of the location of San Andrea makes this point eloquently: "Good custodianship conserves peace through justice" (*Recta servat custodia*

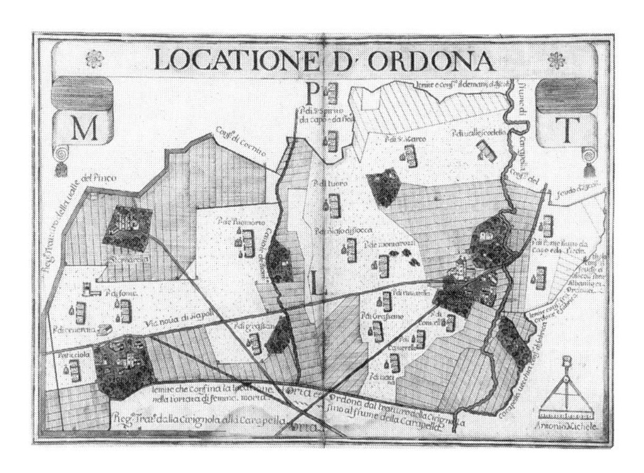

Fig. 1.12 Antonio Michele, Map of the location of Ordona. ASFg, *Dogana*, serie I, fas. 20 (1686) nn. 24

justititia pacem).[57] Simultaneously, the reinstatement of particular locations, which were reserved for large sheepowners and separated them from the mass of small and middling owners, provided a good excuse to map the winter grazing locations. The enforcement of law and order, the shift in winter grazing privileges for large owners, the change in rural credit practices, and the renewed demand from northern Italian wool buyers sparked a recovery in wool production in the 1680s.

Fig. 1.13 Antonio Michele, Map of the location of Arignano. ASFg, *Dogana*, serie I, fas. 20 (1686), nn. 6

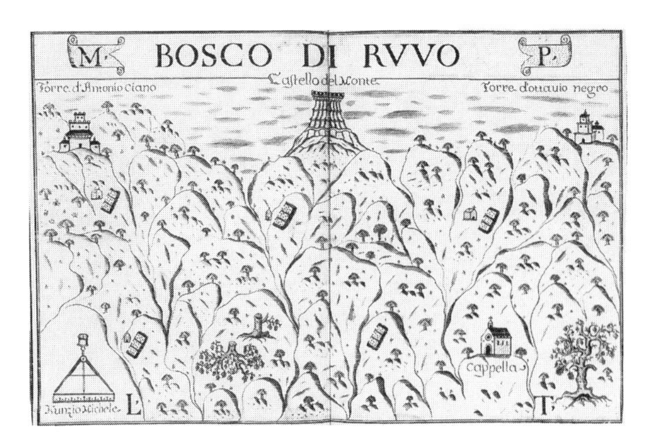

Fig. 1.14 Nunzio Michele, Map of the extraordinary unusual pasture of Bosco di Ruvo. ASFg, *Dogana*, serie I, fas. 29 (1686) nn. 27

The Michele brothers' maps illustrate the boundaries of the winter locations and represent their distinctive features in a celebratory or ideal manner. The location of Ordona (figure 1.12), divided between agricultural plots (36 percent) and pasture (64 percent), lies on the plain and gives an indication of the complex land usage in the sheep customhouse. Note especially the private holdings—like Stornarella and Stornara, property of the Jesuit Collegio Romano—which dot the landscape. The location of Arignano (figure 1.13) lies on the northeastern border of the Tavoliere plain as it edges up the Gargano Mountain flanks. Individual farms and plots are marked, boundaries delimited, rivers emphasized, sheep walks and roads signaled, and towns with their jurisdictions given in schematic or symbolic fashion. Note even the tiny churches, farmhouses, and towers indicated in the plain, especially when associated with specific posts where individual sheepowners were assigned. The map of Bosco di Ruvo (figure 1.14), one of the "extraordinary unusual" pastures assigned to sheepowners as increased sheep numbers pushed winter grazing ever farther into the marginal zone, emphasizes the rugged topography of the Murge just south of the Tavioliere. Frederick II's famed hunting retreat, Castel del Monte, crowns the rolling hills, which are dotted by trees, plots, and buildings (note the two private towers that flank the focal center), but no roads or sheep walks.

For my final example of the Michele maps, I have chosen a view of the general location of Castiglione, which included associated locations of San Iacovo, Fontanelle, and Motta di San Nicolà—all reserved for large holders (figure 1.15). Note that the center map is framed in the corners by four maps of noncontiguous pastures: Palmora grande and piccolà, Versentino, Civitate, and Visciglito, which we have already seen in a 1736 planimetric view. The map orientation is turned almost 90 degrees, so that we are looking up at the city of Foggia from the east. Overall, the maps emphasize how private and public, agricultural and pastoral interests had to be balanced in the quest for good government.

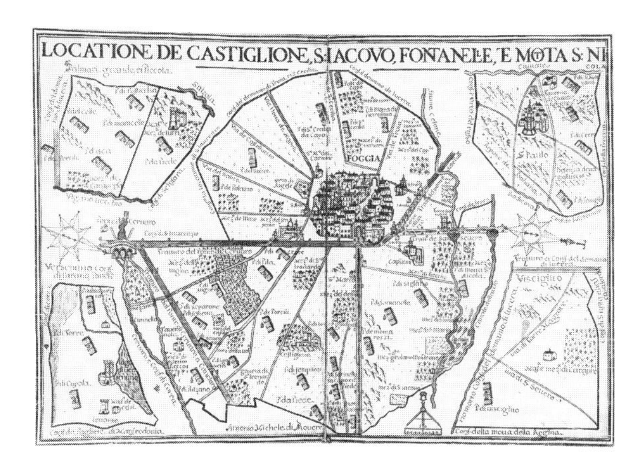

Fig. 1.15 Antonio Michele, Map of the location of Castiglione and associated particular locations. ASFg, *Dogana*, serie I, fas. 20 (1686) nn. 9

Who commissioned the 1686 location maps, or why did the Michele brothers design them? Antonio and Nunzio Michele were officers of the customhouse, employed as *agrimensori* or *compassatori*—literally, "land surveyors" or "measurers with compasses." The office dates from a 1574 decree by the viceroy. Cardinal Granvelle:

> Item. Because it is right, that in the Tribunal of the Royal Sheep Customhouse, there is *so much* decided by the measurements, judgment, and written reports of land surveyors, and since they frequently serve for cases touching upon the Royal Fisc and since the number of these land surveyors is large and *so many of them unqualified;* thus we order and command that in this customhouse no more than six land surveyors be permitted, that we elect them, and so I have ordered the Magnificent Doganiero [head of the customhouse] to notify and advise all those whom this may concern.[58] [Emphasis mine]

The viceroy's emphasis on the large number of unqualified (*inexperti*) surveyors roaming about underlines the demand for their services. The decree's intent was to license or verify the surveyors' competence, and to ensure their integrity by paying them state wages. One year later, the order capping their number at six was amended to allow for employment of more surveyors to whatever number might be convenient.[59]

The following item in both the 1574 decree and its 1575 revision helps clarify the state's intent.[60] The number of lawyers and procurators was also limited for the same reason. But this order was completely rescinded in 1575, perhaps because defendants wanted the right to name their own counsel rather than work with customhouse appointees, or, more likely, because lawyers were a more powerful guild than surveyors and resisted state limitations and appointment by *doganiero.*

Surveyors, nevertheless, had a certain prestige. Their office was said to have descended from the ancient Egyptians who needed to manage the Nile floods.[61] Their skill depended upon a mastery of the principles of geometry and was confirmed by ancient authorities like Herodotus and Plato. This knowledge meant that customhouse surveyors were paid "not only for their physical labor, but like judges and lawyers, for their genius [*ingegno*]."[62]

Their salaries reflected this honor. Established during the 1574–80 tenure of Fabrizio di Sangro as *doganiero,*[63] surveyors' salaries depended on the specific task and varied from ½ to one ducat per day. At about two hundred ducats per year, surveyors' salaries equaled the average for customhouse officers and were well above the high range of "reasonable" salaries mentioned by Fernand Braudel for sixteenth-century Naples.[64] By way of comparison, Mario Cartaro, the royal mapmaker in Naples, received eleven ducats per month in November and December 1590 and thirty ducats per month in 1591.[65]

The duties of the customhouse surveyor were well delineated in four of the forty-nine items of the 1668 pragmatic.[66] In addition to measuring sheep walks and winter pastures, they "examined and measured all of the agricultural plots rented by the customhouse or others, fields both planted and in fallow."[67] Their survey determined both the boundaries and the area of pastoral and agricultural territories.

The focal point of all this countryside activity was the bureaucratic seat in Foggia. Details of Foggia from the customhouse surveyors' maps (figure 1.16) emphasize the town's distinctive feature of the town as the single roadway bisecting it. Via Arpi ran east-west through the old town, past the customhouse palace (until it was transferred after the 1731 earthquake) and out to an open square where grain depositories held harvested reserves in deep-soil pits. In the 1712 map of trail's end of the Aquila-Foggia sheep walk, the town concentration stands in stark contrast to the open expanse of pasture, the wide sheep walk, and isolated churches and buildings (figure 1.16). If we mentally remove the walls, which were added in error, the easy flow between city and country becomes more apparent. Similarly, a charming 1703 view of Foggia invites viewers into a pastoral landscape with gesturing and reclining shepherds in contemplation and disputation (figure 1.17). As late as the eighteenth century, idealized visions of harmony and repose continued the early tradition of town maps emphasizing good government and local pride.

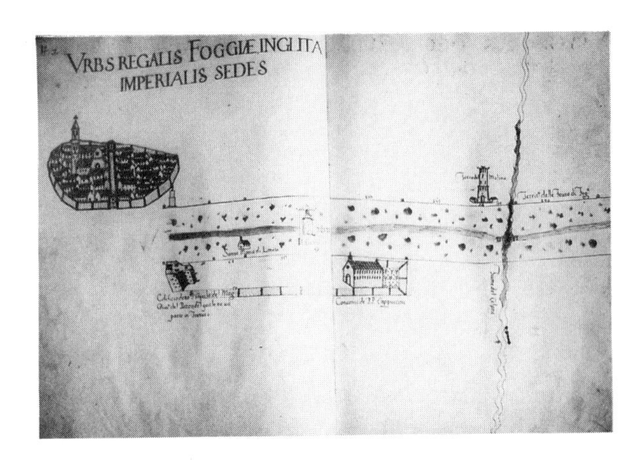

FIG. 1.16 Alfonso Crivelli, Foggia at the trail-end. ASFg, *Dogana,* serie I, fas. 19, cc. 17r–18v (1712)

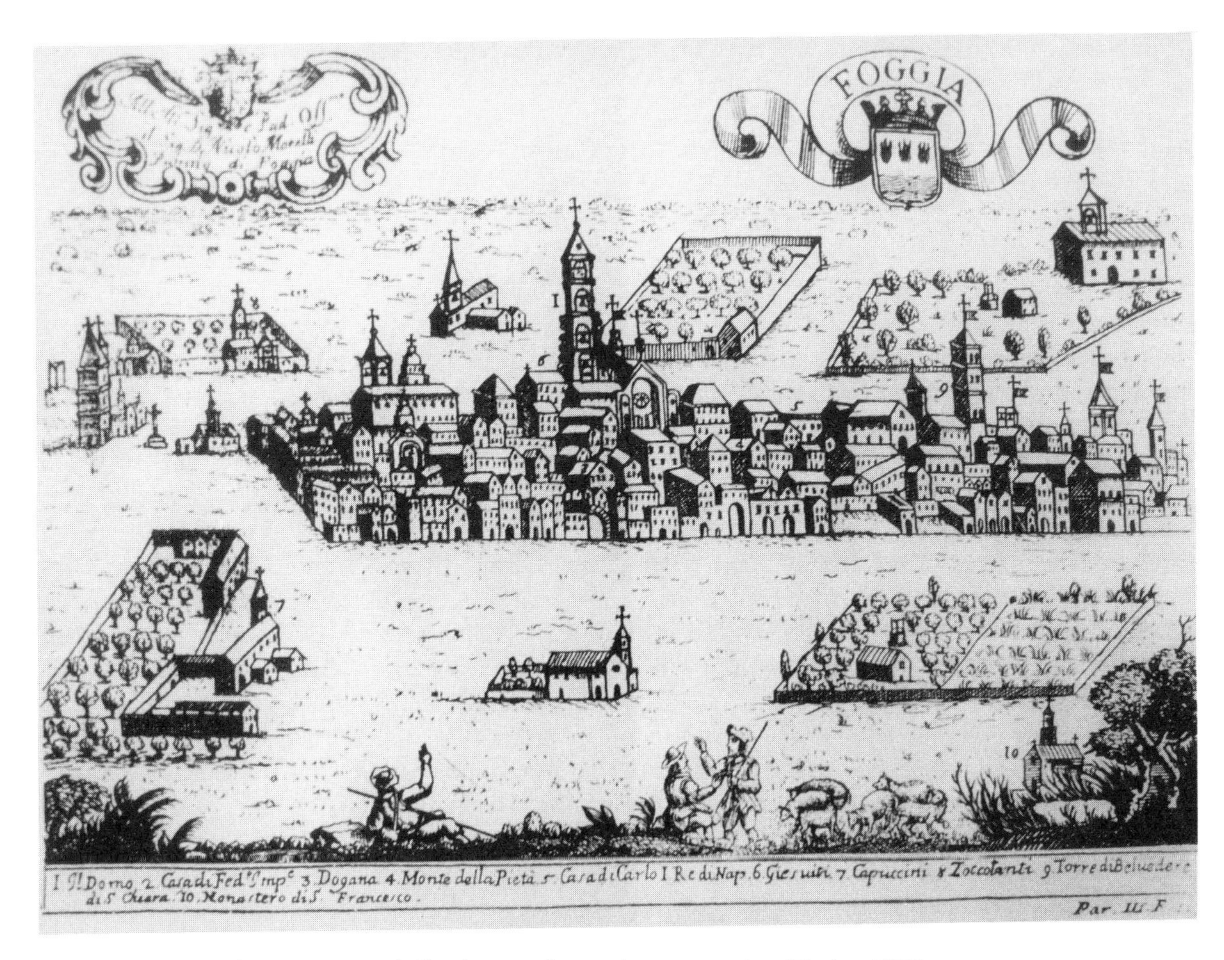

Fig. 1.17 "Foggia" from G. B. Pacichelli, *Il Regno di Napoli in prospettiva,* (Naples, 1703)

Conclusion

As pleasing as this continuity in ideological tone may be, three discontinuities—times of increased mapping production—stand out: the late fifteenth century, the mid-sixteenth century, and the late seventeenth century. Why did maps increase in each of these periods?

The Venetian expansion into mainland Italy in the second half of the fifteenth century first coupled the indigenous district maps of the northern Italian plain with the bureaucratic functions of imperial administration. No mapping culture quickly spread throughout Italy, however, and it seems fruitless to point out any one map as the missing link in the transformation of late medieval maps into modern administrative ones. The development of linear perspective may serve as an intriguing contrast to explain the slow adoption of a mapping mentality. If the revolution in Renaissance painting in the 1420s and 1430s soon captured and dominated Western modes of perception (printing press or not), why did accurate maps representing cities and towns, provinces and principalities appear so late? The answer to this paradox, I believe, lies in the patronage, production, and purpose of maps. Renaissance painting found loyal patrons in the Church, princes, nobility, and merchants, who were all interested in the didactic, commemorative, and/or spiritual functions of art, and the majority of maps from this period followed in this same aesthetic tradition as a kind of decorative art. Until midcentury, I would argue, a kind of mental habit blinded administrators' eyes to the bureaucratic potential of map use. Numbers, accounting and double-entry bookkeeping, the nuts and bolts of Renaissance control over finances both personal and patrimonial, guided the bureaucratic imagination more than spatial design.[68] The accountants' eye for the bottom line, not the drawn line, ruled.

By the mid-sixteenth century, however, two new variables had transformed the Italian states and help to explain the proliferation of maps. First, rational bureaucracies and government offices established increased control over their domains. We have already found that in Florence and Milan institutions overseeing borders, roads, water, and buildings all took on new life in the sixteenth century. And for some reason these renewed offices became the chief users of maps. Almost 75 percent of the ten thousand maps in the Florentine state archives are preserved in only two collections: that of the captains of the Guelph party after 1538 and that of territorial possessions from the mid-seventeenth century. Similarly in Venice, almost one-half of the state archives' ten thousand maps were commissioned by only one agency, the Office of Rural Lands, founded in 1566.[69] The foundation of these offices reflected the second variable, the upturn in the economy often called the Italian Indian summer. The end of the Italian wars with the peace of Cateau-Cambresis in 1559 allowed for increased demand, credit, and eventually production in the cities of the North. Maps were among the goods produced for these revived markets. Not unlike their sister industry, the book publishers, mid-sixteenth-century map publishers began to employ new experts in map composition—surveyors instead of artists.

A new geographic interest, almost a travel-guide mania, grew up by midcentury. Leandro Alberti's 1550 *Descrittione di tutta Italia* may have been the most influential, with eight Italian editions within twenty-five years, two Latin editions in 1566 and 1567, and revised and updated editions from Venice after 1581.[70] His immensely popular and widely imitated topographical, historical, and biographical material ranged through the Italian states province by province, but without any maps. Originally inspired by the ancients like Strabo and moderns like Flavio Biondo, this early antiquarian compendium found a ready market in late sixteenth-century guidebooks, which soon incorporated its contents with bird's-eye views of towns and illustrations to aid travelers.

There can be no doubt that the long gestation period generated by portolan charts and overseas exploration, artistic innovation and scientific inquiry, and the intensive theorizing and practice of warfare and statecraft gave rise to a new spatial mapping mentality. The extensive diffusion of maps in state bureaucracies followed the fortunes of the economy that built the map market, employed surveyor-cartographers, and filled state treasuries with the means to purchase such products. The third burst of mapping intensity in Italian administrative affairs, corresponding to the recovery from the long economic crisis of the seventeenth century, confirms that this periodization is not coincidental. The political and economic climate in mid-sixteenth-century Italy tapped the received tradition of the budding mapping mentality and employed that vision to carry out the reorganized functions of the absolutist state.

Notes

1. *Carte Nautiche e geografiche,* in the Archivio di Stato, Florence (hereafter ASFi), for example, includes two maps of the Mediterranean from the fourteenth century and five from the fifteenth. Roberto Almagià, *Planisferi carte nautiche e affini dal secolo XIV al XVII esisenti nella Biblioteca Apostolica Vaticana,* vol. 1 of *Monumenta Cartographica Vaticana,* 4 vols. (Vatican City: Biblioteca Apostolica Vaticana, 1944–55), reproduces seven maps from the fourteenth and seven from the fifteenth century in part 1.

2. Egnazio Danti (1536–86) was a mathematician, cosmographer, and architect who worked on the fifty-three maps commissioned by Cosimo I for the Guardaroba of the Palazzo Vecchio from 1563 to 1575. The work was completed by Stefano Buonsignori from 1575 to 1584. Danti's designs for the forty maps adorning the Gallerie delle Carte Geografiche in the Vatican Palace were executed by his brother Antonio Danti between 1580 and 1583. For the geographical murals in the Vatican, see Almagià, *Le pitture murali della galleria delle carte geografiche* and *Le pitture geografiche murali della terza loggia e di altre sale vaticane,* vols. 3 and 4 of *Monumenta Cartographica Vaticana.*

3. P. D. A. Harvey, *The History of Topographical Maps: Symbols, Pictures, and Surveys* (London: Thames & Hudson, 1980), 86.

4. Jurgen Schulz, "Jacopo de' Barbari's View of Venice: Map Making, City Views, and Moralized Geography before the Year 1500," *Art Bulletin* 60 (September 1978): 425–74.

5. Harvey, 58–83.

6. Harvey, 58, also cites evidence for two lost maps, Padua (1359) and Lombardy (1379).

7. Ibid., 74.

8. Ibid., 72.

9. Ibid., 60. The order of 1460 applied to all territories, cities, and castles under Venetian rule.

10. Ibid., 60, and pl. 30.

11. Ibid., 60–61. On the Terraferma empire, see Nicolai Rubinstein, "Italian Reactions to Terraferma Expansion in the Fifteenth Century," in *Renaissance Venice,* ed. J. R. Hale (London: Faber & Faber, 1973), 197–217.

12. Archivio di Stato, Venice (hereafter ASV), collects maps in twelve *fondi: Miscellane Mappe, Rason Vecchie, Savi Esecutori Acque, Provveditori camera dei confini, Genio Civile, Dispacci Ambassadori, Provveditori Sanità, Fortezze Disegni, Senato: Dispacci Rettori, Beni Inculti, Relazioni Investigatori Processi,* and *Carte Poleni.*

13. Harvey, 78, 164.

14. ASV, *Rason Vecchie,* 350, dis. 1112, Caorle (pertinenze), "Tracciato relativo ai confini relativi al territorio veneziano di due boschi Harzere and Cavazole(?)" (1505).

15. ASV, *Provveditori dei Confini,* B. 132, c. 127r, Aussa (fiume), "Mappa di terre comprese tra Strasoldo, Cervignano, Aquileia, parte del corse del fiume Aussa e Malisana al confine tra lo Stato Veneto e l'Austriaco" (20 January 1538).

16. ASV, *Beni Inculti, Vicenza,* not. 234, maz. 49, dis. 1, c. 15v, Angarano di Bassano (pertinenze), "Settore dalle Valli S. Michele e Tripona fino al Brenta" (24 November 1564).

17. ASV, *Beni Inculti, Processi,* B. 2, dis. 5/bis, Angarano (Vicenza), "Corso del torrente Silano in contrada S. Michele nella podestaria di Marostica" (10 March 1564).

18. ASV, *Miscellane Mappe,* dis. no. 766, Brenta, "Mappa del fiume Brenta da Angarano a nord di Nove, con prospetti di case" (12 June 1515).

19. ASV *Beni Inculti, Verona,* 66, 58/B, dis. 5, Adige (comprensorio), "Territorio tra Valle Serego e Beccacivetta, congli abitati di Bionde, Ronco, Albaredo, Beccacivetta" (1558).

20. Ibid., rotolo 50, maz. 45/B, dis. 1, Adige (fiume), "Zona in territorio Veronese non specificato, Gruppo possessioni del supplicante e di V. Morosoni tra l'Adige ed il canal di S. Caterina" (1564).

21. ASV, *Miscellane Mappe,* dis. n. 1402, Carpi (Verona), "Mappa del territorio dei signori Barbarigo" (26 March 1563).

22. ASV, *Savi Esecutori Acqua, Laguna,* dis. 6 sui sostegni, Zona Barenosa, "Valle di Fogolana e laguna sud-occidentale, tra il canale di corte e di Siocho. Il Brenta nuovo ed il Bachiglione, il litorale dal porto di Chioggia a quello di Malamocco" (9 September 1542).

23. Ibid., dis. 13, Altino, "Mappa topografica zona tra il fiume Sile, il Canal di S. Maria, e la laguna: Compresso il canal del Sioncello" (26 April 1556).

24. Frederic C. Lane, *Venice: A Maritime Republic* (Baltimore: Johns Hopkins University Press, 1973), 16.

25. ASV, *Senato: Dispacci Inghilterra,* f. 1, c. 38, Calais, "Schizzo del luogo dove si incontrano per le trattative, Inglesi e francesi ed Imperiali alla presenza del Rev^{mo} Legato" (25 May 1555; from the dispatch of 1 June 1555).

26. ASFi, *Acquisto Gonneli* (18th-19th century), 60 maps; *Cetosa di Firenze* (18th–19th century), 43 maps; *Piante dei Capitani di Parte Guelfa* (1538–1773), 2,500 maps; *Piante delle Possessioni* (mid-seventeenth century to 1822), 4,650 maps; *"Piante Antiche" dell'Archivio dei Confini,* 121 maps; *Carte Nautiche e geografiche* (14th–18th century), 22 items; *Piante della Direzione Generale dell'Ufficio delle Acque e Strade* (17th–19th century), 100 maps; *Piante delle Fabbriche* (1548–1851), 600 maps; *Piante Miscellane* (16th–19th century), 650 maps; *Regie rendite* (18th–19th century), 150 maps.

27. ASFi, *Piante dei Capitani di Parte Guelfa* 8:7, "Podere di Montechio" (1530s); 8:8, "Bottega ad uso di Montechio" (1557); 14:12, "Meta del podere posto nella Badia di Zuola Capitanato di Castrecaro" (1538); 19:47, "Sbozzo di Pianta del podere della Colombaja" (1564); carte sciolte 38 manca, "Pianta della Toscana fatta da Gerolamo Bellamato" (1536); *"Piante Antiche" dell'Archivio dei Confini* 9 (1551) "Pianta del corso della Chiana e suoi affluenti dal Ponte di Carnaida fino allo sbocco in Arno al Ponte a Briano."

28. R. Burr Litchfield, *Emergence of a Bureaucracy: The Florentine Patricians, 1530–1790* (Princeton, N.J.: Princeton University Press, 1986).

29. Ettore Verga, *Catalogo ragionato della raccolta cartografica e saggio storico sulla cartografia milanese* (Milan, 1911), 23–30, 72–73, 83–144.

30. *Guida generale degli archivi di stato italiani,* 3 vols. (Rome, 1981–), 2:891–991.

31. Roberto Almagià, *Documenti cartografici dello Stato Pontificio editi dalla Biblioteca Apostiolica Vaticana* (Vatican City: Biblioteca Apostolica Vaticana, 1960), 3.

32. Roberto Almagià, *Carte Geografiche a stampa di particolare pregio o rarità dei secoli XVI e XVII esisenti nella Biblioteca Apostolica Vaticana,* vol. 2 of *Monumenta Cartographica Vaticana,* v.

33. Numa Broc, *La géographie de la Renaissance (1420–1620)* (Paris: Bibliothèque Nationale, 1980), 12–13.

34. R. V. Tooley, "Maps in Italian Atlases of the Sixteenth Century," *Imago Mundi* 3 (1964): 12–47.

35. M. A. Martullo Arpago, et al., eds., *Fonti cartografiche nell'Archivio di Stato di Napoli* (Naples: Ministero per I Beni Culturali e Ambientali, 1987), 21. Note that this catalog was checked against a 1930 inventory to compensate for the World War II archival destruction.

36. Archivio di Stato, Foggia (hereafter ASFg), *Dogana delle pecore,* serie I, fas. 17 bis in two copies, transcribes the 1533 reintegration from Archivio di Stato, Naples (hereafter ASN), *Dipendenze della Sommaria,* serie II, 57/130, fas. 14 with later copies in fas. 15, and ASN, *Sommaria Diversi,* serie I, fas. 103 I for the 1548–53 reintegration; and fas. 17 for the early sheep-walk surveys.

37. Biblioteca Nacional, Madrid, MS. 2857, "Itinerario del Reyno de Napoles di tutto lo circuito del Regno cominciando dalla prima terra di marina, et circuendo tanto il mare come la terra fatto l'anno 1559."

38. Harvey, 62, 146, cites a map associated with the Humanist Giovanni Pontano (1422?–1503), signatory of the map.

39. Ernesto Mazzetti, ed., *Cartografia generale del Mezzogiorno e della Sicilia,* 2 vols. (Naples: Edizioni Scientifiche Italiane, 1972).

40. Biblioteca Angelica, Rome, *Bancone stampe,* n.s. 56. The Lafreri map is reproduced in Cesare de Seta, *Storia della città di Napoli dalle origini al Settecento* (Naples: Laterza, 1973), 178, tav. 37.

41. Vladimiro Valerio, "The Neapolitan Saxton and His Survey of the Kingdom of Naples." *The Map Collector* 18 (Mar. 1982): 14–17, explains how Cartaro worked with Stigliola from 1591.

42. Roberto Almagià, "Note sulla cartografia dell'Italia nei secoli xv e xvi," *Rendiconti della Classe di Scienze Morale, Storiche, e Filologiche,* ser. 8, 6 (January–February 1951), abstract, 1–8.

43. Mazzetti, 2: tav. 11.

44. Carmelo Colamonico, "La più antica carta regionale della Puglia," *Iapigia* 10, no. 2 (1939): 145–85. Mazzetti, 1:124.

45. Pasquale Villani, *Numerazioni dei fuochi e problemi demografici del Mezzogiorno in età moderna* (Naples: Istituto Grafico Italiano, 1973).

46. John A. Marino, "Creative Accounting in the Age of Philip II? Determining the 'Just' Rate of Interest,"

forthcoming, Archivo General de Simancas, *Estado, Nápoles*, leg. 1066, ff. 88, 89.

47. Cesare de' Seta, ed., *Cartografia della Città di Napoli*, 3 vols. (Naples: Edizioni Scientifiche Italiane, 1969), 2: tav. 7.

48. Giovanni Antonio Magini, *Italia* (Bologna, 1620); facs. ed. F. Maranelli (Amsterdam: Theatrum Orbis Terrarum, 1974). The original Cartaro maps, dated 1613, are preserved in Biblioteca Nazionale di Napoli, MS. XII. D. 100, c. 8, 10.

49. John A. Marino, *Pastoral Economics in the Kingdom of Naples* (Baltimore: Johns Hopkins University Press, 1988).

50. I have found only four early maps in the Neapolitan copy of the 1548–53 reintegration; see n. 36, above. ASN, *Sommaria Diversi*, serie I, fas. 103 I, after ff. 450 are sketch maps of specific plots. Only one is dated 1553.

51. Della Croce's 2000-×-1950-mm map was destroyed in World War II, but is poorly reproduced in Andrea Gaudiani, *Notizie per il buon governo*, ed. Pasquale Di Cicco (Foggia: Editrice Apulia, 1981), 147.

52. ASFg, *Dogana*, fas. 18.

53. ASFg, *Dogana*, fas. 28, c. 479.

54. ASFg, *Dogana*, fas. 16.

55. ASFg, *Dogana*, fas. 20. The volume is reproduced in Antonio and Nunzio Michele, *Atlante delle locazioni della dogana delle pecore di Foggia* (Lecce: Capone Editore, 1984).

56. Raffaele Colapietra, "Ambiente e territorio della dogana di Foggia a fine Seicento attraverso l'atlante Michele," *Studi e Ricerche Geografiche* 1 (1985): 96.

57. ASFg, *Dogana*, fas. 20, nn. 3. Marino, *Pastoral Economics*, 164–166.

58. Francesco Nicola De Dominicis, *Lo stato politico ed economico della Dogana della Mena delle pecore di Puglia*, 3 vols. (Naples: Vincenzo Flauto, 1781), 1:358 (30 July 1574), item 22. Translation mine.

59. Ibid., 1:370 (1 June 1575).

60. Ibid., 1:358, 370, item 23.

61. Stefano Di Stefano, *La ragion pastorale*, 2 vols. (Naples: Domenico Roselli, 1731), 2:211.

62. Ibid., 2:215.

63. Ibid., 2:216.

64. Fernand Braudel, *The Mediterranean and the Mediterranean World in the Age of Philip II*, trans. Siân Reynolds, 2 vols. (New York: Harper & Row, 1972), 1:456–58.

65. Mazzetti, 1:88, 143.

66. Alessio De Sariis, *Codice delle leggi del Regno di Napoli*, 9 vols. (Naples, 1794), 4:99.

67. Di Stefano, 2:210.

68. Richard A. Goldthwaite, "Schools and Teachers of Commercial Arithmetic in Renaissance Florence," *Journal of European Economic History* 1 (1972): 418–33; and Paul F. Grendler, *Schooling in Renaissance Italy: Literacy and Learning, 1300–1600* (Baltimore: Johns Hopkins University Press, 1989).

69. Andrea da Mosto, *L'Archivio di stato di Venezia: Indice generale, storico, descrittivo, ed analitico* (Rome: Biblioteca d'Arte Editrice, 1937), 168–74. In 1545 an inspection of rural lands recommended the creation of this permanent office, the Provveditori sopra Beni Inculti. Its major concern was water for agricultural production, both juridical issues over its possession and technical problems of its distribution.

70. Leandro Alberti, *Descrittione di tutta Italia* (Bologna: Anselmo Giaccarello, 1550). For an evaluation, see Eric Cochrane, *Historians and Historiography in the Italian Renaissance* (Chicago: University of Chicago Press, 1981), 305–8; n. 34, 558.

T W O

England I: Pageantry, Defense, and Government: Maps at Court to 1550

Peter Barber

The political and administrative use of maps by English monarchs and ministers seems to stretch back well into the Middle Ages. The creation of the so-called Gough map of England, now in the Bodleian Library, Oxford, has been plausibly attributed to the needs of mid-fourteenth-century government. Its creation certainly argues a degree of map consciousness in official circles at that time, and its sojourn among English state papers until the eighteenth century suggests that it was initially retained for administrative purposes.[1] The survival of a handful of fifteenth-century maps, or of sixteenth-century copies of earlier maps, of localities such as Dartmoor may indicate that on occasions certain departments of state, and particularly the Exchequer and the Duchy of Lancaster, spasmodically commissioned, utilized, or referred to maps before 1500.[2] If such traditions of map use did exist, however, they were extremely weak by the time Henry VII fought his way to the throne in 1485.

A far more vital tradition was the use of world maps, or *mappaemundi*, and of allegorical representations of the universe as symbols and props in court pageantry and propaganda. This stemmed from the *mappamundi*'s role in medieval society as an expression of history, religion, and legend as well of geography. Henry III is recorded as having had a *mappamundi* on the wall behind his throne in the Painted Chamber at Westminster. Other decorative schemes, such as that of the audience chamber of Countess Adela of Blois, sister of William the Conqueror, had the *mappamundi* on the floor, with scenes from biblical, classical, and contemporary history around the walls and the heavens on the ceiling—a still more striking example of the way in which maps, painting, and architecture served to place the monarch, or his relatives, in proper context, in a form of *tableau vivant*, when they appeared in public.[3] This tradition manifested a new vitality in the ceremonial of the fifteenth-century Burgundian court, which made a marked impact on the court of Charles the Bold's brother-in-law Edward IV, and it was further enriched by the pseudoantique pageants of the courts of Renaissance Italy, which, by the early sixteenth century, were being imitated throughout western Europe.[4]

I would like to thank Geoffrey Parker, Sarah Tyacke, and Helen Wallis for reading and criticizing this paper when it was in draft form.

Mapping under Henry VII

In the course of Henry VII's reign a degree of map consciousness can be discerned among two groups that had some contact with king and government: the Bristol merchants and the lawyers. The interest of the former in western exploration seems to have been fed by contemporary Italian portolan charts depicting the "isle of Brasil," "Antilia," and the fabled seven cities to the west of Ireland. Their association with John Cabot and the probable correspondence of at least some of their number with Columbus himself almost certainly involved the consultation of some of the latest maps and charts of the world.[5] On a much more humble level, lawyers and those involved in legal disputes began from about 1400, and in a very limited number of cases, to use sketch plans to illustrate, and when possible to clarify, disputes over rights, ownership, and jurisdiction. Roughly executed, unsophisticated, and undecorative as they usually were, the plans seem to have served their purpose, and eventually some were even called for and used as evidence in courts of law. Thus it was that they came to be preserved among the records of fledgling departments of state. This sort of map, one assumes, would have been relatively familiar to younger lawyers who were to become ministers under Henry VIII, such as Thomas More and Thomas Cromwell.[6]

Henry VII himself was certainly not unfamiliar with maps; a contemporary chronicler reported that John Cabot persuaded him to provide a victual a ship in 1498 for a voyage to Newfoundland and Nova Scotia "by a caart & othir demonstracions reasonable."[7] In a more conventional manner, the library he inherited from Edward IV contained encyclopedic works, such as Corbechon's *Proprieté des Choses,* with world maps among the illustrations. It is also likely that some of the numerous *mappaemundi* that adorned the palaces of Henry VIII in 1547 had been inherited from Henry VII.[8] A form of map was used on at least one occasion as a vehicle of royal propaganda: the festivities celebrating Katherine of Aragon's entry into London in 1501 included a pageant in which, in an effort to compare her betrothed with the sun, a figure representing Arthur, Prince of Wales, was shown enthroned in majesty at the center of the universe.[9] This and the mention of John Cabot's showing Henry a "caart" appear to be the only known examples of any form of active map use at the court of the first Tudor. That he or his ministers ever made much use of maps for government or administration is, indeed, extremely unlikely. For despite some contact with Renaissance figures and a belated grasp of certain modern political notions, Henry's administrative and governmental methods were traditional. King and country were rather backward by German and Italian standards in respect of the new geography and cartography in the early sixteenth century, and the regular use of maps would have been out of keeping with the essentially conservative attitude of Henry and his ministers.[10]

Developments under Henry VIII

The reign of Henry VIII proved a watershed in the history of map consciousness and map use by king and government in England, but such a development would have seemed improbable at the time of his accession. The young king had, in the words of one of the closest students of his reign, Garrett Mattingly, "learnt his geography and politics mostly out of Froissart and Malory,"[11] and his enthusiasm for maps and plans became apparent only very gradually. In May 1512, Sebastian Cabot was paid twenty-six shillings and eight pence "for making of a carde of Gascoigne and Guyon [Gascony and Guienne]"; a couple of months later, the marquess of Dorset led an abortive English attempt to invade Guienne from San Sebastian.[12] In

1514 England itself was threatened by invasion from, and was contemplating the invasion of, France. It may have been at this time that a long and detailed pictorial map of the north coast of Kent was painted.[13] It was almost certainly produced locally on the basis of information supplied by a pilot. There can be little doubt, however, that it was intended for and possibly commissioned by the government, in order to illustrate the local state of vulnerability and preparedness. The nature of the shore, and whether it was favorable or otherwise for landings, is indicated, but so too are existing precautions against invasion like a local beacon and church towers, and a barrier of ships at Whitstable. Indeed, it has been suggested that the map was also partly intended to persuade the king to select Faversham as a port of departure for an invasion of France: notes give soundings for the ship-bearing capacity of the river, and what is possibly a powder mill with a crane for loading is also shown.[14]

Against this rather slight evidence of map use by the government has to be set the absence of any other surviving maps that clearly served military or administrative purposes,[15] and, more tellingly, Henry VIII's and Cardinal Wolsey's treatment of Sebastian Cabot, undoubtedly the most talented cartographer then in the employ of the English crown. The king had scarcely availed himself of his skills before, in September 1512, Cabot was allowed to leave Henry's service for that of Spain, where he eventually rose to be *piloto mayor* in the Casa de Contratación in Seville.[16] Although Spain and England were fairly steady allies during the early part of Henry VIII's reign, and although Cabot seems to have been a difficult character, it is surprising that with his geographical knowledge of Europe and the world beyond, and its commercial and political implications, he should have been allowed to depart so easily. It argues, at the very least, the king and ministers' lack of understanding of the value of maps and cartographical information for government.

The New Generation of Statesmen

The next decade saw the beginnings of a change in attitude. In part this was due to the change of generations. In the 1520s the reins of power began to pass to men born after 1480, who had been at their most impressionable in the 1490s and 1500s. Copies of Ptolemy's *Geographia* were then coming off the presses of Europe, illustrated printed books were appearing in greater numbers, and knowledge of the discoveries in America, Africa, and India was spreading throughout Europe—though for some time there were few generally available printed maps to illustrate those discoveries.[17] In the same period the English court was visited by those champions of the Northern and Southern Renaissance Erasmus (1499) and Castiglione (1503),[18] while involvement as mercenaries in the Italian wars enabled younger and often poorer Englishmen such as Thomas Cromwell to become familiar with the Italian environment and the Italians' novel administrative methods.[19] These included the use of maps, which had been employed by the Venetians since at least the middle of the preceding century.[20] These men, then, who in the 1520s generally occupied second-rank positions in government, had been brought up in a world where maps were becoming more commonplace. A few of this generation probably unconsciously developed a manner of thinking graphically, which was not shared by contemporaries such as Cardinal Wolsey, who were just a few years older but who had been adults in the 1490s.

The case should not be overstated, for we are talking only of a handful of intellectually curious, better educated, or widely traveled men.[21] And cartography does not seem to have been a passion or of any special importance for most of them. Still less were most of them, unlike the Bristol merchants, particularly concerned with accurate measurement or technical skill. But they were familiar with two-dimensional representations of the outside world, whether they called them "plattes," "plottes," "images," "peinctings," or "mappes," and had some inkling of their possible uses.

Members of the older generation were, to some extent, swept along with the new tide, though one suspects without the natural understanding of maps that characterized some of the younger men. Wolsey (1475–1530), who was fitfully interested in overseas exploration and toyed with the idea of reemploying Sebastian Cabot in 1520–21 on an expedition in search of the Northwest Passage and the wealth of Cathay,[22] is known to have designed and supervised the creation of the magnificent tent intended for the use of Henry VIII at the Field of the Cloth of Gold in June 1520 by way of a "plot" that he retained afterward as a souvenir.[23] More significantly, he instituted a mathematics lecture at Oxford, in recognition of the newly appreciated importance of a subject that, in the coming decades, was to enable mapmaking to become more precise and less pictorial.[24]

Thomas More (1744/8–1535), who as lord chancellor was briefly one of Henry VIII's chief ministers, took a lively intellectual interest in the great discoveries.[25] There is no evidence to suggest that this extended to an interest in maps. The only maps indubitably associated with him are those adorning early editions of his best-known work, *Utopia*, but these may well have been concessions made by him to the demands of his map-conscious European publishers and to the thirst of European readers for illustrated travel accounts (a tradition that extended back over several centuries). For, as More undoubtedly realized, the measurements he gave for the island of Utopia defied the rules of mathematics and thus the possibility of accurate cartographic representation.[26] While More may or may not have been interested in maps, however, several members of his close circle were:[27] notably his brother-in-law John Rastell (d. 1536), Thomas Elyot (c. 1490–1546), George Lily, the son of one of More's earliest friends, William Lily (1468?–1522), and the German astronomer Nikolaus Kratzer (1468–1550[?]), who was recruited by the king as court astronomer and horologist in 1519.[28] Kratzer's interest in land measurement and mapping is well documented, and he came from southern Germany, an area that had been associated since the 1480s with the production and publication of relatively sophisticated maps and may have had a tradition of local mapping that extended back even earlier than that.[29] It could well have been Kratzer who awakened an awareness of the utility of maps in the circle associated with More and in the king himself.

In the 1520s Henry probably commissioned the depictions of his early successes in France, notably the portrayal of the meeting with Emperor Maximilian during the siege of Thérouanne (1513), which still hangs in Hampton Court Palace (color plate 1). Although in technique these are paintings, in their particularized topographical content and in the method of perspective employed they are in effect highly polished and embellished picture maps, and their commemorative purpose is identical to that of many of the printed and manuscript battle plans that begin to appear in profusion in Europe from the 1520s.[30] The map consciousness that the paintings reveal was to become ever more pronounced as the years proceeded.

Meanwhile, court pageantry continued to employ maps in the medieval manner. "Cosmological ceilings"—depictions of the heavens and/or of the four elements painted onto the ceilings of temporary structures, whether tents or banqueting chambers—seem particularly to have been associated, as Sydney Anglo has pointed out, with grand diplomatic occasions such as the meeting of Henry VIII and Charles V in Calais in July 1520 or the entry of Charles V into London two years later, when the propaganda of which they were a part became a tool of policy.[31] Even here, however, the 1520s saw a significant change in the type of map that was employed.

The change can be attributed to the organizer of the pageants, John Rastell. Rastell, a printer, lawyer, and man of wide-ranging enthusiasms and activities, was particularly interested in cosmology and geography.[32] This interest extended far beyond the knowledge of medieval theories exemplified in the court pageants of the early 1520s. In 1517, with some degree of royal patronage, he had tried to lead an expedition to colonize North America, and though it got no further than the Irish coast before ending in mutiny and chaos,[33] his interest in the "new geography," and in modern maps as a means of propaganda for it, remained lively. In 1519 he produced an entertainment, the "New Interlude and Mery of the iiij elements," in which the characters displayed maps and instruments on stage. It was, as Helen Wallis has pointed out, "a method of popular (and probably royal) education" in cartography as well as, more generally, a means of encouraging the audience to explore the world and publish scientific works in English.[34] The "interlude" may have been written with Kratzer's assistance, and in 1527 they collaborated again—this time with additional help from Hans Holbein and the king's lesser-known painter, Vincenzo Volpe—in organizing festivities in Greenwich to celebrate the formal conclusion of peace between France and England. The purpose served by the pageantry was much the same as ever: glorification of the king, more or less tactfully, at the expense of his fellow monarchs. The use and style of the cartography, however, seem to have marked a sea change and suggest that maps were finally becoming familiar to the king and his ministers.

The festivities included two maps painted by Holbein that, judging from contemporary descriptions, were very different from the allegorical maps and paintings of past pageants. First there was—somewhat surprisingly in view of the occasion—a "payncting of the plot of Tirwan . . . in grete," showing Thérouanne under siege by the English in 1513, with "the very manner of every man's camp connyngly wrought"—clearly based on an existing, detailed "plot" of the type that was to become very familiar. Second, there was a ceiling "conninglie made by the Kinges Astronimer . . . [with] the hole earth environed with the sea, like a very Mappe or Carte"—or, in the words of the Venetian ambassador, a "mappamondo" but one unlike the earlier examples with their generalized landscapes, in which could be read "tutti li nomi de le provincie principali."[35]

The grand "mappe" had clearly become established on the walls and ceilings of Henry VIII's palaces, and it was not long before the smaller "platte" found its place in the studies of the king and his ministers. The 1520s saw the publication of Machiavelli's *Arte della guerra* and Castiglione's *Il cortegiano*,[36] both of which laid some emphasis on the necessity for maps in defense and warfare in a tradition that, they correctly argued, extended back to Ptolemy and Strabo, and indeed beyond to the campaigns of Alexander the Great.[37] The books made a considerable impact.[38] In 1531 Sir Thomas Elyot published *The Boke named the Governour,* which was clearly influenced to some extent by Castiglione and other Italian and English writers of the Renaissance but contained much that was fresh and independent. The book was intended to provide guidelines for the education of future "governors," or ministers and other servants of the Crown, and on the qualities required in them. Elyot could speak with experience, having until recently served as secretary to the Privy Council under Wolsey. Significantly, he was also a close friend of Thomas More and was acquainted with Kratzer, Rastell, and Holbein, who drew his portrait.[39]

In his work Elyot pays considerable attention to the value of maps to ministers at both a personal and a business level. Concentrating his advice in the sections dealing with painting and cosmography, he adopted a tone similar to Machiavelli and Castiglione's and noted that "by the feat of portraiture or painting a captain may describe the country of his adversary, whereby he shall eschew the dangerous passages with his host or navy; also perceive the places of advantage, the form of embattling of his enemies, the situation of his camp for his most surety, the strength or weakness of the town or fortress which he intendeth to assault."[40] After all, as Castiglione noted, these were "matters, the which though a manne were liable to keep in mynde (and that is a harde matter to doe) yet can he not shewe them to others"[41] without a map or painting. Elyot offered the example of Alexander the Great, who "caused the countries whereunto he purposed any enterprise diligently and cunningly to be described and painted, that beholding the picture he might perceive which places were most dangerous and where he and his host might have most easy and convenable passage. Semblable did the Romans in the rebellion of France and the insurrection of their confederates, setting up a table openly, wherein Italy was painted, to the extent that the people looking in it should reason and consult in which places it were best to resist or invade the enemies." Indeed, the fate of Cyrus of Persia, Crassus, "and divers other valiant and expert captains which have lost themselves and all their army" illustrated the dangers of not using maps for warfare.[42]

But maps, for Elyot, were also means for the private edification and pleasure of the governor. In words that anticipate Robert Burton's in the *Anatomy of Melancholy* (1621), and seem to describe the beautiful planispheres that were being produced in Italy and Portugal and were soon to be made in Dieppe, he eulogized

> the pleasure . . . in one hour to behold those realms, cities, seas, rivers and mountains that unneth [scarcely] in an old man's life cannot be journeyed and pursued; what incredible delight is taken in beholding the diversities of people, beasts, fowls, fishes, trees, fruits and herbs: to know the sundry manners and conditions of people, and the variety of their natures and that in a warm study or parlour, without peril of the sea or danger of long and painful journeys: I cannot tell what more pleasure should happen to a gentle wit than to behold in his house everything that within all the world is contained.[43]

Maps were not, however, meant only for war or quiet pleasure. First—and earliest in the future governor's life—they were "most commodious and necessary," once he had reached the age of fourteen and had learned the lessons of the classical poets and rhetoric, "to prepare the child to understanding of histories, which, being replenished with the names of countries and towns unknown to the reader, do make the history tedious or else the less pleasant, so if they be in any wise known, it increaseth an inexplicable delectation. It shall be therefore, and also for refreshing the wit, a convenient lesson to behold the old tables of Ptolemy, wherein all the world is painted, having first some introduction into the sphere, whereof now of late he made very good treatises, and more plain and easy to learn than was wont to be."[44] Elsewhere he elaborates on this. For as regards geometry, astronomy, and cosmography, "called in English the description of the world . . . a man shall more profit in one week by figures and charts well and perfectly made then he shall by the only reading or hearing the rules of that science by the space of half a year at the least; whereof the later writers deserve no small commendation which added to the authors of those sciences apt and proper figures."[45]

Once the governor had grown to adulthood this use of maps could be adapted "to the administration of other serious studies and business" great and small.[46] Wars abroad were one obvious field, as we have seen, but—and here Elyot appears to be breaking new ground—he also advocates that "in visiting his own dominions [the governor] shall set them out in figure, in such wise that at his eye shall appear to him where he shall employ his study and treasure, as well for the safeguard of his country, as for the commodity and honour thereof, having at all times in his sight the surety and feebleness, advancement and hindrance of the same."[47] Here, surely, he is speaking as an experienced "governor" himself, as he is when, in discussing the role of maps and plans in short-term planning, he mentions the

> pleasure and also utility . . . to a man . . . to express the figure of the work that he purposeth according as he hath conceived it in his fantasy[.] Wherein by often amending and correcting he finally shall so perfect the work unto his purpose that there shall neither ensue any repentance nor in the employment of his money he shall be by other deceived. . . . Where . . . that which is called the grace of the thing is perfectly expressed, that thing more persuadeth and stirreth the beholder and sooner instructeth him, than the declaration in writing or speaking doth the reader or hearer.[48]

Elyot's words, expressing as they do the realities of his time and place, were to have an enormous influence on the next and following generations, for his book was reprinted at least seven times before the 1850s. It would be surprising if Elizabeth I and her great minister, Lord Burghley—an eleven-year-old when the book was written—were not familiar with it, particularly in Burghley's case since, as will be seen, his use of maps closely parallels the theories outlined in *The Boke named the Governour.* Shakespeare, too, seems to have known and used it,[49] and it was purchased by the tutors of James VI of Scotland, who was to become James I of England, as one of the books for him to study.[50]

In the context of the time of writing, however, the *Boke* was a prologue to and herald of what has justly been called a revolution in English cartography and the use of maps that accompanied the better-known revolutions of the 1530s in religion, national self-awareness, and government.[51] From 1530, maps begin to appear regularly with other, more traditional types of documents, as tools in the processes of government and administration. Indeed, the revolution in government and the cartographic revolution were linked. For maps were to become one means by which Cromwell's (and later Burghley's) objective of enhancing royal authority and control throughout the land could be achieved, and it was thus that they made their most profound contribution to Tudor government.[52] It is significant that one of the leading figures in the cartographic revolution, Richard Lee, was at the time regarded as being one of Thomas Cromwell's "creatures."[53] It is also fitting that the famous double portrait of Jean de Dinteville, the French envoy to Henry VIII, and his friend Georges de Selve, better known as "the Ambassadors," with its prominent display of globes, astronomical instruments, and mathematical treatises, should have been painted in the English court by Holbein in the course of the decade (April 1533).[54] But while the painting demonstrates the extent to which cartography, as exemplified in globes, had been assimilated into European courtly society, the nature of the revolution in England centered on a quite different sort of map: the humble and frequently inelegant manuscript plat.

The Emergence of Manuscript Plats

The context was the crisis provoked by Henry VIII's divorce of Catherine of Aragon, his break with Rome, and his consequent alienation of Emperor Charles V and, to a lesser extent, of Francis I of France. Since 1485 England had normally been able to rely on an alliance with one of these rulers—generally but not invariably with the Habsburgs—to ward off any threat from the other.[55] England's isolation after 1529 was, therefore, alarming—and particularly so when Charles and Francis appeared to be on the point of resolving their own considerable differences and allying against Henry, as was the case in autumn 1533 and again in the summer, autumn, and winter of 1538–39. England's defenses and its communications with Calais, its last bridgehead on the European continent, had to be looked to. The problems were daunting. Complacency, a lack of central control, local parsimony, and the forces of nature that threatened to silt up vital harbors and channels, such as Dover and Sandwich, had left England ill-prepared to face a challenge from France. As for a challenge from Spain to Ireland and the southwest coasts of England, it seems hardly to have been thought of previously.[56]

In 1530 Henry commissioned a map of Rye and its surroundings from Vincenzo Volpe.[57] Two years later, Dover Corporation, in an attempt to secure government aid in improving the harbor and defenses, paid Vincenzo Volpe twenty-two shillings for preparing a plat showing its proposals for a second, inner harbor.[58] This elegant presentation map, intended for the king's eyes, survives.[59] It was the first of a long line. For although these particular plans were not acted on, the attention of Henry and of Thomas Cromwell did become focused on Dover and on Calais, with the same cartographers at times employed on the mapping of both places.[60] Over the following decade, proposal followed proposal, some realistic, some fantastic, but most accompanied by plats as varied in style and skill as the reports to which they were appended.[61] In the same years the artist gradually gave way, as the creator of plats, sometimes to the foreign-born military engineer, but more often to the native master-mason who was being transformed under the pressure of circumstances from medieval builder-architect into military engineer and surveyor-cartographer.[62]

The crisis of 1533 passed quickly. That of

1538–39 seemed altogether more serious, with a Habsburg-Valois reconciliation in June 1538 being accompanied by the promulgation of a papal bull excommunicating Henry VIII and much loose warlike talk from English exiles in Rome. Fortunately for Henry, the crisis coincided with the dissolution of the monasteries in England, which placed in his hands the enormous revenues of the monastic lands. For the only time between 1485 and 1660 the Crown had significant financial means at its disposal—means that, this once, outdid those of the English nobility and merchants. Henry's response to the challenge could be and was extravagant diplomatically—the alliance with the Protestant princes of Germany symbolized in his marriage with Anne of Cleves—and militarily. Traditional means such as musters were resorted to, but side by side with them and of vastly greater importance was the decision to modernize and enormously expand England's defenses. This was to result in England's largest military building program before the nineteenth century, and the creation of many of the earliest surviving local and regional maps of the kingdom.[63] As a by-product it was also, once and for all, to establish maps and plans as one of the English government's everyday tools in the formulation of policy and in the processes of administration.

There had been governmental calls for surveys of the English coasts in 1519 and again in 1533,[64] but now, backed with the resources of the monastic lands channeled through the Court of Augmentations, action was taken. In February 1539, at Cromwell's instigation, Henry commissioned certain "sadde and expert men of every shire in Ingland [and Wales] beyng nere the see . . . to viewe all the places alongest the secost wher any daunger of invasions ys like to be and to certifie the sayd daungers and also best advises for the fortificacion thereof."[65] At the same time Henry named heads of construction teams for forts along the Thames, in Kent, and in Hampshire—places that had received royal visits in the previous summer. The commissioners must have surveyed and, in certain cases, mapped the coasts, if only roughly, in a mere couple of months, for by April 1539 Thomas Cromwell was able to compile a list, presumably on the basis of the surveys, written and cartographic, of twenty-eight sites "where fortification is to be made."[66] Presentation copies of several of the maps still survive in the Cotton collection in the British Library.[67] They seem to be based on pilots' surveys, local knowledge, and, in certain cases—such as the famous long view of the southwest coast of England (Cotton Augustus I.i.35, 36, 38, 39)—preexisting town views of the kind that had originated in Italy and the Netherlands in the previous century. Purely pictorial, they were not drawn to scale, being intended to focus the attention of the king and his advisers on the areas presenting the greatest dangers of invasion, notably inlets and sandy bays, and on the existing or proposed defense mechanisms such as beacons and church towers. Even so, a few nonmilitary features, such as the location of royal and noble deer parks, were included, giving the maps some potential administrative as well as military function.[68]

These maps were retained for reference, and some were later annotated to indicate the progress made with the proposed coastal forts.[69] They marked only the first stage of the Henrician cartographic revolution, however, and on their traces quickly followed a profusion of plats of particular locations by military engineers. Each operation seems to have spawned a multitude of plats showing elevations and plans of the existing situation, proposals for improvement, and recent developments—most of them with a minimum of written explanation, since they were to be explained orally or accompanied written reports from which they have since become detached. For similar reasons, no differentiation is usually made in individual plats between representations of proposals and of the status quo.[70] Though the skill displayed and methods employed in these plats vary enormously, with different techniques—plan, elevation, bird's-eye view, measured drawing, and estimation—often found on the same sheet,[71] the overall impression is of surprising sophistication when they are compared to the earlier work of cartographer-artists such as Volpe.

It is all the more impressive when the differing backgrounds of their creators are borne in mind. Some, like Richard Lee and John Rogers, were masons by training,[72] others, like John à Borough of Northam in North Devon, were pilots,[73] while Richard Cavendish (Caundish) began his career as a master gunner.[74] Only a few, such as the Italian Giovanni Portinari and the Moravian Stephen von Haschenperg, were trained engineers.[75] None, with the exception of the former pilots, appears to have had any earlier experience of mapmaking.[76] But most made up for this, it would seem, with curiosity about and a readiness to emulate and amend continental maps and cartographical ideas. These were transmitted in a variety of ways. Some were officially communicated by the enemy, such as the map of Boulogne (Cotton Augustus I.ii.82) that was probably supplied to Richard Lee by French commissioners in May 1546.[77] Some, such as the maps of foreign towns, and the picture-views of sieges that decorated Henry VIII's palaces,[78] may have been presented by allies or, if printed, acquired by purchase. By 1546, leading French cartographers such as Nicholas de Nicolay (later to be cartographer to the French king) and Jean Rotz, were working for Henry VIII, and they may have served as examples and been a source of new ideas.[79] A couple of detailed charts of England and Scotland's coasts have survived and give an impression of the greater accuracy obtained by them (figure 2.1).[80] It is clear that Scottish scholars like the Earl of Angus's physician, Dr. Cromer, and the "Redshanks" clerk John Elder, who served Henry VIII from 1543, had some influence also in expanding knowledge of Scotland, which had previously been a terra incognita to the English. Their maps, which seem to have been based on expeditions and voyages made over the previous thirty years by a group associated with Elder and the navigator Alexander Lyndsay, are now lost. From at least 1546, however, they must have been available for study in Whitehall, and they inspired the precociously accurate maps of Scotland by Nicolay, Laurence Nowell, Rotz, and others which were produced in the following decades. Indeed, the original maps—which may have perished in the Whitehall fire of 1698—must have contributed to, and may help to explain, England's notable military successes in Scotland between 1544 and 1547.[81] Last, and throughout the period, there was the informal exchange of ideas between the Italians in French and English service with their "disegnos" and the native English[82]

These influences were undoubtedly the key to the amazing progress in technique and map types that characterized the early 1540s, but there were other factors. There was the sheer weight of the engineers' work load, which led them from place to place and from problem to problem in rapid succession, so that they soon became familiar with most aspects of large-scale mapping. There was the need to

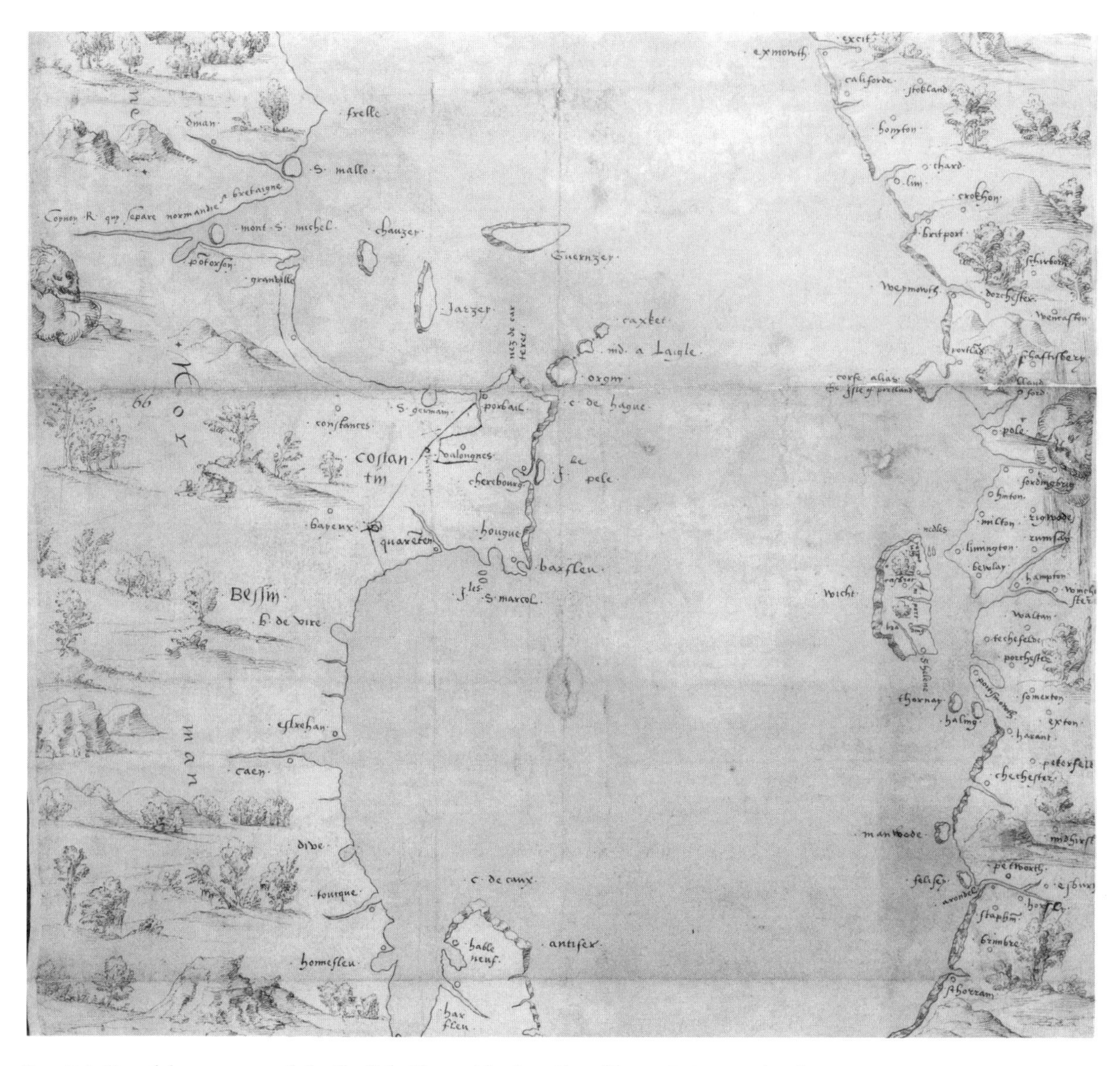

Fig. 2.1 Detail from a map of the English Channel by Jean Rotz(?), c. 1542. British Library (BL) Cotton MS Augustus I.ii, 65, 66

hold one's own in the frequent encounters with the king and his advisers after having been summoned to court to explain one's plats in person; and the king tended to put his views forcibly. Henry and his courtiers and ministers, through their ever-increasing familiarity with maps and plans (it should not be forgotten that Henry's enormous palace-building program spawned a multitude of architectural plans), clearly became ever more aware of their versa-

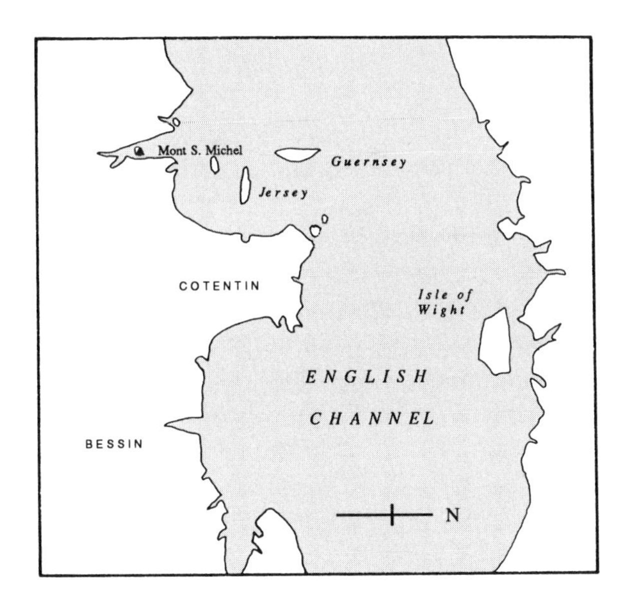

tility and potential for government, which resulted in increasingly diverse demands being made on the engineers by a ruling group that was beginning to elucidate its problems by graphic means.[83]

Thus in the seven years preceding Henry VIII's death, maps were produced to meet a wide variety of purposes other than the simple planning of England's defenses. On repeated occasions, engineers were sent as spies,[84] and several of the maps thus produced survive, like John Rogers's plans of the French fortifications of Outreau opposite Boulogne, Giovanni di Rossetti's plan of the French-held fort of Ardres on the border of the Calais Pale, and the sketchy chart of the Dutch coast secretly executed by John à Borough in September 1539, when Henry was contemplating the possibility of bringing his future bride Anne of Cleves to England by sea from Guelderland.[85] There were maps of colonization, such as the "plaict" for a proposed settlement within the Pale of Calais which was sent by Henry VIII to his commissioners in September 1541.[86] By the spring of 1546, engineers were attached to the English team that negotiated over the future of Boulogne and the Boulonnois with the French, and in this capacity they produced maps of the area to serve as a basis for discussion and to keep the king informed of the course of the negotiations (figure 2.2).[87] Among the plans is one that must rate as among the earliest of the detailed "frontier" maps. It shows Boulogne and its neighborhood, and on it Rogers indicated what were probably the maximum wishes of the English commissioners in the form of a line of red dots.[88] By this time Henry and his advisers had become sufficiently sophisticated to criticize one of the maps by Rogers (probably Cotton Augustus I.ii.75) on the grounds that by appearing to accept the French interpretation of the course of the River Liane it was likely to cost England territory.[89]

By the early 1540s maps were being demanded to illustrate all manner of written texts, in the manner approved by Elyot. Sketch plans were enclosed in letters describing sieges.[90] Plans were required for administrative purposes from subjects far removed from the political and cartographic front line. In 1541, in pursuance of the act of abolishing many of the places of sanctuary that had survived from the Middle Ages, the mayor of Norwich was ordered to produce a plat showing the bounds of Norwich as a sanctuary. It is significant that he was successful in commissioning the map required, which still survives in the Public Record Office.[91]

The period 1539–47 also saw the first appearance in England of symbols and of explicit scales on large-scale maps of small areas.[92] These were of immense value to absent decision-makers when reaching decisions that depended on an accurate knowledge of size, distance, terrain, and patterns of settlement. John

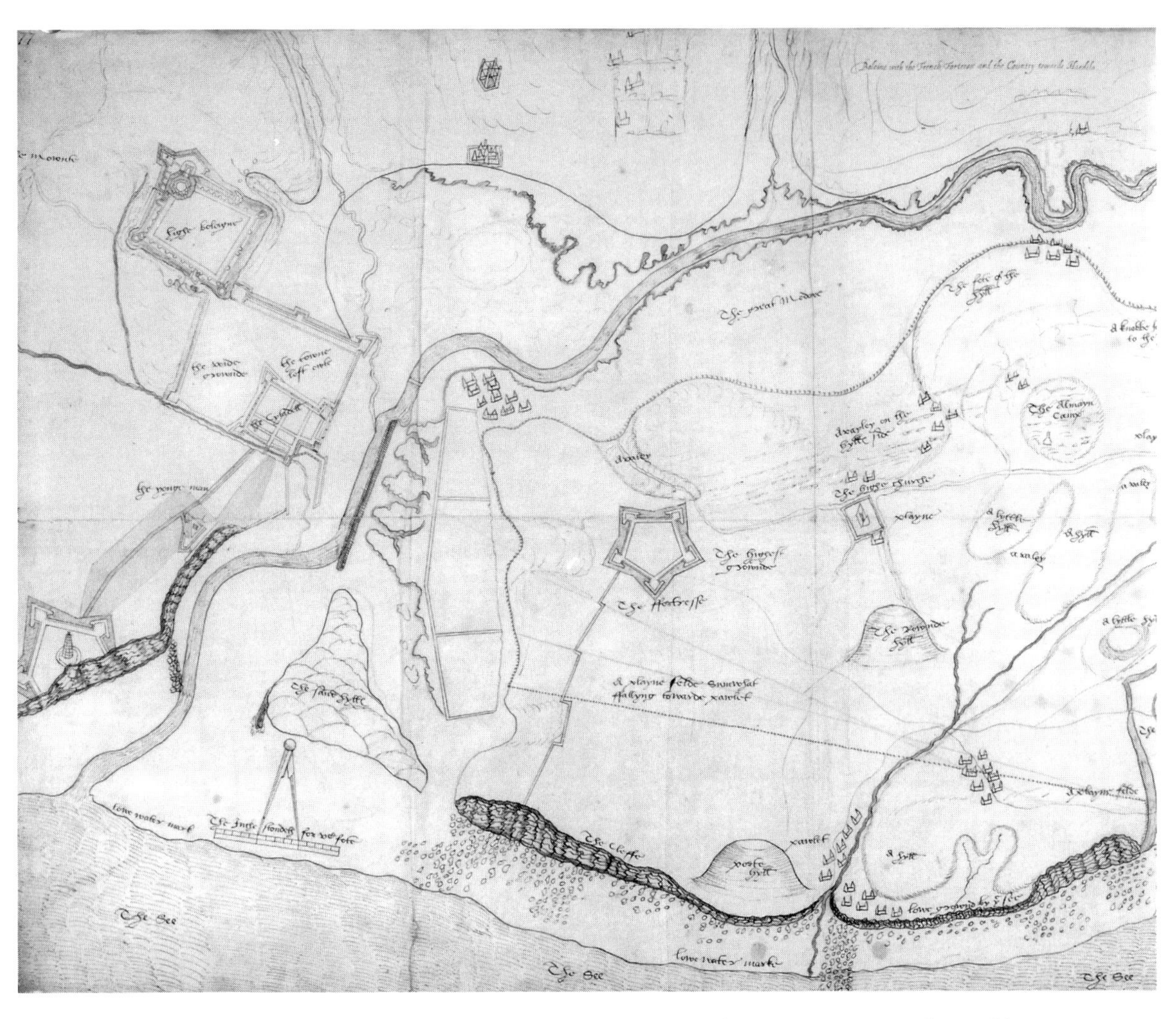

Fig. 2.2 John Rogers, "Boleine [Boulogne] with the French Fortresse and the Country towards Hardilo," c. 1546. BL Cotton MS Augustus I.ii, 77

Rogers referred to the innovation of scale with barely suppressed pride in his correspondence with secretaries of state.[93] But it should be borne in mind that the sort of precision to be found in the scale maps was not often required by decision-makers, who could make do perfectly adequately on most occasions with rough-and-ready picture or position maps, lacking scale or standardized conventional signs. These continued to be produced and used for decades—and centuries—to come,[94] even if within a few years some bureaucrats, like the clerk who annotated Richard Lee's map of Orwell Haven with a meaningless scale bar, felt shamed by them.[95]

There was one major area where maps

might, perhaps, have been expected to have made a significant contribution but did not. This was in the survey of the monastic lands themselves. Not that they were left unsurveyed; the surveys were, however, in accordance with medieval tradition, entirely written and probably not based on measurement.[96] The first English textbook on geometrical land-measurement and surveying had, in fact, been written by a canon of the Augustinian priory of Merton, Richard Benese, and published in 1537.[97] Benese was later to become a surveyor of works for the royal palaces at Hampton Court, Oatlands, and Nonesuch (1539–41), but he seems not to have had drawn plans in mind.[98] The first measured estate plans appeared in England only some forty years later, though they were preceded, after about 1540, by a growing number of unmeasured plans.[99]

Kings and ministers displayed varying degrees of enthusiasm about all this mapping activity. Although Henry seems to have shown no marked interest in cartography until the 1530s, he had long been fascinated by the related disciplines of mathematics, navigation, and astronomy.[100] From at least the late 1530s, his contact with maps was intense and a source of great pleasure and excitement. During one of the Turkish wars, in 1538, the French ambassador found the king poring over "a map of the very place where the Armada of the Levant was."[101] In the next year he was reported by the Lord High Admiral as being "marvellously inflamed" with à Borough's "Carte" showing the proposed passage of his future wife, Anne of Cleves, from North Holland to England—"supposing many things to be done thereon."[102] A handsome map or atlas combining the latest cartographic techniques with colorful charts embodying the latest discoveries and depictions of people in far distant lands undoubtedly also delighted Henry, despite his lack of real commitment to overseas exploration, and there can be little doubt that he welcomed the magnificent *Boke of Idrography* presented to him in 1542, as his testimonial, by the Franco-Scottish hydrographer Jean Rotz.[103]

Henry's excitement at maps and ability to draw the most from them is best seen, however, in the context of military maps. It should not be forgotten that Rotz was employed not to execute exotic charts of distant parts of the globe, but to map the coasts and ports of England for defensive purposes.[104] In his robust way, Henry was in the habit of commissioning the drafting of plats of fortifications, harbors, and other defensive systems, embodying his own ideas and based on the evidence of his own eyes, and of combating with further plats the ideas contained in plats produced and explained to him by his engineers. In the process of this dialectic the original ideas were refined, in the manner outlined by Elyot, though generally in accordance with the king's views.[105] Henry also seems by the end of his reign to have tried to supervise the conduct of sieges from afar on the basis of information contained in maps, as was apparently the case with Landrecy in 1543.[106] There are suggestions, in places, that at times the king actually drafted some plans himself—something Elyot rather frowned on as too plebeian for a governor to practice openly.[107] But if any such plans were executed, none have so far been identified.

Maps and Ministers

Few if any of Henry's ministers were as enthusiastic about maps as their master. Sir William Paget, secretary of state in Henry's closing years—a post he continued to hold under Edward VI and Mary—was clearly, however, quite interested. Comments on maps are to be found in his letters, as for instance in May 1546, when he was moved to describe a French plat of Boulogne (tentatively identified as Cotton Augustus I.ii.82) as the fairest and best that he had ever seen,[108] and it is interesting to learn that in February 1553 one of the walls of his London home was adorned with a "mappe" of the French king's house—possibly a precursor of the painted maps of French palaces and their surrounding estates of about 1600 that are still to be seen in the Galerie des Cerfs at Fontainebleau.[109] There are also signs of interest among lesser ministers and generals. Henry Manners, second earl of Rutland, who was to serve as lord warden of the Eastern and Middle Scottish marches under Edward VI, retained the numerous plats of Scottish castles and their surrounding country, executed for him partly by a certain William Hunt, which are still housed in the ancestral home, Belvoir Castle.[110] A similar degree of interest on the part of Lord Fitzwilliam, Henry's Lord High Admiral in the 1530s, can be deduced from his letters, though surviving cartographic evidence is lacking.[111] The attitude of Henry's most important minister after 1532, Thomas Cromwell, toward maps and plats seems to have been businesslike. Sufficient references exist in the *Letters and Papers* of Henry's reign to make it certain that Cromwell was very familiar with them. He does not seem to have annotated any of the plats that have survived, however, and no examples of the specific ways in which he used maps or of his comments on them have yet come to light.[112]

A leading member of the younger generation of Henry's advisers was Edward Seymour, earl of Hertford and brother of the king's third and best-loved wife, Jane Seymour. After 1540 he held important offices in English-occupied France, and in 1544 he led the English army that invaded Scotland. Contemporary letters and accounts show him constantly evaluating and commissioning plats for the information of his master. In the process he came into close contact with the cartographers themselves.[113] Like most English administrators, he had differences of opinion with the capable but independent-minded John Rogers, who by the mid-1540s was the king's favorite engineer.[114] Richard Lee seems to have been more amenable, and the earliest surviving picture map of Edinburgh may have been commissioned from him by Hertford to commemorate its capture in 1544.[115]

Mapping under Edward VI

On Henry's death in January 1547, Hertford became "protector" to his nephew, the young Edward VI, and promptly had himself created duke of Somerset. The three years of his regency mark a continuation of Henry's reign as far as foreign policy and map use are concerned. Following his stunning victory over the Scots at Pinkie on 10 September 1547, Somerset used maps and plans to exploit his advantage politically and psychologically as well as militarily. In 1548 he chose to display a now lost plat of England's impressive new fortification of Haddington in Scotland prominently on the wall of the chamber in which he was entertaining the French ambassador.[116] The ambassador took note of this polite warning, but the French king chose to disregard it and two years later had his military and cartographic revenge. In 1550 Henri's triumphal entry ito Rouen was accompanied by banners containing landscapes of the Scottish lowlands, including Haddington, which had been won back by French arms from the English.[117]

The map of Haddington was a possibly more polished example of several. For after

1547 English and Italian engineers, like Richard Lee and Giovanni di Rossetti, Thomas Petyt and Gian Tommaso Scala, accompanied their creation of Italianate forts and fortifications in the Scottish lowlands and northern England, which were some of the most technically advanced in Europe, with architectural plats and large-scale maps of the border country.[118] Some of the plats merely showed the immediate surrounds of the fort, actual or proposed.[119] Others depicted the terrain of a whole district and also identified the homes of potential friends and foes.[120] This type of map was thus able to serve as an aid for the tactical planning of a campaign, as a guide for the siting of strong points and forts, and as a source in the formulation of policy. Far more than simple "military" or "fortification" plans, they were of a kind that was to become very familiar in the reign of Edward's sister, Elizabeth. Still other maps were commissioned to illustrate the English and Scottish claims to the aptly named "Debatable Lands" on their border.[121] These played a part in achieving the French-mediated agreement of 1552, though tension contained long after, resulting in several more surviving maps of the same area.[122]

The reign of Edward VI saw the creation of similar plats of the Boulonnois and the Pale of Calais, for it became clear that the French intended to seize the first opportunity to expel the English from their last continental toeholds, so that Somerset and his colleagues were compelled to look ever more closely at the areas' defenses.[123] These maps were often produced by the same men who had been active in Scotland, such as Thomas Petyt, who, in 1550, was probably responsible for a detailed plat of the marches of Calais. This shows bulwarks and dykes, defensive lines and mills, plans of the town of Calais itself and of the forts on the edges of the Pale—almost everything the government needed to know for its purposes.[124]

By this time, too—unlike the case some thirty years earlier, when Sebastian Cabot had been allowed to take service with the Spanish king seemingly without obstacles being raised—English leaders realized that, if maps in the right hands could be an aid to conquest and defense, in the wrong hands they could facilitate invasion and defeat. The hydrographer Jean Rotz had to resort to some complicated subterfuge in order to effect his return to French service in 1547. And, as in the case of his compatriot Nicholas de Nicolay, his maps and plans of England and Scotland's coastal fortifications served as his passport.[125] The English ambassador in France, Wotton, was only too aware of the consequences, warning Somerset a few months later, in March 1548, that by means of these "pictures" the French "May land their men that go to Scotland easily."[126] The French invasion of the Scottish lowlands did soon follow,[127] and by December 1551, a chance visit to Portsmouth by a French ambassador en route to Scotland in the company of an engineer-mapmaker was sufficient to alarm the English authorities into ordering the refortification of its castle.[128]

The traffic in cartographers was not totally to England's disadvantage, however. In 1545 Diogo Homem, distinguished scion of one of the most talented families of Portuguese hydrographers and son of the Portuguese royal hydrographer Lopo Homem, was welcomed to England, apparently without difficulties over the murder that had precipitated his flight.[129] In 1547 he was followed by Sebastian Cabot, whose worth as a cartographer, navigator, and expert on the northern regions was finally appreciated in the country of his youth,[130] and later by several other Portuguese pilots who were ipso facto excellent cartographers.[131] Their knowledge of the cartography of the non-European world was to be indispensable to the English merchants, intellectuals, and seamen who were responsible for the English overseas enterprise that was, at last, getting under way.[132] The government's tolerating the immigration of these cartographers was thus of historical importance.

Somerset and his colleagues were almost certainly aware of the advantages that would flow from this immigration. For by 1550 the court as a whole, and not simply the ministers, seems to have been cartographically sophisticated. Edward VI's tutor, Sir John Cheke, was one of the great mathematicians of the day. He took a particular interest in the practical applications of geometry, to the extent not only of purchasing the appropriate literature for his young charge but also of designing an astronomical quadrant for him in 1551, and he enjoyed close contacts with other young dons, scientifically inclined gentlemen, and seamen who were alive to the necessity of improving the standard of English navigation and cartography.[133] In large part their motivation was patriotic, but the precise influence Cheke and his friends had on Edward cannot be judged. Yet his schoolboy passions for fortifications and foreign wars and sieges and his almost obsessive interest in the detailed itineraries of Continental armies, which spring forth from almost every page of his diary, were certainly fed by a steady supply of plats and probably by an increasing number of printed commemorative battle plans from Germany and Italy.[134]

Edward was himself almost permanently surrounded by maps and plats, as is clear from the surviving inventories of the contents of the royal palaces made in 1547 and 1549 following Henry VIII's death.[135] These demonstrate the extent of map consciousness and map use at court by the middle of the century. It is significant that maps, globes, and plats were to be found exclusively—but in great numbers—in the palaces that were most extensively used for grand receptions and for the business of government: above all Whitehall, but also at Greenwich, Hampton Court, and "St James hous nighe Westminster," which then served as "a subsidiary royal palace close to the centre of government."[136] Maps and plats were not objects like books and tapestries that might be found anywhere. By 1550 they were particularly associated with places intended for pageantry, propaganda, and government.

There were a few relics of previous centuries, like the "rownde mappae mundi"[137] in the window of the Long Gallery at Hampton Court and the traveling "small . . . mappae mundi in parcheament" which formed part of "the Removing Guarderobe . . . attendaunt at the Corte upon the Kinge's most Roiall parsonne where the same for the tyme shall happen to be,"[138] and which may have been intended for the same use that Henry III's *mappaemundi* had been put to. The "rounde mappe" that stood in the window of "the closet over the watersteire" at Greenwich[139] might also have been a *mappamundi,* while "the discription of the citie of Jherusalem of stayned clothe" and "mappe of the holye lande of stayned clothe set in frame of woode" in Whitehall may also have been medieval.[140] The majority of the maps were clearly no more than a generation old, however, and were intended for official show or for direct use in government.

Many are described as framed or as being backed with or painted on cloth for hanging. The framed maps and some of those on cloth were meant for public display in galleries such as the Privy Gallery in Whitehall, where several of the maps listed in the inventories of 1547/9 are known still to have been on show in the reign of James I.[141] And there could not have been a better location, for the Privy Gallery was a place where "courtiers lounged, strolled or waited to catch the eye of King or Ministers,"[142] as also did foreign envoys and distinguished foreign visitors prior to their audiences. For most of the time they had little else to do than look and, it was hoped, be impressed by the king's knowledge of his realms, by depictions of his victories (or those of his allies), and by examples of his wise patronage of the best astronomers, artists, and cartographers of the age. Despite its charm and artistry, then, the display was every bit as much a part of the infrastructure of state power as the law courts or the coastal forts.[143]

Many of the maps meant for display were probably gifts from abroad, and several showed famous modern cities and commemorated military victories. Thus with maps of Europe and of Italy,[144] and plan/views of Paris, Jerusalem, Antwerp, Venice, Constantinople, Ancona,[145] "the castell of Millayne," and Florence, "with sondrye tounes ioynynge theirunto"—and sometimes more than one example of each—mostly of "stayned clothe" or "stayned . . . paper . . . pasted upon cloth," there were depictions of the siege of Vienna (1529), "the seige of Pavie when [to Henry's delight[146]] ye Frenche kynge was taken"[147] (1525), "the discription of Naoples being biseged with Monsier (*sic*) de Londraith," and what was probably a picture map of an episode from Charles V's campaign in Tunis in 1535.

In the same company, there were Henry's own victories such as "the discription of the siege and wynnynge of Bolloigne"—almost certainly that of September 1544 rather than his father's conquest of 1492—and, in memory of the achievements of his young, "a large mappe of the siege of Guines," accompanying the painted "tables" of Henry's other military and diplomatic triumphs of 1513, which can still be seen in Hampton Court. The King also displayed depictions of his by then numerous palaces[148] (as did other rulers) and of the realms over which he ruled or aspired to rule. There was a "mappe of Englande" in the lobby of a gallery at Greenwich,[149] small-scale "plattes" of England and Wales, but also Normandy and Scotland, at Hampton Court,[150] while Whitehall housed a "little mappe of Englande, parte of Scottlande Ierland and Brytayne [Brittany] on parchment."[151] An accomplished, if still essentially medieval, map of Great Britain and Ireland on parchment of about this period, which is now in the British Library, may also have been originally intended for this sort of display.[152]

Modern world maps and globes were the other category of "public" maps. These included a "mappe or a card maryne of thoole [*sic*] worlde off parchment set in a frame of white woode with the Kynges Armes crowned supported by His Grace's beasts," which sounds tantalizingly like a lost masterpiece of Rotz, probably broadly resembling planispheres like the Harleian *mappemonde* of about 1547 (British Library Add. MS 5413) or the Desceliers world map of 1550 (British Library Add. MS 24065), which were both executed for royal or noble patrons by cartographers from Rotz's hometown of Dieppe.[153] Another world map, described simply as of "paper pasted uppon borde,"[154] may have been a recently printed wall map from Germany, Flanders, or Italy. Last, there were globes, like the "rounde Globe of thole worlde standinge uppon a foote of woode coulored grene being brokin" at St.

James.[155] These items served the purposes of political propaganda much as did the maps of the realms in Europe claimed by the king, and from 1549 they were joined in the Privy Gallery by a new, probably woodcut world map, which was regularly to be described by visitors for the next century and a half.[156] Designed by Sebastian Cabot and cut by Clement Adams, the map (which probably perished in the great Whitehall Palace fire of 1698) showed the Northwest Passage and had inscriptions pinpointing the other North American discoveries that Cabot claimed to have made while in English service at the beginning of the century. Thus the map did much more than simply demonstrate Cabot's prowess as a navigator and cartographer. As R. A. Skelton and Helen Wallis have observed, it served also to justify the claims of England's right to colonize North America that were then beginning to be formulated.[157]

The public galleries were not the only parts of English palaces with maps by the mid-sixteenth century, however. "The little study called the new Librarye" in Whitehall was the kingdom's nerve center, where the king worked and consulted his ministers and his records. This is clear from the description of its contents in the 1547–49 inventories.[158] Here were dispatches "from sondrie places beyond the sea," copies of acts of parliament, "treatises and commissions for the peace . . . wrytynges and specialties belonging to the Lorde legatte [Wolsey, whose London home Whitehall had originally been]." But here also were "a black coffer covered with fustian of Naples full of plattes . . . a cuppbourde full of tilles [drawers] viz. [in] the lower parte sondrie wrytinges concerning rekoninges with plates and petygrees. The next rowme [drawer] above the same, sondrie plattes and wrytinges, twoo tilles next above the same with sondrie wrytinges in them . . . the thirde having bookes declaring thordre of battell . . . upon two shelves paternes for Castles and engynes of warre . . . a case for a platte covered with leather." This then seems to have been the original home of the smaller manuscript maps and of the architectural plans, all generically called "plats," which had been created in the 1530s and 1540s. They were filed away, with other visual aids such as, in the words of a contemporary endorsement, "figures, drafts, plats and images"—a clear enough indication, incidentally, of the pictorial nature of these functional early governmental maps and of their users' inability or refusal to distinguish between different sorts of visual aids.[159]

Plats were not the only cartographic items in the New Library. There was also "a great globe of the description of the worlde."[160] In Sir Anthony Denny's keeping[161] were, moreover, a few items that but for their size would also have been kept in the New Library, since they were, almost certainly, too politically and militarily sensitive to be exposed to public view. For they related to the defense of the kingdom. The "description of Rye and Winchelseye" has been identified as the map commissioned by Henry VIII from Vincenzo Volpe in 1530.[162] The "discriptions" of "parte of Suffolke, Essexe and London," of "parte of Hampshire with the Isle of Wight . . . painted upon clothe" and of "the towne of Callice of parchment," sound very much like companion pieces to the surviving surveys of 1539–40, such as the long view of the coasts of Devon and Cornwall.[163]

The inventories of 1547–49 provide evidence that as well as having a multitude of maps at hand, the king and his ministers had the means available to them for fully exploiting even the most technically advanced. In addition to "sondrie wrytinges concerning rekoninges," the New Library contained "a rule of boxe a fote and a halfe longe," presumably for use with plats that were drawn to scale, and "ix compasses of iron, one partely gilte." Indeed, with "two squre tables to knowe the distance of a grounde" and "a piece of brassell to trye the lengthe of greate pieces," an English king or minister would have been able to go on surveying expeditions himself.[164]

The impression from the inventories that by 1547 ministers had a wealth of plats at their disposal is supported indirectly. There can be little doubt that in succeeding years most of the plats mentioned in contemporary letters were lost. There was no official repository for state papers until early in the next century. Many plats must have been used and annotated to the point of destruction. Yet in the 1580s and 1590s Elizabethan statesmen were able to lay their hands on Henrician plats with relative ease when they wanted to consult them, and today the number that survive in the British Library, the Public Record Office, Hatfield House, and Belvoir Castle is impressive, even when the efforts of later antiquaries and archivists to preserve them are taken into account.

Between 1530 and 1550, then, maps were enrolled into the service of English government, and by the middle of the century they had become a fully integrated aid in the formulation of policy and an instrument of administration. The next half century was to see further elaboration of, but no essential change in, these roles.

Notes

1. E. J. S. Parsons, *The Map of Great Britain circa 1360 Known as The Gough Map: An Introduction to the Fascimile* (Oxford, 1970), 2, 15.

2. The plan of Dartmoor, now in the Royal Albert Memorial Museum in Exeter, is reproduced and discussed in *Local Maps and Plans from Medieval England*, ed. P. D. A. Harvey and R. A. Skelton (Oxford, 1986), 293–302. The same volume contains other examples of early maps with official provenance, such as plans of the Isle of Ely and of Inclesmoor in Yorkshire from the archives of the Duchy of Lancaster (89–98; 147–62) and of Chertsey in Surrey from the archives of the Exchequer (237–43); see also *Catalogue of Maps and Plans in the Public Record Office*, vol. 1, *British Isles, c. 1410 to 1860*, ed. P. Penfold (London, 1967), where the government department can be identified from the suffix originally attached to the map's file number (e.g., "E" = Exchequer).

3. David Woodward, "Reality, Symbolism, Time, and Space in Medieval Maps," *Annals of the Association of American Geographers* 75, no. 4 (1985): 510–21. Pamela Tudor Craig, "The Painted Chamber at Westminster," *Archaeological Journal* 114 (1957): 92–105, esp. 100–101. I am grateful to my colleague Andrew Prescott for this reference.

4. See G. Kipling, *The Triumph of Honour: Burgundian Origins of the Elizabethan Renaissance* (London, 1977). Roy Strong, *Art and Power: Renaissance Festivals, 1450–1650* (Woodbridge, Suffolk, 1984).

5. James A. Williamson, *The Cabot Voyages and Bristol Discovery under Henry VII* (Cambridge, 1962). Helen Wallis, "Globes in England up to 1660," *Geographical Magazine* 35, no. 5 (September 1962): 267. Kenneth R. Andrews, *Trade, Plunder, and Settlement: Marine Enterprise and the Genesis of the British Empire, 1480–1630* (Cambridge, 1984), 41–50. The recently discovered "Aslake" world map shows that portolan charts were known in England, and probably at court, as early as the later fourteenth century (P. M. Barber, "Old Encounters New: The Aslake World Maps" in *Géographie du monde au Moyen Age et à la Renaissance*, ed. M. Pelletier (Paris, 1987), 67–88.

6. P. D. A. Harvey, *Topographical Maps: Symbols, Pictures, and Surveys* (London, 1980), 93–96. William Ravenhill, "The Plottes of Morden Mylles, Cuttell (Cotehele)," *Devon & Cornwall Notes and Queries* 35, no. 5 (Spring 1984): 165–74, 182–83. Sarah Tyacke's introduction to *English Map-Making, 1500–1650: Historical Essays*, ed. Sarah Tyacke (London, 1983), 16–17. Several early plans of this sort are to be found in Harvey and Skelton's *Local Maps and Plans*. After 1500 the number of such survivals increases dramatically. Several are to be

found in the Public Record Office's *Catalogue*, vol. 1, their departmental provenance ascertainable from their original file numbers. Others are to be found among the British Library's Cotton manuscripts (e.g., Cotton Augustus I.i.65: a plan of waterways in Nottinghamshire). For a plan that got no further than a Yorkshire lawyer's notebook, see BL Add. MS 62534 f. 180, and for another plan of about 1550 that was explicitly intended for display at the assizes in Northampton, BL Add. MS 63748. As Harvey has pointed out, however, this was a continent-wide phenomenon, and, if anything, English lawyers seem to have lagged behind their European counterparts.

7. Williamson, *Cabot Voyages*, 220. Helen Wallis has suggested that the "demonstracions reasonable" may have included a globe that Cabot was reported as having constructed, with a world map, in December 1497 (Helen Wallis, "Some New Light on Early Maps of North America, 1490–1560," *Wolfenbütteler Forschungen* [Munich, 1979], 94).

8. *Catalogue of Western Manuscripts in the Old Royal and King's Collections*, ed. G. F. Warner and J. P. Gilson (London: British Museum, 1921), 2:175–76. D. H. Turner, *The Hastings Hours* (London, 1983), 99–100. The Corbechon from Edward IV's library (more properly a translation by him of Bartolomaeus Anglicus's *Liber de Proprietatibus Rerum*, copied in Bruges in 1482 by Jean du Ries) is now BL Royal MS 15.E.III. The *mappamundi* appears on folio 67 and is reproduced and discussed in P. M. Barber, "The Manuscript Legacy: Maps in the Department of Manuscripts," *Map Collector* 28 (September 1984): 18–21.

9. Sidney Anglo, *Spectacle, Pageantry, and Early Tudor Policy* (Oxford, 1969), 77, 82.

10. For Henry VII, see R. L. Storey, *The Reign of Henry VII* (London, 1968), and G. Kipling, "Henry VII and the Origins of Tudor Patronage," in G. F. Lytle and S. Orgel, *Patronage in the Renaissance* (Princeton, 1981), though there is no evidence of Henry's acting directly as a patron of the cartographic arts. For the lack of interest before 1550 in the new geography and the world beyond Europe of most Englishmen (with the obvious exception of the Bristol merchants): Andrews, *Trade, Plunder, and Settlement*, 57; E. G. R. Taylor, *The Mathematical Practitioners of Tudor and Stuart England* (Cambridge, 1954), 16–18; henceforth referred to as *Mathematical Practitioners*.

The forthcoming chapter by Sir John Hale, "Warfare and Cartography in the Renaissance, c. 1450–c. 1600," for *History of Cartography*, ed. J. B. Harley and D. Woodward, vol. 3, and the contributions to this work by David Buisseret and Geoffrey Parker suggest that there was little to choose between England and the great western European monarchies of France and Spain in respect of the use of maps for war and government as of 1509; all three lagged considerably behind northern Italy and southern Germany.

11. Garrett Mattingly, *Catherine of Aragon* (London, 1971), 114. More generally for the "right-about-turn to the Middle Ages" at the beginning of Henry VIII's reign, see David Starkey, *The Reign of Henry VIII: Personalities and Politics* (London, 1985), 37–51.

12. Williamson, *Cabot Voyages*, 281, referring to BL Add. MS 21481, f. 92. Mattingly, *Catherine*, 114–15; J. J. Scarisbrick, *Henry VIII* (London, 1972), 50–52. C. G. Cruickshank, *Army Royal* (Oxford, 1969), is an account of the campaign.

13. BL Cotton Roll xiii.12. Reproduced in part in S. Tyacke and J. Huddy, *Christopher Saxton and Tudor Mapmaking* (London, 1980), opposite p. 16. For the war scare, see Scarisbrick, *Henry VIII*, 78–79. It is possible, however, that the map may be connected with a survey of England's shores ordered in 1519; see Howard Colvin, ed., *The History of the King's Works* (London, 1951–82), 4:367. It may even date from the late 1530s. BL Harley MS 590, f. 1, a map of the same area showing the marshland manors of Lord Cobham and Sir Thomas Wyatt, is so similar in style as to be possibly by the same hand, but it must date from after 1537, when Wyatt was knighted.

14. Private communication from Hugh Perks of Faversham, 19 January 1981.

15. There is no evidence in C. G. Cruickshank, *The English Occupation of Tournai* (London, 1971), or in Colvin, *King's Works*, 3:375–82, of maps or plans' having been commissioned or utilized in connection with the refortification of England's principal conquest from the 1512–14 war, Tournai, during its seven years of English occupation (1513–20), in contrast with the prodigious use made of them for similar works at Boulogne thirty years later. Certainly no plans seem to survive, but for Tournai's importance, see Starkey, *Reign of Henry VIII*, 50. It can be inferred that the originals of the commemorative picture-maps of engagements in the 1512–14 war, now in Hampton Court, were made by Germans serving with Emperor Maximilian I (see O. Millar, *Tudor, Stuart, and Early Georgian Pictures* [London, 1963]). Thus they belong to a separate tradition, of which a surviving, even

earlier example of about 1505, commemorating a campaign of 1499, is illustrated in Harvey, *Topographical Maps*, 100–101. And see L. Bagrow and R. A. Skelton, *History of Cartography* (London, 1964), 93.

16. D. B. Quinn and A. M. Quinn, eds., *New American World: A Documentary History of North America to 1612* (New York, 1979), 1:123, item 92. Williamson, *Cabot Voyages*, 146–47, 171–72.

17. Generally see Tony Campbell, *The Earliest Printed Maps, 1472–1500* (London, 1987), and W. H. Stahl, *Ptolemy's Geography: A Select Bibliography* (New York, 1953). Karl-Heinz Meine, ed., *Die Ulmer Geographia des Ptolemäus von 1482: Zur 500 Wiederkehr der ersten Atlasdrucklegung nördlich der Alpen* (exhibition catalog, Ulm, 1982), 26, lists seven editions between 1470 and 1500 and contains a recent bibliography on the subject. Hartmann Schedel's profusely illustrated *Nuremberg Chronicle*, first published in 1493, was printed in such numbers in German and Latin that it soon became well known throughout western Europe. See also Horst Kunze, *Geschichte der Buchillustration in Deutschland: Das 15 Jahrhundert* (Leipzig, 1975); Karen S. Pearson, "The Emergence of the Illustrated Geography Book in Renaissance Germany," *Technical Papers of the American Congress on Surveying and Mapping Convention, Denver, March 14–20, 1982* (Falls Church, Va., 1982), 85–91.

18. Neville Williams, "The Tudors: Three Contrasts in Personality," in *The Courts of Europe: Politics, Patronage, and Royalty, 1400–1800*, ed. A. G. Dickens (London, 1977), 153. Margaret Mann Phillips, *Erasmus and the Northern Renaissance* (London, 1949), 40–41.

19. G. R. Elton, *England under the Tudors* (London, 1955), 127.

20. Harvey, *Topographical Maps*, 58–61. See also, however, John Marino's chapter in the present volume.

21. Andrews, *Trade, Plunder, and Settlement*, 57; Taylor, *Mathematical Practitioners*, 17–18. This group was familiar with Latin, often with Greek, and frequently with Italian and French. Its members thus had direct access to the latest writings of the leading European theorists, unlike the vast majority of their countrymen, who had to wait several years—and often decades—for the same ideas to be expounded in English.

22. Quinn and Quinn, *New American World*, 1:171–79.

23. Anglo, *Spectacle*, 141, 144. For a surviving design, see BL Cotton MS Augustus iii, f. 18 (reproduced in Richard Marks and Anne Payne, *British Heraldry from Its Origins to Circa 1800* [exhibition catalog, British Museum, London, 1978], cover).

24. Taylor, *Mathematical Practitioners*, 13, 165. Nikolaus Kratzer was apparently the first lecturer, and he was particularly interested in the application of mathematics to mapping (see also John D. North, "Nicolaus Kratzer—the King's Astronomer," *Science and History: Studies in Honor of Edward Rosen, Studia Copernicana* 16 [Wrocław: Polish Academy of Sciences, 1978], particularly 217–22, 225–26).

25. R. W. Chambers, *Thomas More* (London, 1935, 1963), 116, 131, 134–35.

26. Gillian Hill, *Cartographic Curiosities* (London, 1978), 21. More was, however, interested in the related field of astronomy (North, "Nicolaus Kratzer").

27. For the association of the following with More, Chambers, *Thomas More*, 48, 49 (Rastell), 275 (Elyot), 72–73 (William Lily), 101, 103 (Kratzer). Also *"The King's Good Servant": Sir Thomas More, 1477/8–1535*, ed. J. B. Trapp and Hubertus Schulte Herbruggen (exhibition catalog, National Portrait Gallery, London, 1977), nos. 187, 188 (Kratzer), 14 (William Lily), 269 (George Lily), 279, 280 (Elyot).

28. North, "Nicolaus Kratzer," 206, 213ff., 225–26, 230–32. North, thinking primarily of Kratzer's work as an instrument maker, stresses his role in bringing German science and technology to the court of Henry VIII. And see Taylor, *Mathematical Practitioners*, 13–14, 165. These studies supersede the article in the *Dictionary of National Biography.*

29. Bagrow and Skelton, *History of Cartography*, 147–50. Harvey, *Topographical Maps*, 88.

30. Oliver Millar, *The Tudor, Stuart, and Early Georgian Pictures in the Collection of H.M. the Queen*, nos. 22, 23. The general resemblance between these and some of the earlier picture-maps in the British Library's Cotton Collection (e.g. BL Cotton MS Augustus I.i.43) should be noted. It is the picture of Henry VIII's meeting with Maximilian that seems to be referred to in the inventory of 1547 (BL Harley MS 1419A f. 447v) as the "Table wherein is conteined the seginge of Torney and Turwyn" at St. James's Palace. Millar, *The Queen's Pictures* (London, 1977), 225, n. 9, accepts that the painting of the Battle of Pavia (1525), now in the Tower of London, "is clearly by the same hand," and this too seems to be listed in the 1547 inventory (f. 130v) as "A Table of the Siege of Pavie." Although by 1977 he had changed his mind, in 1963 Millar still held that the painting of the Battle of the

Spurs, now in Hampton Court, was from Henry VIII's time, and probably by an Italian. Its appearance (perhaps even more than the painting of the meeting with Maximilian it resembles a picture-map executed in perspective) strongly suggests that the original from which it must have been copied was of the 1520s and probably, like the painting of the meeting with Maximilian, ultimately derived from a now-lost German commemorative plat (see n. 35 below). Both paintings relate to engagements where German troops were present or close at hand. Roy Strong has suggested (*Holbein and Henry VIII* [London, 1967], 25–26) that the better-known paintings at Hampton Court of Henry VIII's journey to the Field of the Cloth of Gold (Millar, *Tudor, Stuart, and Early Georgian Pictures,* nos 24, 25), may be Elizabethan copies of lost Tudor frescoes of the 1520 or 1530s. In this case the original, pictorial commemorative plats would presumably have been English. The 1547 inventory lists several other paintings, such as one showing the siege of Guines (1513), which may well have also been commissioned in the 1520s to commemorate the king's youthful feats (for such early commemorative military picture-plats, see E. Pognon, "Les Plus Anciens Plans de villes gravés et les événements militaires," *Imago Mundi* 20 [1968]: 13–19; *Wien 1529: Die erste Türkenbelagerung* [exhibition catalog, Historisches Museum der Stadt Wien, Vienna, 1979], nos. 147–50, 156–57). When discussing the possible existence of these maps, R. A. Skelton's warning words (in *Maps: A Historical Survey of Their Study and Collecting* [Chicago, 1972], 28–35) about the low survival rate of early maps should be borne in mind.

31. Anglo, *Spectacle,* 163, and also 140–41, 160, 196–97.

32. Ibid., 166, 219. H. Wallis ("Some New Light," 99–100) particularly emphasizes his activities as a map publisher. Taylor, *Mathematical Practitioners,* 312.

33. Andrews, *Trade, Plunder and Settlement,* 54–55. Chambers, *Thomas More,* 132.

34. Taylor, *Mathematical Practitioners,* 13, 312. Wallis, "Some New Light," 99.

35. Anglo, *Spectacle,* 211–24 (the English quotations are from Hall's chronicle). Millar, *Tudor, Stuart, and Early Georgian Pictures,* no. 22. The "payncting" may well have been the prototype for the picture of the meeting of Henry VIII and Maximilian now at Hampton Court and discussed in n. 30 above. Its details tally closely with the description in Hall's chronicle of the "payncting." The "plot of Tirwan" on which the "payncting" was based was probably a commemorative picture-map of the occasion by a German in the suite of Maximilian. The same "plot" almost certainly served as the model for the woodcuts, completed in 1515, of the same scene by Dürer and others for Maximilian's Triumphal Arch (1518) (see Eduard Chmelarz, Adolf Hokhausen, et al., eds., *Maximilian's Triumphal Arch: Woodcuts by Albrecht Dürer and Others* [New York, 1972], introduction and plate 20). The similarity of the groupings in the Hampton Court painting and the woodcut is striking (indeed, it could be that the "payncting" was derived from the "plot" through the woodcut, with which Dürer's correspondent Kratzer, as well as Holbein, might have been familiar).

In 1529, as Wallis has discovered ("Some New Light," 99), Rastell referred, in *The Pastyme of the People,* to a "quarto Mappa mundi," presumably printed by himself. It is tempting to wonder whether this was a scaled-down version, intended for general circulation, of the "mappamondo" prepared for the Greenwich festivities which had so caught the Venetian ambassador's eye. The so-called Münster-Holbein world map printed in Basel in 1532 (Rodney Shirley, *The Mapping of the World: Early Printed World Maps, 1472–1700* [London, 1983], 74–75, no. 67) may also be related. As on the map displayed at Greenwich, the names of the provinces of the world are one of its most prominent features.

36. The years 1521 and 1528, respectively.

37. They took their arguments from the late fourth-century writer Vegetius, who in his *Epitoma rei militaris* had urged the necessity of generals' providing themselves with maps (translated by John Clarke in *Roots of Strategy* [Harrisburg, 1940], 133). I am most grateful to Sir John Hale for this reference.

38. For a recent discussion of Castiglione's influence, see David Starkey, "The Court: Castiglione's Ideal and Tudor Reality: Being a Discussion of Sir Thomas Wyatt's Satire Addressed to Sir Francis Bryan," *Journal of the Warburg and Courtauld Institutes* 45 (1982): 332–39.

39. S. E. Lehmberg, *Sir Thomas Elyot: Tudor Humanist* (Austin, 1960), and n. 27 above. In *Holbein and the Court of Henry VIII* (exhibition catalog, Queen's Gallery, Buckingham Palace, London, 1978), 63–64, the drawing (Royal Library 12203) is reproduced and discussed.

40. Sir Thomas Elyot, *The Book Named the Governor* (Everyman edition with an introduction by S. E. Lehmberg; London, 1962), bk. 1, chap. 8, pp. 23–24.

41. Baldassare Castiglione, *The Book of the Courtier from the Italian of Count Baldassare Castiglione Done into English by Sir Thomas Hoby Anno 1561* (London, 1900), 91.

42. Elyot, *The Book,* bk. 1, chap. 11, p. 36.

43. Ibid., 35.

44. Ibid. As David Woodward has recently pointed out ("Reality, Symbolism, Time," 510–21), this idea of a map as a guide to history is also found in medieval times.

45. Elyot, *The Book,* bk. 1, chap. 8, p. 24.

46. Ibid., p. 26. As he stated in chap. 11 (p. 36) "it may not be of anywise denied but that cosmography is to all noblemen, not only pleasant, but profitable also, and wonderfully necessary."

47. Ibid., 23–24.

48. Ibid., 24.

49. Lehmberg has noted that the passages on political theory in *Henry V* and *Troilus and Cressida* include "remarkably close verbal echoes" of the opening chapters of *The Book Named the Governor* (introduction, vii). See also E. Grether, *Das Verhältnis von Shakespeares "Heinrich V" zu Sir Thomas Elyot's "Governour"* (Berlin, 1938).

50. Lehmberg, introduction, vii.

51. Hale, "The Defence of the Realm, 1485–1558," in *King's Works,* ed. Colvin, 4:373–74.

52. The possession of maps imbued monarchs and ministers with considerable confidence and independence in their dealings with lieutenants in regions distant from court, on whose advice they might otherwise have been almost totally dependent. Thus, with plans in hand, Henry VIII was able to disregard local administrators and noble-born generals when supervising the refortification of Hull and Boulogne after 1540, preferring to rely instead on the plebeian mason-engineer John Rogers, and he even disregarded Rogers over the fortification of Ambleteuse near Boulogne in 1546 (L. R. Shelby, *John Rogers, Tudor Military Engineer* [Oxford, 1967], passim and esp. 76–78). See also Colvin, *King's Works,* 4:373–74, 391–92. These consequences are implicit in the passage from Elyot quoted in the introduction to this paper and probably constitute the most profound contribution made by maps to Tudor government and statecraft. At the same time I would not wish to exaggerate the role of maps as instruments for enhancing royal authority (given the peripatetic court, one can hardly speak of *central* authority), at the expense of the other technological, administrative, social, personal, and political factors that have traditionally been discussed in this context.

53. Colvin, *King's Works,* 3:14, 352. The prestige the humble-born Lee consequently enjoyed as surveyor of Calais is amply demonstrated in Muriel St. Clare Byrne, ed., *The Lisle Letters,* abridged ed. (Chicago, 1985). He was even allowed to marry the daughter of the high marshal of Calais, Sir Richard Grenville—herself a grandniece of Edward IV's bastard son Viscount Lisle, the lord deputy of Calais (ibid., 375–77). Another mason-engineer, John Rogers, secure in the king's favor, at times treated his noble-born superiors with an insouciance bordering on contempt, to their—impotent—disgust (Shelby, *John Rogers,* passim).

54. Independently of each other, North ("Kratzer," 227–28) and Wallis ("Some New Light," 97) have shown that the globes and mathematical instruments depicted were actually owned by Kratzer. The painting thus provides another example of his key role as a popularizer of cartography in court circles.

55. R. B. Wernham, *Before the Armada: The Growth of English Foreign Policy, 1485–1588* (London, 1966), passim.

56. Colvin, *King's Works,* 4:367–71; also 484 (Plymouth and the West), 731 (Dover); Byrne, *Lisle Letters,* 65, 69, 71–72; Colvin, *King's Works,* 3:353–54 (defense of Calais 1533); Shelby, *John Rogers,* 111–15 (Sandwich).

57. Colvin, *King's Works,* 4:418.

58. Ibid., 731–32.

59. B. L. Cotton MS Augustus I.i.19.

60. E.g., BL Cotton MS Augustus I.i.22, 23 and I.ii.70: plans of Dover and Calais of 1538, both in the same style. This is presumably the hand of Richard Lee, the surveyor of Calais, who certainly executed the plan of Orwell Haven (BL Cotton MS Augustus I.i. 56) and almost certainly the fragmentary map-view of Edinburgh (BL Cotton MS Augustus I.ii.56) in 1544. M. Biddle and J. Summerson (in Colvin, *King's Works,* 4:744–45) attributed the plan of Dover to John Thomson, the surveyor there, but the plans of Dover and Calais are so close in style, and Lee so (relatively) well known by contemporaries as a cartographer, that he seems far likelier to have been the maker.

For the plans produced in Calais, see Colvin, *King's Works,* 3:345–55, 401–4. The earliest "plait" (1531), by the surveyor William Lambert, is now lost, and few survive from earlier than about 1550. For the sequence of sixteenth-century plans of Dover, see W. Minet, "Some

Unpublished Plans of Dover Harbour," *Archeologia* 62 (1922): 185–225; A. Macdonald, "Plans of Dover in the Sixteenth Century," *Archeologia Cantiana* 49 (1937): 108–26.

61. R. A. Skelton and J. Summerson, *A Description of the Maps and Architectural Drawings in the Collection Made by William Cecil, First Baron Burghley; Now at Hatfield House* (Oxford: Roxburghe Club, 1971), 46, no. 36 (CPM.1.64); henceforth Hatfield House Maps. For the proposals and work at Dover to 1542, see Colvin, *King's Works,* 4:731–50, 764–67.

62. Colvin, *King's Works,* 4:378–79, 392–93. Shelby, *John Rogers,* esp. 5–8, 127–44, 151–57. A good example is provided in South England, where the artist Vincenzo Volpe (Dover, 1532) gave way to the Moravian engineer Stefan von Haschenperg (Camber, c. 1540) and the English mason Richard Lee (c. 1538–43).

63. Colvin, *King's Works,* 4:367–69.

64. Ibid., 367.

65. "Remembraunces" of Thomas Cromwell (with amendments in his hand), probably early February 1539 (BL Cotton MS Titus B.i.ff.473–74); *Letters and Papers, Foreign and Domestic, of the Reign of Henry VIII,* ed. J. S. Brewer et al. (London, 1862–1932), 14, pt. 1, 400; henceforth referred to as L&P). This would seem slightly to antedate the "Devyce by the King" (L&P 14, pt. 1, 398; actually drafted by Cromwell) of early 1539, in which the heads of construction teams and commissioners are named, consisting predominantly of leading local gentry with some noblemen. For all of this and the remainder of the paragraph, I am indebted to Sir John Hale's contribution in Colvin, *King's Works,* 4: 369–71.

66. "Remembraunces" of Thomas Cromwell, April 1539 (L&P 14, pt. 1, 655).

67. E.g., BL Cotton MSS Augustus I.i.83 (Hull and environs). For this dating see Skelton and Harvey, *Local Maps,* 353–54. Augustus I.i.35, 36, 38, 39 (Exeter to Land's End and St. Ives); Augustus I.i.18 (Brighton); Augustus I.i.57 (Essex and Suffolk from the Naze to Bawdsey, probably by Richard Cavendish); Augustus I.i.31, 33 (Dorset), Augustus I.i.8 (north coast of Somerset). BL Royal MS 18.D.III ff. 9v–10 (Poole Harbor, Dorset), later owned by Lord Burghley, and BL Cotton Roll xiii. 12 (Feversham and the River Swale in Kent) may have been created at this time.

68. The parks were, of course, of military importance as sources of meat and timber for the forces, and the administrative use for the map was potential rather than actual.

69. E.g. BL Cotton MS Augustus I.i.35, 36, 38, 39 with comments such as "half made" by sketchy depictions of castles.

70. This is particularly well illustrated in the case of John Rogers's plans of the early 1540s of Guines (BL Cotton MS Augustus I. Supp. 14; I. Supp. 2; I.ii.12; I.ii.51; I.ii.52), of Hull (BL Cotton MS Augustus I. Supp. 4; I. Supp. 20; I. Supp. 1; Augustus I.ii.11; I.i.84; I.ii.13), Ambleteuse (BL Cotton MS Augustus I.i.59; I.ii.68; I.ii.73; I.ii.83; Hatfield House CPM 1.54) and Boulogne (BL Cotton MS Augustus I. Supp. 5; I.ii.77; I.ii.53; I.ii.82), for which see Shelby, *John Rogers,* and Hatfield House Maps, 63, no. 92. By the early 1550s some draftsmen began to distinguish between existing and proposed works on their plats; see, for instance, BL Add. MSS. 69824, an anonymous plan of Dover Harbor dating from 1552.

71. E.g. BL Cotton MS Augustus I.ii.7 (Tynemouth by Gian Tommaso Scala, 1545), and see Hatfield House Maps, 63, no. 92 (CPM 1.54: plan of Ambleteuse by John Rogers, 1547).

72. For Lee, see Colvin, *King's Works,* 3:13–14 4:410–11 and passim, which correct and supplement accounts in the *Dictionary of National Biography* (henceforth DNB) and A. H. W. Robinson's pioneering *Marine Cartography in Britain: A History of the Sea Chart to 1855* (Leicester, 1962), 145. For Rogers, see Shelby corrects Robinson in *John Rogers,* 147.

73. Taylor, *Mathematical Practitioners,* 167; Alwyn Ruddock, "The Earliest Original English Seaman's Rutter and Pilot's Chart," *The Journal of the Institute of Navigation* 14 (1961): 416. Colvin, *King's Works,* 4:454, 747, 751.

74. Colvin, *King's Works,* 4:380, 617, 630, 689, 743, 746–47; Taylor, *Mathematical Practitioners,* 169. Robinson, *Marine Cartography,* 146–47, should be treated with caution.

75. For Portinari: Colvin, *King's Works,* 4:394, 400, 409. For Haschenperg: B. H. St. John O'Neill, "Stefan von Haschenperg, an Engineer to King Henry VIII, and His Work," *Archeologia* 91 (1945): 137–55. Colvin, *King's Works,* 4:378, 667–68.

76. It is possible that à Borough may himself have created at least some of the maps that he owned (and is recorded as having lost) as early as 1533 (Taylor, *Mathematical Practitioners,* 167).

77. Shelby, *John Rogers,* 102–3.

78. BL Harley MS 1419 A & B.

79. Jean Rotz, *The Maps and Text of the Boke of Idrography Presented by Jean Rotz to Henry VIII*, ed. Helen Wallis Oxford: Roxburghe Club, 1981), 9–16. Rotz's manuscript treatise on the natural sciences, which he also presented to Henry VIII, contains one of the earliest explanations in English of triangulation (W. Ravenhill, "Mapping a United Kingdom," *History Today* 35 [October 1985]: 28).

80. On the grounds of handwriting, style, and similarity of content in the depiction of the French coast with charts in the *Boke of Idrography*, it is probable that a hitherto little-noted chart of the French and English coasts lining the English Channel (BL Cotton MS Augustus I.ii.65, 66, described in F. Madden and J. Holmes, *Catalogue of the Manuscript Maps, Charts, and Plans and Topographical Drawings in the British Museum* [London, 1844, 1962], 1:55) may be by Rotz and date from 1542–44, when Henry was contemplating war with France and would have needed such information about the French coasts (Scarisbrick, *Henry VIII*, 548ff.; and Rotz, *Boke of Idrography*, 12–13). I hope to discuss the chart at greater length in a future article.

Apart from its depiction of cliffs and bays, the chart is significant for its explicit scale of leagues; this may well have impressed and influenced Rotz's English counterparts, who seem generally to have been little acquainted with portolan charts and their rudimentary scale bars. Rotz and Nicolay, who was recruited into English service in 1546 by John Dudley, Viscount Lisle (D. Moir, "A History of Scottish Maps," in Royal Scottish Geographical Society, *The Early Maps of Scotland to 1850*, 3d ed. [Edinburgh, 1973], 1:20; Rotz, *Boke of Idrography*, 15–16), learned much from the work of the English also, however, and particularly from Scottish cartographers. Nicolay displayed this in his 1583 map of Scotland and acknowledged his debt in the text of the map's cartouche. It had already stood Rotz and him in good stead when they decided, on Henry VIII's death, to return to French service.

81. For Henry's own recognition of his ignorance of Scottish geography and his request for a map in 1542, see Scarisbrick, *Henry VIII*, 565. For Scottish cartographers and cartography and the links with Henry VIII's court, see Moir, "History of Scottish Maps," 92–93, 144, 146, 163–65; A. B. Taylor, "Name Studies in Sixteenth-Century Scottish Maps," *Imago Mundi* 19 (1965): 81–99; and A. B. Taylor, *Alexander Lindsay: A Rutter of the Scottish Seas, circa 1540*, Maritime Monographs and Reports 44 (Greenwich, 1980). For a hint that these detailed large-scale maps showing fortifications were used by the English in Scotland, see the reference, on Rossetti's map of Broughty (1547), to "la carta del paese" which showed two forts there (Hatfield House Maps, 60, referring to CPM Supp. 12).

Nicolay described on several occasions how in England in 1546 he had seen Lyndsay's rutter and map and copied them. After preparing manuscript copies of the map in c. 1547 for French forces invading Scotland and subsequently for private patrons such as the Cardinal de Lorraine (1559: BL Harley 3996, text only), he engraved it in 1583 (*La Navigation du Roy d'Ecosse Jacques cinquiesme . . . autour de son Royaume*). Lynday's and Elder's works must also have been studied in the early 1560s by Laurence Nowell. They provided the basis for his beautiful and detailed maps of Scotland (now BL Cotton MS Domitian A. xviii, ff. 98–99, 104–5, 106–7). They also provided the basis, several decades later, for the Scottish charts in the anonymous manuscript "Book of the Sea Carte called the Rutter . . . [of] the whole of Britanye," now BL Add. MS 37024. The family resemblance between all these maps and their common derivation from Lyndsay's now lost original was noted by A. B. Taylor as long ago as 1961 ("Some Additional Early Maps of Scotland," *Scottish Geographical Magazine* 77, no. 1 [1961]). I am most grateful to Geoffrey Parker, then of St. Andrew's University, for reminding me of the importance of the Scots!

82. Colvin, *King's Works*, 4:393. Marcus Merriman, "Italian Military Engineers in Britain in the 1540s," in *English Map-making*, ed. Tyacke, 57. Both Hale, in *King's Works*, vol. 4, and Merriman point out that there is little evidence that Italian engineers actually exerted the cartographic influence formerly attributed to them (by, e.g. Shelby, *John Rogers*, 143).

83. Shelby, *John Rogers*, and Colvin, *King's Works*, vols. 3 and 4 passim amply demonstrate these points. For Henry's palace-building program, see *King's Works*, 4:1–364.

84. Shelby, *John Rogers*, 50; Ruddock, "Earliest Original," 415–16.

85. BL Cotton MS Augustus I.ii.77; I.ii.53 (compare I. Supp. 5; Outreau); I.ii.74 (Ardres, and see Colvin, *King's Works*, 4:378, 383, 393); Cotton MS Augustus I.ii.29 (and see Ruddock, "Earliest Original," passim).

86. BL Cotton MS Augustus I.ii. 69 (Colvin, *King's*

Works, 3:374 and pl. 29).

87. Shelby, *John Rogers*, 94–101.

88. BL Cotton MS Augustus I.ii.77.

89. Shelby, *John Rogers*, 99–100.

90. E.g. BL Cotton MSS Augustus I.i.49, I.i.50: two plans of the Anglo-Imperial siege of Landrecy (1543).

91. Tyacke, *English Map-Making*, 16. The map (now in two parts) is Public Record Office MPI 221. Another early map in the Public Record Office, of York (MPB 49 & 51) may also have been made for the same purpose (see John Harvey's article in *Local Maps and Plans from Medieval England*, ed. Harvey and Skelton, 343). Such prodding by central government may have helped to make local corporations themselves more map-minded: the earliest surviving printed map of an English town was William Cunningham's map of Norwich illustrating his *Cosmographical Glass*, published in 1559, by which time Norwich may already have had something of a mapping tradition. As Harvey has remarked (*Topographical Maps*, 156), "It may be that the impetus to the spread of map-making came from the use of maps for official purposes."

92. P. D. A. Harvey, "The Portsmouth Map of 1545 and the Introduction of Scale Maps into England," *Hampshire Studies Presented to Dorothy Dymond* (Portsmouth, Hampshire, 1981), 33–49. See also his *Topographical Maps*, 160–62; Merriman, "Italian Military Engineers," 57–67; S. Tyacke and J. Huddy, *Christopher Saxton and Tudor Map-Making* (London, 1980), 12–14; Shelby, *John Rogers*, particularly 151–57.

93. E.g. to Sir William Paget, 21 February 1547, quoted and reproduced in Shelby, *John Rogers*, 174–75, pl. 27.

94. Hatfield House Maps, 28–29; R. A. Skelton, "The Military Surveyor's Contribution to British Cartography in the Sixteenth Century," *Imago Mundi* 24 (1970): 79; Harvey, *Topographical Maps*, 156, 158; Tyacke and Huddy, *Christopher Saxton*, 15–18.

95. BL Cotton MS Augustus I. i.56.

96. E.g. the written and finely decorated survey of Colchester Abbey's lands, 1540 (BL Egerton MS 2164).

97. *The maner of measurynge al maner of land* (Taylor, *Mathematical Practitioners*, 168, 312).

98. Tyacke and Huddy, *Christopher Saxton*, 18–19; Colvin, *King's Works*, 4:181, 206.

99. For this now see P. D. A. Harvey's forthcoming essay "Estate surveyors and the spread of the scale-map in England 1550–80." Harvey draws a distinction between unmeasured plans of manors or districts, of which several are known from before 1580 (e.g. the plan of manors in North Dorset of 1569–74, now BL Add. MS 52522), and measured estate plans, drawn to a consistent scale and based on geometrical survey, the earliest surviving examples of which first appeared in 1579–80 (e.g. Israel Amyce's survey of Edmund Tirrell's lands of Essex—now BL Lansdowne MS 6697—and Ralph Agas's map of Toddington, Bedfordshire, of 1581, now BL Add. MS 38065). Margaret Conlon of the Public Record Office has discovered a measured plan datable to 1571.

100. Scarisbrick, *Henry VIII*, 32; Starkey, *Reign of Henry VIII*, 133.

101. J. Hale in Colvin, *King's Works*, 4:374, and, more generally, Starkey, *Reign of Henry VIII*, 13.

102. Ruddock, "Earliest Original," 420. Ruddock assumed that this related to BL Cotton MS Augustus I.ii.29, a rough sketch by John à Borough, of the Zuider Zee of September–October 1539, accompanying a written rutter. She now agrees that it seems more likely to refer rather to BL Cotton MS Augustus I.ii.64, a far larger chart clearly based, in its coverage of the Zuider Zee, on the rough sketch but also showing the proposed passage of Anne of Cleves to England in a royal ship while a line of English ships between Dunkirk and Dover prevents French or imperial interference. Though based on à Borough's work and ideas, the draftsman may have been the king's painter Anthony Anthony, who was skilled in depicting ships. See the British Museum's *Catalogue of Manuscript Maps*, 3:195.

103. Rotz, *Boke of Idrography*. It is true that Henry lent some support to the New World voyages of Rastell (1517), Rut (1527), and Hore (1536), and to Cabot's projected voyage in 1521, which was largely frustrated by the lack of interest shown by the London merchants, the main backers. On the other hand, apart from his offhand treatment of Cabot, he signally failed to respond to the proto-imperialist arguments of the Bristol merchants Robert Thorne and Roger Barlow, even after 1536, when, with money from the dissolved monasteries, he would have had the means and—by the early 1540s, when the threat of invasion had receded—the opportunity (Andrews, *Trade, Plunder, and Settlements*, 52–56; H. Wallis et al., *Raleigh and Roanoke* [exhibition catalog, Raleigh, N.C., 1985], 21, 24–26; Joyce Youings, *Sixteenth-Century England* [Harmondsworth, 1984], 233–34).

104. Rotz, *Boke of Idrography*, 12–16.

105. Hale in Colvin, *King's Works*, 4:367–94.

Shelby, *John Rogers*, passim, utilizing L&P.

106. Hale in Colvin, *King's Works*, 4:375–76.

107. Ibid., 376–77. Starkey, *Reign of Henry VIII*, 13. Elyot, *The Book*, 1, chap. 8, pp. 23–24. Castiglione, on the other hand, felt that "our courtyer ought in no wise to leave . . . out . . . the cunning in drawing" (*Book of the Courtier*, 91). I, with Hale, incline to the view that while Henry instigated many plans he did not actually draft them. The equipment in his New Library and the elaborate hourglass-clock with two compasses, designed by Holbein as Sir Anthony Denny's New Year's gift for 1544 to the king (illustrated in Starkey, *Reign of Henry VIII*, 62), testified to Henry's interest in maps but were surely intended to facilitate his use of them, not to assist him in mapmaking itself.

108. Quoted by Shelby, *John Rogers*, 102–3.

109. "Inventory of all manner of stuff remaining in Paget Place, London," 20 February 1552/3. Greater London Record Office, Acc. 446/H/1 (Paget MSS). I am grateful to my former colleague J. Alsop for this reference. Items in a two-volume collection of plans and views of French châteaus by Jacques Androuet du Cerceau dating from about 1570, now in the British Museum (K. Top. 56, 87), may also resemble Paget's "mappe," as the similar views at Whitehall listed in the 1547/9 inventory of the contents of Henry VIII's palaces (BL Harley MS 1419 A. f. 127) may also have done.

110. See Marcus Merriman's contribution in Colvin, *King's Works*, 4:698. The plans in Belvoir Castle all relate to the Scottish campaigns of 1547–50. They are a major and as yet little exploited source for the military and cartographic history of the period and are particularly valuable given the scarcity of this type of map.

111. Ruddock, "Earliest Originals," 415–16, 420.

112. Nevertheless, he did draft the "Devyce" and "Remembraunces" of early 1539, which envisaged the creation of maps for the specific purpose of coastal defence (see notes 65, 66 above). For other suggestions of his familiarity with maps: Ruddock, "Earliest Original," passim, and Tyacke, *English Map-Making*, 16.

113. Shelby, *John Rogers*, 75ff.; Colvin, *King's Works*, 4:391, 398, 698–99; Tyacke and Merriman in *English Map-Making*, 19, 22, 59, 60, 63, 66.

114. Shelby, *John Rogers*, 87–90.

115. Colvin, *King's Works*, 3:357. The map is now BL Cotton MS Augustus I.ii.56. For its authorship, see n. 60 above. Indicative of the rising status of the engineer—and, indirectly, of his plats—was the knighthood conferred on Lee following the taking of Edinburgh.

116. Merriman in *English Map-Making*, ed. Tyacke, 59. Somerset's brother, the reckless Sir Thomas Seymour, also used maps as a political aid and a psychological weapon. In 1549, during his trial for treason, one of the witnesses mentioned how, when plotting against his brother, the Protector, Seymour "would divers times look upon a chart of England . . . and declare . . . how strong he was and how far his lands and dominions did stretch . . . and what shires and places were for him. And when he came to Bristol he would say this is my Lord Protector's and of others that is my Lord of Warwick's" (cited by Kevin Sharp, "Crown, Parliament, and Locality: Government and Communication in Early Stuart England," *English Historical Review* 101 [April 1986]: 331). I am most grateful to Geoffrey Parker for this revealing reference. That Seymour could use a map of England in this way and that his meaning could be understood by others suggests the extent to which maps as geographical aids and as symbols for reality had become familiar at court and in court circles thirty years—a whole generation—before the publication of Mercator and Saxton's wall maps.

117. Strong, *Art and Power*, 47. Jean Rotz was presented and was ennobled on this occasion, presumably in recognition of his services in supplying Henri with plans of Scotland's forts and harbours (*Boke of Idrography*, 18). It was another sign, with Lee's knighthood (see n. 115), of how highly cartographers were coming to be prized by their political masters.

118. Merriman in Colvin, *King's Works*, 613–726. Apart from two maps in Hatfield House, of Castlemilk (CMP II 27) and Broughty (CPM Supp 12 alias CP 205/78), one in the Public Record Office (MPF 257), and two maps possibly of Eyemouth in the British Library (Cotton MSS Augustus I.i.60; I.i.76), surviving maps and plans of Scottish forts—six in all—are in Belvoir Castle. For the total number of maps generated during the campaign, one would, of course, have to multiply this number many times over. And see Merriman in *English Map-Making*, ed. Tyacke, 59; Royal Commission on Historical Manuscripts, *Report on the Manuscripts owned by the Duke of Rutland at Belvoir*, 1:38. 4 vols. (London, 1888–1905)

119. E.g. Rossetti's plan of Broughty of 1547 (CPM Supp. 12), discussed by Skelton, Hatfield House Maps, 60, no. 85, reproduced in Tyacke, *English Map-Making*, pl. 28. A slightly earlier example from North England is Scala's map of Tynemouth of 1545 (BL Cotton MS Au-

gustus I.ii.7), reproduced in *English Map-Making*, pl. 24.

120. Notably the "platte of Castlemilk" (Hatfield House CPM II 27, reproduced in *English Map-Making*, ed. Tyacke, pl. 20), and see Marcus Merriman, "The Platte of Castlemilk 1547," *Transactions of the Dumfriesshire and Galloway Natural History and Antiquarian Society* 44 (1967): 175–81; Hatfield House Maps, 59–60, no. 82.

121. PRO DL 44/764 (anonymous but probably dating from Henry VIII's reign); PRO MPF 257 by Henry Bullock (d. 1561), another master-mason by training like John Rogers. This map is reproduced and discussed in Tyacke and Huddy, *Christopher Saxton*, 14–15. Laurence Nowell is recorded as having made yet another, now lost, map of the Debatable Lands in the 1550s (T. Randolph to Sir W. Cecil, St. Andrews, 25 April 1562, *Calendar of State Papers Foreign, 1561–1562* (London, 1866), 63, no. 1051).

122. Tyacke and Huddy, *Christopher Saxton*, 14–15. Also PRO SP 59/5 ff. 44–45 (Liddesdale, 1561); MPF 285 (1590). BL Royal MS 18.D.III f. 76 (map 33; 1590); Cotton MS Titus F.xiii f. 201 (1592). Hatfield House CPM I.3 (1597).

123. Shelby, *John Rogers*, 72, 76, 86ff., 102–111 (Boulogne), 121–26 (Calais). Colvin, *King's Works*, 3:369–70, 370–75, 401–2 (Calais and Pale), 383–93 (Boulogne). Edward VI discusses them in his journal: *The Chronicle and Political Papers of King Edward VI*, ed. W. K. Jordan (London, 1966), 35 (13 June 1550).

124. BL Cotton MS Augustus I.ii.71, discussed in Colvin, *King's Works*, 3:358, n. 3, 4:397, and Hatfield House Maps, 63, no. 93 (a propos of CPM I.55, a map of the defenses of Calais, possibly by Robert Lythe). It should be noted that Petyt's map must be based on several now-lost surveys of single forts in the Calais Pale—implicitly suggesting, once again, the high rate of loss of these early plats. For Petyt's earlier activity in Scotland, see *King's Works*, 3:357, 4:700 and passim. For maps of Calais executed while he was surveyor there, earlier under Henry VIII (1546–47), see BL Cotton MS Augustus I.ii.57b. Other engineers who served in France and Scotland (often seemingly commuting between the two) were Richard Lee, John Rogers, and Giovanni di Rossetti.

125. *Boke of Idrography*, 13–16. Moir, "History of Scottish Maps" 21. And, for the Scottish maps, see notes 80 and 81 above.

126. *Calendar of State Papers: Foreign, 1547–1553* (London, 1861), 70.

127. In June 1548: Colvin, *King's Works*, 4:397.

128. Edward VI, *The Chronicle and Political Papers of King Edward VI*, ed. W. K. Jordan (London, 1966), 97 (26 December 1551).

129. Armando Cortesao and Avelino Texeira da Mota, *Portugaliae Monumenta Cartographica* (Lisbon, 1960), 2:5, 7; henceforth PMC. By 1547 the Portuguese king, John III, was offering Diogo Homem a free pardon if he returned to Lisbon within two months. This magnanimity may have been due to pressure from Lopo Homem, and may reflect doubt as to his culpability (he had jumped bail while awaiting trial), but it surely also reflects his value to his countrymen.

130. *Boke of Idrography*, 15. Andrews, *Trade, Plunder, and Settlement*, 64–65, 101, 136–37, gives 1548 as the year of his return. A. A. Ruddock, "The Reputation of Sebastian Cabot," *Bulletin of the Institute of Historical Research* 47 (1974): 95–98.

131. The most notable was Simão Fernandes (for whom see chap. 3 in this book, n. 59), but there were others such as Antonio Anes Pinteado, Francisco Rodrigues, and Bartolomeu Bayão (D. Waters, *The Art of Navigation in England in Elizabethan and Early Stuart Times* [London, 1958], 82, 89–90, 113, 120–21, 151, 496, 535–36) Andrews, *Trade, Plunder, and Settlement*, 106–7, 138–39, 167). For the Portuguese pilots used by Drake during his "Famous Voyage": Helen Wallis, "The Cartography of Drake's Voyage," *Sir Francis Drake and the Famous Voyage 1577–80: Essays Commemorating the Quadricentennial of Drake's Circumnavigation of the Earth*, ed. N. Thrower (Berkeley, Calif., 1984), 131.

132. Andrews, *Trade, Plunder, and Settlement*, 64ff. England's dependence—beyond court circles—on foreign cartographers was made explicit in the official judgment on Diogo Homem's attempt to secure payment for one of his charts in the English High Court of Admiralty in April 1547. The high price he had asked (one hundred gold ducats) was accepted as being justified "having regard . . . to the wante and lack of expert lernyd men in that facultie of makying of cartes or mappes and the scarcyte and price of such cartes withein this realme of England" (PMC, 2:7).

133. Taylor, *Mathematical Practitioners*, 12. 18, 20 (pl. 2), 168. The quadrant made by Cheke for Edward is now owned by the British Museum (M&LA 58. 8–211). T. A. Birrell has pointed out that, probably at the same time, Cheke purchased for the king Robert Recorde's *Pathway to Knowledge . . . First Principles of Geometrie*

. . . *applied unto practice, bothe for the use of instruments . . . also for projection of plattes* (London, 1551). The copy now forms part of the "Old Royal Library" in the British Library. See T A. Birrell, *English Monarchs and Their Books from Henry VII to Charles II* (London, 1987), 15.

134. E.g. *Chronicle and Political Papers,* 140–41. E. Pognon, "Plus Anciens Plans."

135. BL Harley MS 1419 A & B, partly edited—but only for the grander paintings—as W. A. Shaw, *Three Inventories . . . of Pictures in the Collections of Henry VIII and Edward VI* (London, 1937). Helen Wallis has utilized them in her article "The Royal Map Collections of England," *Revista da Universitade de Coimbra* 28 (1980): 461–68, and I am grateful to her for drawing my attention to them.

136. Colvin, *King's Works,* 4:2, and see also 4:1–366 passim.

137. BL Harley MS 1419 A, f. 246.

138. F. 414.

139. F. 59.

140. F. 135. The inventory lists these items under "Westminster," but a fire in 1512 "evidently terminated the history of the palace [of Westminster] as a royal residence" (Colvin, *King's Works,* 4:287), and in 1536 Cardinal Wolsey's former palace was officially renamed the "new palace of Westminster" or "the Kyng's Paleys at Westmynster," of which the old palace formed "a member and parcel," though from 1542 the new palace was colloquially called Whitehall (*King's Works,* 4:301).

141. The journal of Johann Ernst, duke of Saxe-Weimar (1613), in *England As Seen by Foreigners in the Days of Elizabeth and James I,* ed. William Brenchley Rye (London, 1865), 159–62, 164.

142. Colvin, *King's Works,* 4:17.

143. George Kish has perceptively noted how decorative, painted "maps and perspective views were favourite . . . means of symbolizing knowledge and power" in sixteenth-century Italy ("The 'Mural Atlas' of Caprarola," *Imago Mundi* 11 [1953]: 52); and see Victor Morgan, "The Cartographic Image of 'The Country' in Early Modern England," *Transactions of the Royal Historical Society,* 5th ser., 29 (1979): 141.

144. The following are listed in BL Harley MS 1419 A, ff. 133–35. It is possible, given its age and general appearance, and the provenance of other maps in the Cotton Augustus series, that the "description of Italie of parchment sette in a frame of woode" listed in the inventory may be the fifteenth-century Italian map "Italie provincie modernus situs" surviving as BL Cotton Roll xiii. 44.

145. An odd companion for the other, major, cities, but those prolific later fifteenth- and early sixteenth-century chartmakers Grazioso and Andrea Benincasa and Conte Ottomano Freducci were patricians of Ancona, and, judging from the wording of the authorship inscriptions on their work, they were proud of it (and see M. Emiliani, "Le Carte Nautiche dei Benincasa Cartografi Anconitani," *Bolletino della Reale Società Geografica Italiana* 8, no. 1 [1936]: 2–27). Could one of them have produced this map-view? The other views of famous towns may have been intended as a subtle counterpoint to the famous actions also depicted in the Privy Gallery. Jurgen Schulz has argued that this was the case with the display of town views in certain Italian palaces at the turn of the fifteenth and sixteenth centuries ("Jacopo de Barbari's View of Venice," *The Art Bulletin* 60 (1979): 465–67.

146. Scarisbrick, Henry *VIII,* 184.

147. This is a "mappe," not the painting, also listed in the inventory, which Sir Oliver Millar has identified as the one that is now in the Royal Armories, Tower of London (*Queen's Paintings,* 225, n. 9), and is reproduced in Starkey, *Reign of Henry VIII,* pl. 32.

148. F. 127: with them, in a manner recalling Paget's London home (n. 109 above), Henry had "fygures" of the royal French châteaus of Amboise, Cognac, etc.

149. F. 58.

150. Ff. 245–46.

151. F. 133. It may be significant that its coverage must have been almost identical to Laurence Nowell's "general description of England & Irelande with the costes adioyning" of c. 1564 (BL Add. MS 62540). And see chap. 3 in the present volume.

152. BL Cotton MS Augustus I.i.9 (reproduced in Tyacke and Huddy, *Christopher Saxton,* 8).

153. F. 135. It seems still to have been on display in the Privy Gallery in 1613 (Rye, *England As Seen by Foreigners,* 164). It has been suggested by Skelton that the Harleian *mappemonde* was produced in England, since it shows the fictitious "Sea of Verrazano" which was probably unknown to the Dieppe mapmakers, and that it was later on view in the Privy Gallery in Whitehall during Elizabeth I's reign (R. A. Skelton, "Raleigh as a Geographer," *The Virginia Magazine of History and Biography* 71 [1963]: 135). For a more recent discussion see Wallis, "Some New Light," 101–2.

Skelton's speculations fail to explain why, if made in England, the chart should display the arms of Henri II as king of France, as he was from 1547, with his arms as dauphin; why Elizabeth should wish to display a chart with a French king's arms in her Privy Gallery when there was already a very similar one with her father's arms on show there; and why, if it was displayed, it was not remarked on by visitors, such as a French ambassador, who might be expected to have noticed. It seems more reasonable to suppose that the map was produced in France at the time of Henri's accession (31 March 1547) and the return to France in June of Rotz and Nicolay, who may well have seen the map Verazzano had presented to Henry VIII in 1528 or globes based on it.

154. F. 134.

155. F. 447. In 1584 Richard Hakluyt recorded the existence of another globe in the Privy Gallery apparently of "Verarsanius makinge" and probably presented to Henry in about 1530. It showed the Sea of Verazzano, the supposed existence of which acted as a great spur to English maritime enterprise under Elizabeth I, through offering the possibility of a shorter passage to Cathay (Wallis, "Some New Light," 101ff.).

156. E.g. Humphrey Gilbert (c. 1566), Hakluyt (mentioned in *Principal Navigations,* 1589), Purchas (c. 1615). The map may also be included among "Sebastian Gabots Maps" mentioned in a catalog of maps in the Royal Library of 1661–66 and in a list (20 April 1685) of maps from the Royal Library transferred by Samuel Pepys to Lord Dartmouth (R. A. Skelton, "The Royal Map Collections of England," *Imago Mundi,* xiii [1956], 181–83; R. A. Skelton, "The Royal Map Collections," *British Museum Quarterly* 21 [1962]: 4–6, particularly n. 5; Wallis, "Some New Light," 102–5).

157. The only surviving world map by Sebastian Cabot, of 1544, is known in two examples, of which the most generally familiar is in the Bibliothèque Nationale in Paris (Rodney W. Shirley, *The Mapping of the World* [London, 1983], 92–93, no. 18). It does not show the Northwest Passage, but Cabot was a complex and secretive character (Wallis, "Royal Map Collections," 464, and the articles by Skelton on the same theme cited in the preceding note).

158. BL Harley MS 1419 A, ff. 186–88. Toward the end of Elizabeth's reign Robert Beale, Francis Walsingham's secretary, remembered that "heretofore there was a chamber in West[minster] where such things towardes the latter end of K. Hen. 8 were kept and not in the Secretarie's private Custodie" ("A Treatise of the Office of a Councellor and Principall Secretarie to her Ma[jes]tie," in Conyers Read, *Mr Secretary Walsingham and the Policy of Queen Elizabeth* (Cambridge, 1925), 1:431). For the identity of Westminster and Whitehall, see n. 141 above. As late as 1660 the diarist John Evelyn recorded seeing a room in Whitehall that sounds remarkably similar to the New Library (Skelton, "Royal Map Collections," BMQ, 5).

159. The "plat" so endorsed is an architectural plan of the Exchequer at Calais (BL Cotton MS Augustus Supp. 7) which has been dated to 1534 (Colvin, *King's Works,* 3:349–50). The endorsement was deleted some decades later. For the users' changing perception of maps in sixteenth-century England—and Europe—as, initially, a type of painting and later as scientific objects, see J. B. Harley, "Meaning and Ambiguity in Tudor Cartography," in *English Map-Making,* ed. Tyacke, 25–26, and note how in the 1520s and 1530s, Castiglione and Elyot discuss maps under the headings "Painting" and "Cosmography," whereas from the 1550s mapmaking tends to be discussed as a manifestation of applied mathematics.

160. Possibly Henry's globe showing the 'Sea of Verazzano' which might have been considered too politically sensitive a feature to be displayed publicly in 1547 (and see n. 157 above).

161. For Denny's suitability as a guardian of this material, see Starkey, *Reign of Henry VIII,* 132–36. As keeper of Whitehall Palace and chief gentleman of the Privy Chamber, he was "the key figure in the struggles over Henry's death-bed." He was also associated with the "new learning," with Elyot, and with George Lily (as a youth), and was a patron of several leading scholars under Elizabeth. He would, therefore, have appreciated the value of the maps in his keeping from the standpoint of technique as well as of politics.

162. Colvin, *King's Works,* 4:418.

163. BL Harley MS 1419 A, ff. 134–35.

164. BL Harley MS 1419 A, ff. 186–87.

THREE

England II: Monarchs, Ministers, and Maps, 1550–1625

Peter Barber

Court, Country, and Cartography

The middle years of the sixteenth century were marked by social, economic, and political crises that coincided with a change in generation of the English leadership. "The great year of rebellions," 1549, brought unrest to large parts of the realm.[1] The underlying economic and social problems were worsened by the drying up of the traditional European markets for English wool and by financial troubles that since the closing years of Henry VIII's reign the Crown had sought to avoid by successive debasements of the currency. To make matters still worse, in 1549–50 the French recaptured Boulogne and the Boulonnois, and soon afterward Franco-Scottish forces drove the English from the Scottish Lowlands and across their own borders. Eight years later, England's participation in yet another war against France, at the side of Mary's husband Philip II, king of Spain and titular king of England, brought the loss of Calais, England's last Continental possession and its last link with what were perceived as the glorious years of Edward III and Henry V.

The period also witnessed change in England's leadership, as the generation that had emerged in the 1530s and earlier gradually disappeared. This should not, of course, be exaggerated. Some veterans such as William Paulet, marquess of Winchester, and Sir Ralph Sadler, who had entered royal service in the 1520s and 1530s, were still advising the Crown in the 1570s and 1580s. Moreover, the accidents of death, of Edward VI in 1553 and of Mary in 1558, would have brought change regardless of the shift of generations and internal and external crises. Nevertheless, in the course of the 1550s the political and intellectual leadership of the country passed into the hands of men (one thinks primarily of William Cecil, Lord Burghley and of John Dee) who were still, in many cases, to be in control in the closing years of the century.

These changes brought important developments in the relationship between cartography and government. The first and most negative was the ending of the fortification-building program that had provided the spur for the enormous amount of official mapmaking since the 1530s.[2] The results were drastic for several of the engineers concerned. In May 1551 John Rogers found himself recalled from southern Ireland, where he had been making plats of Cork, Kinsale, and Baltimore, and five years elapsed before he was fully employed again.[3] Other engineers were still less fortunate and found themselves forced, like Giovanni Portinari, to take service abroad or starve.[4] Throughout the remainder of the century and for much of the next, the Crown would not again have the means at its disposal to undertake direct patronage of cartography on the scale that Henry VIII had after 1539.

There was, however, no danger that maps would be neglected and eventually disappear as instruments of government. For with the new leadership came a growing sophistication regarding maps in the country as well as at court. By 1550 and still more so by 1580, as the research of Victor Morgan, Paul Harvey, and others has shown, map consciousness—the ability to think cartographically and to prepare sketch maps as a means of illuminating problems—was becoming ever more widespread.[5] In the same period, the advances made by English mathematical theorists, popularizers, and practitioners, which were first analyzed by E. G. R. Taylor, and the appearance of native mathematical instrument-makers like Humphrey Cole and Augustine Ryther (who also worked as map engravers) made possible a great improvement in the mathematical precision of English mapmaking, with the maps beginning to resemble the modern layperson's idea of what a map should look like.[6] The expansion of English overseas enterprise, which also occurred after 1550, acted as a stimulus, with these factors, to the production of more accurate maritime charts by Englishmen.[7] By the second half of Elizabeth's reign, these were being manufactured in some number and on a regular basis.[8] Earlier still, not only military engineers but also people of other, widely differing backgrounds, such as clergymen, antiquaries, university dons, and even members of the gentry, had begun to draft maps to the highest mathematical standards to assist in the study, or propagation, of matters of particular interest to them.[9]

Crown and ministers were not immune to these changes. Ministers came to expect a greater precision in maps than had their predecessors, and several became more sophisticated in their evaluation of, and their awareness of the potential uses of, maps for government. And the Crown's means were not so straitened that it could not to varying degrees encourage developments in mapping and charting which were felt to be useful to the state. Moreover, the foreign and domestic crises that punctuated the reign and became ever more menacing after 1583 continued to generate a need for maps of almost all conceivable types and of all conceivable geographical areas. To meet these needs, the Crown was now able to call not only on the services of its own often highly accomplished servants but also on the patriotism of an increased number of other citizens who were also skilled mapmakers or appreciated the value of good maps and were prepared to commission them. Thus, while there was a decline—absolutely as well as relatively—in the role of the Crown, though not of ministers as semiprivate individuals, in the field of direct cartographic patronage, the later sixteenth century saw the extensive use of maps—and often of maps of considerable sophistication and precision—by government.

Nevertheless, after 1550 most maps, charts, and plans were produced not for the Crown but for those groups in society that were better able to pay for them and now grasped their importance: predominantly merchants and members of the landed gentry. By the mid-1550s London merchants, through the Muscovy Company, were employing mariners of the caliber of Stephen and William Borough, John à Borough's sons, and encouraging their collaboration with theorists such as Robert Recorde, Richard Eden, and John Dee, to produce scientifically constructed charts to guide their ships in search of much-needed new markets.[10]

By the 1580s the merchant William Sanderson made possible the accurate mapping of the Carolina Outer Banks by Thomas Harriot and John White, through his support for the colonizing efforts of Sir Walter Raleigh.[11] The gentry were less active in financing chart and mapmaking than the merchants.[12] However, from the 1570s the desire to protect their lands (an increasingly precious commodity) in courts of law, and to manage them efficiently, and motives of family honor and pride of possession induced many country squires to become very active patrons of the emerging estate surveyors and of their plans—drawn to a consistent scale and laden with pictorial and heraldic images.[13]

There was, however, no dichotomy between Crown and country. The Crown was also genuinely interested in the expansion of England's overseas trade and influence.[14] In order to be able to encourage and, from a distance, control and direct this enterprise, the Crown needed to be kept informed of its progress and of the environment in which it occurred, through maps and charts. Ministers and courtiers acted as a bridge between Crown and city, for while they were executors of royal policy, individually they were also often closely involved with the chartered companies and, as investors, with particular voyages. Moreover, the queen herself occasionally played the merchant and gave covert assistance to Drake's and Raleigh's overseas enterprises.[15] Ministers were also often great landowners and as such shared the gentry's concerns and their interest in estate cartography. Common intellectual interests resulted in some ministers' exchanging ideas with the leading theorists by letter or in person. Thus, when discussing the patronage of cartography in late sixteenth-century England, one is talking not of royal/ministerial *or* merchant/gentry patronage but of a continuum extending from, in a few cases, complete and direct royal patronage to a few cases of patronage by merchants or country gentlemen alone.

Patronage and Provision

Direct Support

Despite the economies forced on government after 1550, the Crown continued to be responsible for the defense of the realm, and for its administration. Maps were used in both areas, and the government found itself the paymaster of skilled platmakers such as Richard Popinjay, surveyor of Portsmouth from 1563 to 1590;[16] the artist-architect Robert Adam (the son of Sebastian Cabot's engraver, the artist-schoolteacher Clement Adams), who briefly served also as surveyor of the Queen's Works;[17] Paul Ivey or Ives, an engineer who from the 1570s until his death in Kinsale in 1604 was to serve in the Netherlands, Dover (1584), the Channel Islands (1593–95), Portsmouth (1597), Ostend (1596–99), and Ireland (1602–4), with briefer periods in Falmouth and the Cinque Ports;[18] and Rowland Johnson, who was, by contrast, active only in Berwick in the 1560s.[19] Probably the most significant pattern to be discerned in the period is the increasing reliance on native talent. Italian engineers, such as Giovanni Portinari and Jacopo Aconcio, also known as Contio, disappeared from the scene after 1566, so that the employment of Federico Genebelli between 1585 and 1602 was very much of an exception, though it has been suggested that Robert Lythe may have been of Flemish origin.[20]

By the 1560s, however, the government was no longer dependent on military engineers, or indeed on its own official servants, for the provision of maps and plans, even in the military and defense spheres. It could, for instance, turn to well-educated and patriotic country

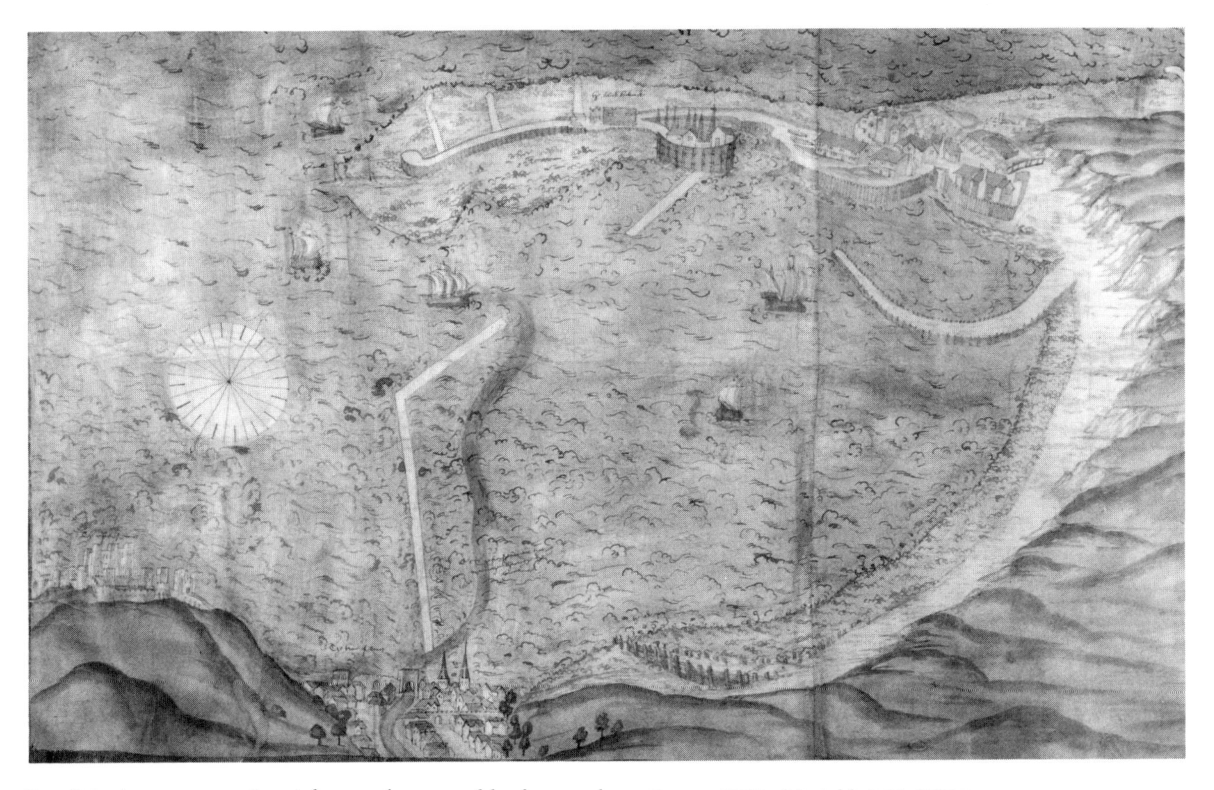

FIG. 3.1 Anonymous pictorial map of proposed harbor works at Dover, 1552. BL Add. MS 69824

squires like Thomas Digges and Sir Edmund Yorke. Digges was a distinguished mathematical theorist in his own right who probably had nothing to learn from the engineers in terms of cartographic skills.[21] In the 1580s and 1590s he was employed in advising on and making plats of yet more of the innumerable proposals for improving the defences of Dover (figure 3.1).[22] Yorke, the son of Sir John Yorke, master of the mint under Edward VI and Mary, had a crude but effective technique, and plats accompanied his military and diplomatic career in Norfolk, Suffolk, Ireland, and France from about 1588 to his death in 1592.[23] Even generals and admirals such as Sir John Norris and Sir Martin Frobisher showed at the unsuccessful siege of Brest in 1594 that they were quite capable of producing adequate sketch maps to illustrate their letters.[24]

Occasionally distinguished cartographers better known for their work in other areas were also temporarily utilized by government. Thus we find Robert Norman and William Borough charting the positions of sandbanks and channels at the mouth of the Thames, the key to London, at times of national crisis in 1589 and 1596,[25] and Ralph Treswell the Elder, who is best known for his estate and building surveys, acting as mapmaker to Sir John Norris during his expedition to Britanny in 1594, and producing a competent map of the province while he was there.[26] Nevertheless, the most representative work by Borough, Norman, and Treswell was done for private and corporate patrons, who sometimes offered greater challenges and possibly better pay.[27] Several leading cartographers and mathematical theorists such as John Dee, John White, Thomas Harriot, William Bourne, and Edmund Gunter received no official employment at all, even if some of them received favors from ministers acting in a private capacity.[28] By contrast, government seems to have shown a growing appetite for printed maps, which were cheaper, increasingly plentiful, and less prone to scribal errors in transmission than their manuscript counterparts. By the end of the sixteenth century and at the beginning of the next, increasing use was being made of them, to illustrate sieges, and more subtly, for instance, to accompany muster lists.[29] Most printed maps, however, seem to have been regarded as ephemera and were used roughly, with the result that few with clear links with government survive.[30]

FIG. 3.2 Detail from an anonymous map of Leix and Offaly, Ireland, c. 1563. BL Cotton MS Augustus I.ii, 40

The Crown's greatest cartographic achievement in the second half of the sixteenth century is to be found in Ireland. Building on the work done by Robert Dunlop at the beginning of the twentieth century, J. H. Andrews and Gerard Hayes-McCoy and their students have studied the matter so thoroughly in recent decades that little needs to be said here.[31] The challenges facing English governments in Ireland after 1550, enormous, varied, and everchanging as they were, called for some idea of the island's geography if ministers in London were even going to begin to confront them. As of 1550, the English had little knowledge of Ireland beyond the Dublin Pale.[32] By 1610, by contrast, ministers were familiar with the physical and political geography of the kingdom—in places in considerable detail.[33] The government and its representatives in Ireland played a leading part in initiating and supervising the mapping of the country that took place in the interim, even if, as might be expected in an area that could be regarded as the Tudors' "moving frontier,"[34] the mapping activity itself was piecemeal and tended to go by fits and starts.[35] By the middle of James I's reign the results were nonetheless impressive, with Robert Lythe's surveys of Munster and Leinster (1568–71), the two John Brownes' survey of Connaught (1580s), Richard Bartlett's survey of Ulster (1597–1603), and Francis Jobson's work throughout the island providing a geographical basis for its government and administration (figure 3.2). Cartographers were also involved in the more detailed mapping of plantations with which the Crown was increasingly involved as Elizabeth's reign proceeded, and in the planning against Spanish invasion and combating of the internal insurrections that characterized Irish history after 1550. By 1590 the need to protect the English position in Ireland created the need for forts in the interior, and their building was inevitably charted in numerous plats. Collectively, this officially sponsored mapping activity helped to bring about the conditions of relative stability necessary for the survival of the plantations. These in their turn created a demand on the part of planters and proprietors for detailed estate maps, which were to be the most common form of cartographic activity in Ireland after 1620. English estate surveyors had begun to move across the Irish Channel at the behest of private patrons as early as 1586, however, and some of their work, such as the fine plan of Sir Walter Raleigh's estate of Mogeely, County Cork, in 1598, is the equal of the best then to be found in England.[36]

Indirect Support

There was one other way in which the Crown was active as a patron of cartography, albeit less obviously. This was as a provider of employment, pensions, or lands for worthy mapmakers who worked largely for private patrons. Thus Humphrey Cole, engraver of the first biblical map to be printed in England, a pioneer of English mathematical instrument-making, and a mathematical theoretician of note, was given employment—and income—as an engraver and die-sinker in the Royal Mint. Similarly, Edward Wright, one of the most brilliant mathematicians of his day and the first to develop and promote the use of the Mercator projection, was, toward the end of his life, appointed tutor and later librarian to Prince Henry, James I's eldest son.[37]

The most obvious examples of this sort of indirect official patronage and support are to be found, however, in the fields of national mapping and of navigation. The importance of both to Crown and people in the sixteenth century need not be emphasized and was clear to contemporaries, but no contemporary government, least of all the impecunious governments of Mary and Elizabeth, had the means to support and sustain thoroughgoing surveys on land or at sea. Varying degrees of official support for private initiatives therefore came to characterize these fields of cartographic activity.

The Mapping of England

As early as 1524, Nicholas Kratzer had talked of producing a new map of England in a letter to Dürer,[38] and Elyot indirectly mentioned the value of such a map in 1531.[39] The first post-medieval non-Ptolemaic printed map of Great Britain originally appeared in Rome in 1546, as an illustration to Bishop Paolo Giovio's *Descriptio Britanniae, Scotiae Hyberniae, et Orchadum.* Its creator was George Lily (d. 1559), the son of Thomas More's close friend William Lily. George Lily served as chaplain to Cardinal Reginald Pole, a cousin of the king and, in effect, the leader of the Catholic opposition to the Protestant leaders surrounding Henry VIII and Edward VI. It seems inconceivable that Lily could have completed the map without the cardinal's full support and approval. The whole project, like the publication of Giovio's treatises, was political in inspiration: a part of Catholic Europe's propaganda war with Protestant groups inside England in Henry VIII's last years and in the wake of his death. Following Mary's accession, Pole, now archbishop of Canterbury and papal legate, may well have assisted Lily (who was to become a canon of St. Paul's in London and of Canterbury Cathedral), in transporting the original plate to England

and in ensuring its publication there, in 1555, by the Flemish-born instrument-maker, engraver, and platprinter Thomas Gemini.[40]

The map itself appears to have been widely used,[41] but it was nevertheless very much the map of an exile. It showed great advances in the depiction of Scotland, which, being a Catholic country, still had close links with Rome.[42] Its depiction of England, however, represented little advance on the maps derived from the medieval "Gough" map. For, like his master, Lily was unable to benefit from the advances made in the depictions of England's coastlines by English engineers since Henry VIII's break with Rome. Moreover, the map's scale limited its usefulness as an instrument of government or administration.

By the 1560s Elizabeth I's principal minister Sir William Cecil, who was to be made Lord Burghley in 1571, was clearly on the lookout for a cartographer to undertake a larger, county-by-county survey of England and Wales. From 1561 to 1563 John Rudd (1498–1579), vicar of Dewsbury and a prebendary of Durham Cathedral, was given paid leave, at the queen's request, to view "divers parts" of England. Rudd, whose experience of mapmaking dated to at least the early 1530s and who had produced a map of England in the 1550s, had been associated with the court from 1540, as clerk of the closet to Henry VIII. It is thus possible that he had been familiar with the many lost plats of parts of England produced by Henry's engineers from 1539. Despite these links with and indirect assistance from the court—and the aid of his assistant, the young Christopher Saxton—Rudd's survey was never completed.[43]

In June 1563, just when the likelihood of this failure was becoming plain, a member of Burghley's household, the antiquarian and

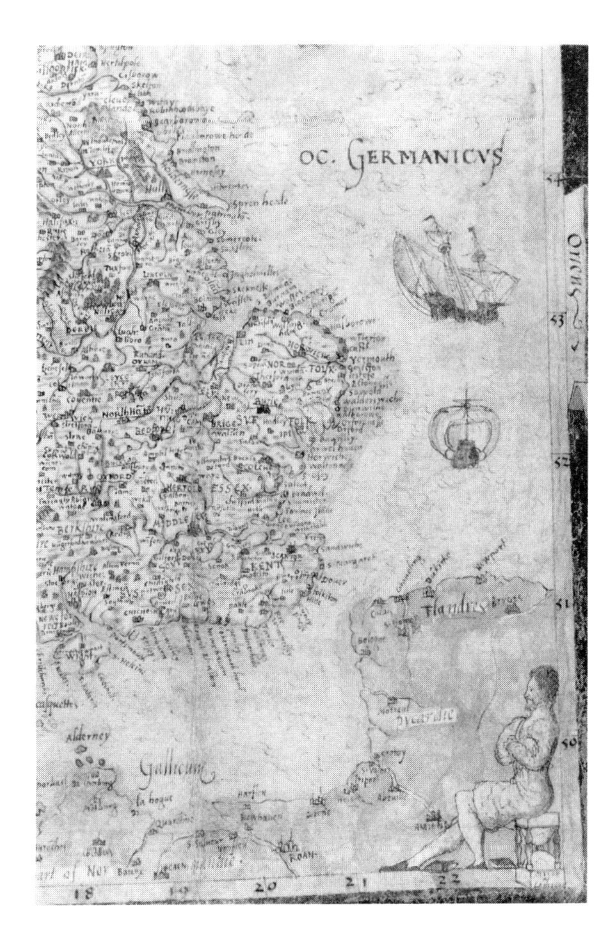

FIG. 3.3 Detail from Laurence Nowell, "A General Description of England and Ireland with the costes adioyning," c. 1564. BL Add. MS 62540 f. 4

philologist Laurence Nowell, pleaded for permission to undertake a similar survey, in a letter containing the earliest surviving reference to Burghley's interest in maps for the purposes of government. Nowell was authorized to produce a specimen map of the realm. The result, a pocket map, his "general description of England and Ireland with the costs adioyning" (1564), is a remarkable achievement by any standards (figure 3.3). But despite his own surveying work in Wales and Ireland (which he is now known to have visited in 1560), he could hardly have obtained all the detailed and accurate information that underpinned the map without recourse to existing official plats and to the libraries of Burghley and the archbishop of Canterbury, Mathew Parker, which he is known to have consulted.[44]

For reasons that have never been satisfactorily explained, Burghley decided against Nowell's proposals, and, superficially, it would appear that with this his interest lapsed. Christopher Saxton's great atlas of maps of the English counties, published in 1579, which provided the basis for his wall map of 1583, was largely paid for by Thomas Seckford, a master of the Court of Requests and surveyor of the Court of Wards and Liveries (1581). The recent researches of Ifor Evans, Heather Lawrence, William Ravenhill, and R. A. Skelton, among others, have demonstrated, however, that Burghley's and the Crown's interest remained as lively as ever. It was reflected in Saxton's selection of the county as the unit of survey—the county being the basic unit of local administration with which the central government dealt, as opposed to the English diocese or hundred. It found expression, too, in the speed with which Burghley acquired proof sheets of the atlas in the years preceding 1579. It is suggested by the close links between Seckford and Burghley, who was master of the Court of Wards and seems to have known Seckford since the early 1540s, when they studied law together at Gray's Inn. It is most convincingly indicated, however, in the indirect official support Saxton received in the form of passes that put at his disposal all the facilities of the local administration, in the printing privilege (1577) that allowed him a ten-year monopoly on the production and sale of his maps, and in the grants of lands and offices made to him by the Crown during and immediately after his great survey.[45]

For all the information it provided, for the first time, in printed form, Saxton's atlas did not meet all the official requirements.[46] Most notably, it did not give equal treatment to all the counties, several of which were crowded together on a single sheet,[47] and it showed neither roads nor (consistently) the divisions of the shires. Within fifteen years, the estate surveyor, topographer, and (probably) religious controversialist John Norden had set out to remedy these omissions in his series of county topographies cumulatively called the *Speculum Britanniae.*[48] The single volumes boasted a number of patrons, some of whom had little connection with government, such as the Countess of Warwick, to whom a copy of the manuscript survey of Hertfordshire was dedicated;[49] another, William Waad, who financed Norden's survey of Middlesex, was a clerk of the Privy Council, like Elyot.[50]. Burghley's and the Crown's interest was not far behind, though. It emerges not only from other dedications in the series[51] and from the words of Norden's *Preparative* (1596), in which he explains the purpose and methods employed in the *Speculum,* but also in Norden's printing privilege (1592) and official passes (1594), which went further than Saxton's in that they also authorized him to search local archives and libraries for relevant material.[52]

There was one other attempt at a national survey before the close of Elizabeth's reign. This resulted in the so-called anonymous series of twelve county maps, printed in the Netherlands in 1602–3, and now known to be the work of the herald and topographer William Smith.[53] These maps, too, show an improvement on Saxton, and Smith may have received some official support. His appointment as Rouge

Dragon Poursuivant in the College of Heralds in 1597,[54] for which ministerial support would have been opportune though not essential, came at a period when royal support for Norden's county series had lapsed (possibly on account of his association with the earl of Essex and the intemperate tone of some of his religious tracts)[55] and when Smith, who already had a reputation as a topographer,[56] may have been beginning work on the series.

Charting the Coasts

There was a similar pattern of semiofficial patronage in the field of hydrography. It was thanks to Mary I's unpopular marriage with Philip II of Spain that, following the success of his Russian voyage of 1556, Stephen Borough was in 1558 able to visit the most advanced mapping establishment in Europe, the navigation school of the Casa de Contratación in Seville, as an honored guest.[57] Borough came away deeply impressed by the Casa's standards, training, and organization and determined that these should be emulated inside England. As a first step he persuaded the Muscovy Company to sponsor the translation of a standard manual of navigation and chart-making used by the Spaniards themselves, Martin Cortés's *Arte de navigar,* from a copy he had brought back. The book, translated by Richard Eden as *The Arte of Navigation,* and published in 1561, has been called "one of the most decisive books ever printed in the English language. It held the key to the mastery of the sea."[58] Borough's longer-term ambition, the appointment of a chief pilot of England who, like the Spanish *"piloto major,"* would supervise the training and ascertain the competence of all English mariners, remained unfulfilled (though in January 1564 Elizabeth I got as far as drafting a commission for Borough as "Cheyffe Pilote of this owr realme of Englande" along the lines he desired). All was not lost, however, for Borough was able to carry through several of his ideas later in the reign as Master of Trinity House (officially Trinity House of Deptford Strand, to distinguish it from the other Trinity houses), which came to be the nearest English equivalent of the Casa de Contratación. The results contributed to the achievements of English navigators after 1570, though as late as the 1580s there was still a considerable dependence on the expertise of foreign-born sailors, such as the Portuguese Simão Fernandes, for charts and navigation in non-European waters.[59]

There was much ignorance within England even as to the coasts of Europe. Hence, in 1585–6, at the suggestion of Lord High Admiral Charles Howard,[60] the Privy Council apparently decided to support the translation into Latin of Lucas Janszoon Waghenaer's *Spieghel der Zeevaerdt,* first published a year earlier in Dutch, the first purpose-made printed atlas of sea charts with accompanying sailing directions and related texts to be published in Europe.[61] The reason for the decision was simple: the atlas contained, in convenient form, much that was essential for the defense of English shores against the expected Spanish invasion, and more than most English mariners already knew. John Dee and others had long complained of the way in which foreign sailors had been furtively charting English coastal waters.[62] The next stage in the process of familiarizing English sailors with Waghenaer's work was its translation into English. It is another sign of the importance attached to this task by the government that, though nominally a private venture, it was undertaken by a clerk to the Privy Council, Anthony Ashley, at the behest of the lord chancellor, Sir Christopher Hatton.[63] Though apparently commissioned in 1587, it was only published a few weeks after the Armada had come and been dispersed in 1588. Nevertheless, its value was soon realized. Burghley secured a copy to join the Latin example he already had, and he may have had a Dutch example in addition,[64] while Waghenaer's name, transmuted into Waggoner, slipped into the English language as a synonym for an atlas of sea charts (color plate 2).

Maps of Distant Shores

In the context of England's overseas expansion after 1550, it would be wrong to talk of any official patronage of cartography of the type to be found in Spain or Portugal or in France before 1560.[65] Yet even here maps played their part in the intercourse between Crown, companies, and individual adventurers. Maps and charts of distant shores would be sent, if only as a matter of politeness, to ministers[66] and could not be refused if requested by them, particularly if, as private individuals, they were investors themselves. This presumably accounts for the couple of explorers' maps to be found in atlases known to have been owned by Burghley (figure 3.4).[67] But the advantage was not all one way. By the presentation of maps subjects could seek to influence royal policy. Thus, John Dee's map of part of the Northern Hemisphere of 1580, containing a justification of English imperialism on its back, was an attempt to persuade the queen to support overseas ventures more actively.[68] Similarly, Francis Drake might well have hoped that the world map showing his discoveries and the lands he claimed for England during his circumnavigation of the globe in 1577–80, which he presented to Elizabeth in October 1580, would persuade her to support the planting of English colonies in South America.[69] In these instances

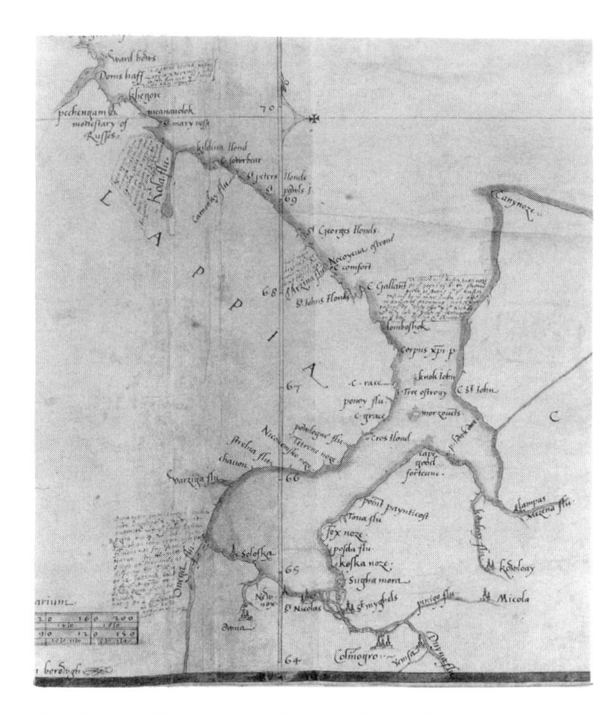

FIG. 3.4 William Borough, "Sayling Plat" of the Muscovy Company's annual route by North Cape and "Wardhows" [Vardo] to "St. Nicholas" [Archangel], c. 1568. BL Royal MS 18.d.III f. 124

the queen was not to be persuaded.[70] However, by 1620, with the Jamestown colony firmly established and Spanish reprisals less feared, Drake's map was displayed, with Sebastian Cabot's map, in the Privy Gallery at Whitehall. There it finally did make propaganda for England's overseas enterprise and made clear royal approval, thus associating exploration with national pride and patriotism.[71]

Maps and Ministers

Although by the 1550s map consciousness was becoming increasingly common and thus unremarkable in educated circles, the interest shown by some ministers in cartography was exceptional. John Dudley, viscount Lisle, earl of

Warwick, and, from 1551, duke of Northumberland, the son of Henry VII's infamous minister and himself the most powerful man in the land between 1551 and 1553, had become familiar with the use of maps for military and administrative purposes while serving as one of Henry VIII's soldier-administrators after 1540.[72] He had also been the person responsible for recruiting the French cartographer Nicholas de Nicolay into the king's service in 1546.[73] During the reign of Edward VI, because of his own close links with John Cheke, he came into contact with the brilliant young polymath John Dee (1527–1608), a pupil of Gemma Frisius, one of the earliest exponents of the theories of triangulation, and a friend of the great cartographer and map publisher Gerhard Kramer, better known as Mercator.[74]

Northumberland's acquaintance with Leonard Digges may have dated back to the early 1540s, when both were involved with attempts to improve the fortifications of Calais and Boulogne. Digges, a country squire of Protestant sympathies from Kent, was one of the prime movers in the popularization of the use of mathematics and scientific measurement for surveying on land and navigation at sea. Like Northumberland, however, he fell foul of the regime of Queen Mary, and following his implication in Wyatt's Rebellion, a protest inter alia against the queen's forthcoming marriage to Philip II, he was attainted.[75] This political act had unfortunate consequences for English cartography, since it placed great difficulties in the way of the publication of Digges's theories. His two great works of popularization, *Prognostication of Right Good Effect* and *Tectonicon*, were, indeed, published anonymously by Thomas Gemini in Queen Mary's reign, but Digges's other works had to await posthumous publication by his son Thomas, in 1571.[76]

Northumberland's involvement with cartography did not end with Dee, Digges, and Nicolay, however. As chief patron, on its foundation in 1553, of the "Companie of the Merchants Adventurers for the discoverie of Regions, Dominions, Islands and places unknown," which was soon to be renamed the Muscovy Company, he came into regular contact with Sebastian Cabot, the company's governor.[77]

Northumberland was the grandfather of Sir Robert Dudley, the creator of *Arcano del Mare* (1646). Before the advent of Dudley, however, the most cartographically interested of Northumberland's immediate family was his son-in-law Sir Henry Sidney. Though Sidney first came to prominence in the early 1550s, he trimmed his sails to the prevailing winds under Mary, to emerge as one of the leading administrators under Elizabeth I, whom he served repeatedly as an energetic and talented lord deputy of Ireland and as lord president of the Welsh Marches. It was he who in 1553 had personally recommended Richard Chancellor to the merchants who financed the northeastern voyage of 1553–54,[78] and who was the main sponsor and dedicatee of Anthony Jenkinson's map of Muscovy, engraved by Nicholas Reynolds and first published in 1562. As lord deputy, as John Andrews has shown, he showed himself particularly cartographically minded, and between 1567 and 1571 he was the inspiration for and ensured the completion of Robert Lythe's survey of the provinces of Leinster and Munster, the first detailed cartographical survey of any part of Ireland.[79] It also seems likely, to judge from the surprising number of surviving Elizabethan maps of Shropshire and the Welsh marches—which unlike Ireland were peaceful if previously ill-mapped—that he was as active in commissioning maps while in Ludlow, the seat of the lord president, as he was when in Dublin.[80]

Of Elizabeth's principal servants, two particularly stand out for their links with cartography: Francis Walsingham, her principal secretary of state from 1573 to 1590, and his immediate predecessor William Cecil, Lord Burghley, who served as secretary of state from 1550 to 1553 and again from 1558 to 1572, and as lord treasurer from 1572 until his death in 1598. Of the two men, Walsingham's involvement is the more difficult to chart; unlike Burghley, he left no sons to safeguard his papers, which were dispersed on his death (some of them passing into the lord treasurer's possession,[81] and partly because he did not have Burghley's passion for annotating the maps and papers that passed through his hands. Nevertheless, the surviving evidence of his interest in and use of maps is impressive enough and suggests that there is much more to be discovered. Married to the daughter of one of the magnates of the Muscovy Company, he was also one of its directors or "assistants" from 1569 and was involved in most of the ventures of expansion that occurred during his period as secretary of state.[82] It has been suggested that the surviving set of John White's watercolors of the Roanoke Colony, including his beautiful map of Raleigh's Virginia, were originally intended for Walsingham.[83] D. B. Quinn has described this as "the most careful detailed piece of cartography for any part of North America made in the sixteenth century, and the first to be based, at least by an Englishman, on a survey made on the ground," the survey having been made by the brilliant mathematician Thomas Harriot.[84]

Walsingham was no less cartographically minded when it came to the affairs of England and Europe. His copy of the first edition of Saxton's atlas of maps of the English counties (1579)—alas, unannotated except for the insertion of his signature—still survives,[85] as does Robert Adams's handsome presentation plan of Flushing, one of the "cautionary towns" in the Netherlands which had been ceded to England in 1585 (figure 3.5).[86] Walsingham's diary of engagements for 1585, containing notes of his meetings with Richard Hakluyt, Thomas Digges, and Ralph Agas in connection with the harbor and fortifications works at Dover,[87] for which plans possibly by Digges still exist,[88] show the extent to which he was involved with maps, cartographers, and people connected with mapmaking on a day-to-day basis. Indeed, such was his enthusiasm for maps that in 1587 his agent in Amsterdam acquired what were probably printed wall maps to adorn the gallery of his town house at the Savoy between the cities of London and Westminster.[89]

Lord Burghley, Maps, and Government

It is William Cecil, Lord Burghley, however, above all other Tudor statesmen, who will be forever associated with the use of maps in government, thanks largely to the research work of R. A. Skelton.[90] The importance in this connection of Burghley's background and youth has not, however, been previously touched on. The Cecil family had been associated with the Tudor dynasty and with the court, in an unspectacular but profitable way, from the start of Henry VII's reign.[91] William's father, who had been present at the Field of the Cloth of Gold, had been close to the person of Henry VIII as a groom of the Wardrobe, even if the position, in court terms, was lowly. William himself, thanks to this connection, had served as a page at court between 1529 and 1534. He probably knew John Rastell and Sir Thomas Elyot and was acquainted with *The Boke named the Gov-*

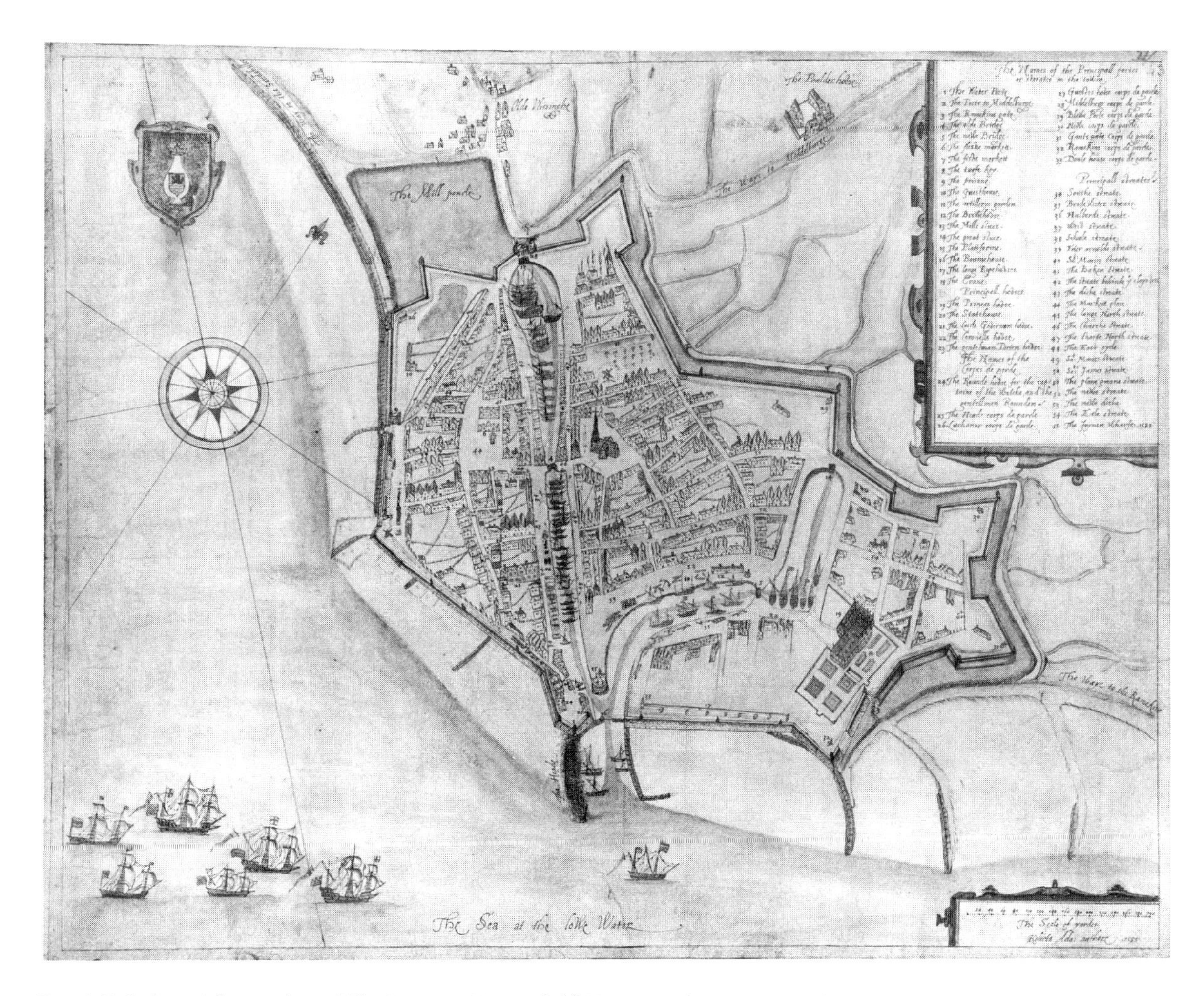

Fig. 3.5 Robert Adams, plan of Flushing, 1585. Hatfield House Archives, CPM II/43

ernour, which first appeared when he was eleven years old.[92] He would also have been familiar with the court during the crisis years of the 1530s. He cannot have known a time when plats were not used and spoken of as a means of communication among monarch, ministers, generals, and administrators, and so his later reliance on them is not surprising. But he could also remember the time when such a use of maps was novel and when their versatility was being ever more clearly demonstrated. Hence, perhaps, the sense of excitement about maps and the high esteem for them that he never lost and on which contemporaries such as Laurence Nowell remarked.[93]

This enthusiasm also grew from his intellectual background. For his student days in Cambridge (1535–41), in mathematics and cartography as in education and religion, "coincided with the establishment of the norms . . .

which guided his world. . . . his experience and study at Cambridge enabled him to absorb the formative ideas of the future."[94] To a large extent, these ideas were being formulated by Cheke and his circle, and Cecil's contact with them could not have been closer, since he married Cheke's sister, Mary. Mary died after barely two years; by this time he was at Gray's Inn, London, where in the course of his legal studies he may well have worked with the sketch maps used to illustrate legal disputes. His links with Cheke were maintained, however, as was demonstrated by his second marriage (1545) to Mildred Cooke, one of the most talented pupils of Cheke's soulmate Roger Ascham, the tutor to Lady Jane Grey and Princess Elizabeth. It was through Cheke that, in 1551, Cecil was introduced to John Dee,[95] with whom he was to remain in contact for the rest of his life and whose 1580 map of the polar regions, illustrating a northwest passage, remains in the possession of his descendants, the marquesses of Exeter, at the seat he had built for the family, Burghley House.[96]

Cecil himself may have had no detailed understanding of geometry,—one of the chief interests of Cheke and his circle—and probably had little skill as a mapmaker,[97] but throughout his career he showed his appreciation of those who did. His private library was rich in Ptolemies and in newer books on geography, as well as containing Ortelius's atlas in two editions (1570, 1595) and an early edition of Mercator's atlas.[98] As the queen's first minister, he was of course bombarded by unsolicited pleas for support from a host of mathematical, cartographical, and geographical theorists. It was through his good offices, as we have seen, that several obtained official posts or direct rewards from the government. Tracts by theorists and practitioners such as William Bourne (1572), Thomas Digges (1573), Edward Worsop (1582), John Blagrave (1584), John Davis (1594), and John Norden (1591, 1594, 1596) were presented to him and thus found their way into his library, which was also enriched by purchases of his own and maps sent by colleagues who knew of his enthusiasm for cartography.[99] His hoard of maps was probably further increased through his contacts with merchants and the chartered companies. For though by no means as extensive as Walsingham's, they were close. As well as investing one hundred pounds in the Muscovy Company, he was actively interested in Frobisher's expedition in search of a northwest passage in the mid-1570s, and in exploration in the vicinity of the Great Fisheries off Newfoundland.[100]

Burghley was not uncritical in his use of maps. He recognized that Ortelius's map of Ireland was inadequate and corrected his copy, now in the National Library of Ireland in Dublin.[101] More usually he added information to maps to increase their utility, as testified by the plentiful examples to be found in the four great surviving repositories of Tudor maps, the Public Record Office, Hatfield House, the British Library, and Trinity College, Dublin (figure 3.6). Occasionally he made sketch-maps for his own use, and while the known examples are far from distinguished, they do demonstrate the extent to which he thought cartographically and his realization of maps' potential as means of clarifying problems with even the slightest geographical connections (figure 3.7).[102] His fascination never left him; in his seventies and responsible for the conduct of war against Spain, he still found time to correspond about surveying techniques with Ralph Agas.[103]

Burghley's interest in maps was, nevertheless, primarily geared to serving the needs of the state. In combination with treatises on

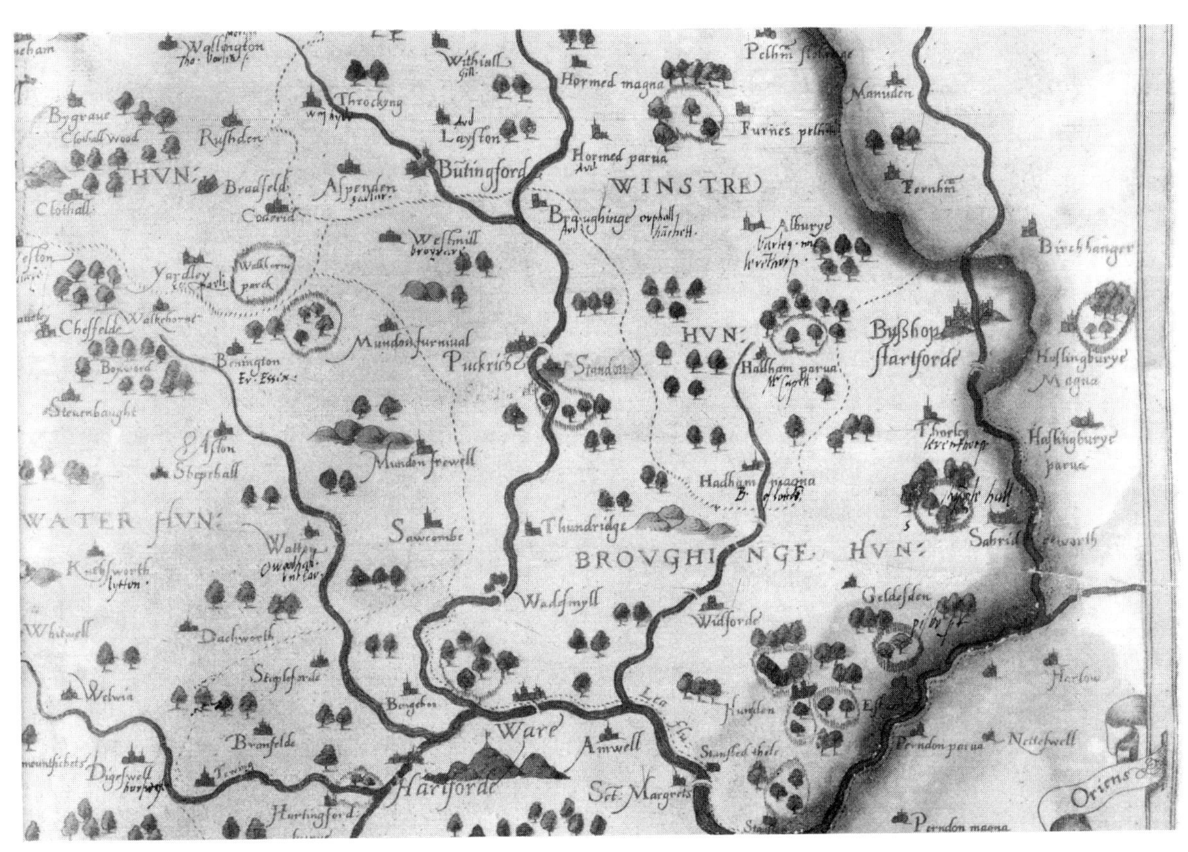

FIG. 3.6 Detail from the proof state of Christopher Saxton's map of Hertfordshire, 1577. BL Royal MS 18.d.III f. 341

finance and government, collections of acts of parliament, treaties and legal precedents, volumes of dispatches from abroad and letters from the provinces, they became vital instruments in the processes of administration and the formulation of policy.[104] His seemingly incurable urge to annotate the maps and papers that passed under his eyes and the chance survival of large parts of his archives have presented posterity with what must be a unique opportunity of seeing in detail how the maps were actually used.

In 1592 Robert Beale (1541–1601), a diplomat clerk to the Privy Council like Elyot, Ashley, and Waad, and secretary to Walsingham, wrote a lengthy "Treatise of the Office of a Councellor and Principall Secretarie to her ma[jes]tie" in which he emphasized the principal secretary's dual functions of gathering intelligence of all sorts and of preserving it in an organized manner in order to dispatch business efficiently. He also gave extensive advice on the sort of information a Secretary should obtain.[105] Included, in a much-quoted passage, was "the booke of Ortelius['s] Mapps, a booke of the Mappes of England, w[i]th a particular

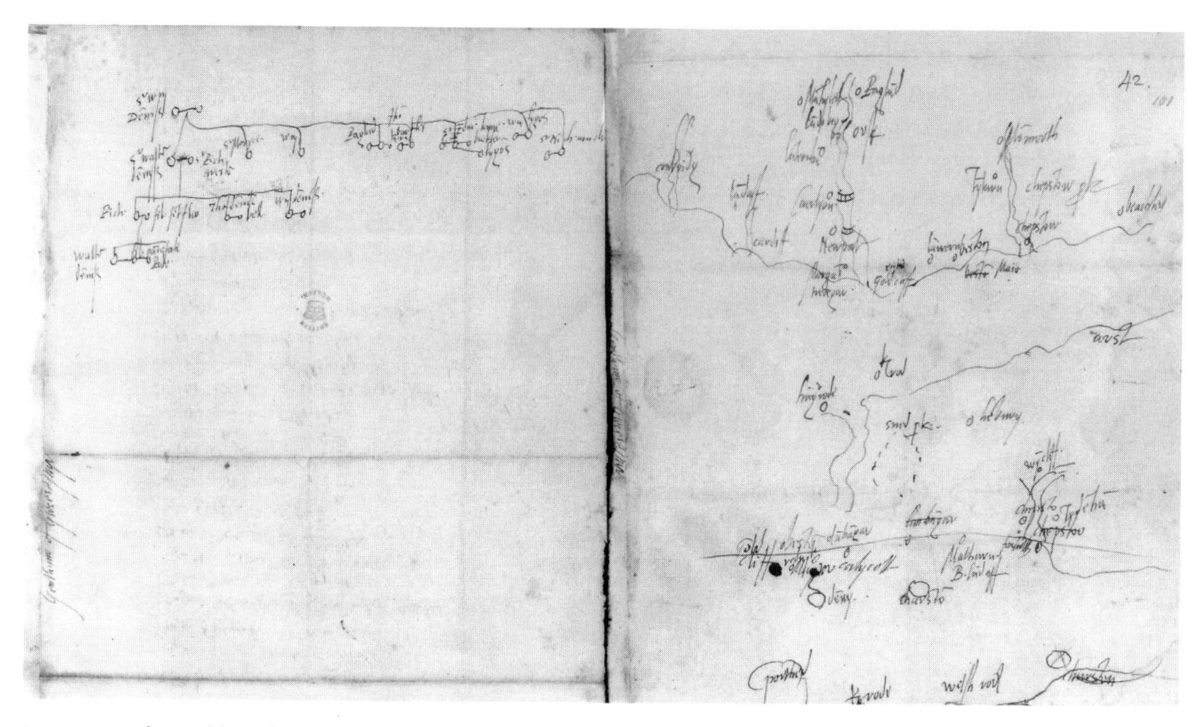

FIG. 3.7 Lord Burghley, sketch-map of the Bristol Channel, and genealogical table, c. 1590. BL Lansdown MS 104, ff. 100–101

note of the divisions of the shires into Hundreds, Lathes, Wappentaes, and what Noblemen, Gent[lemen] and others be residing in every one of them; what Citties, Burrows, Markett Townes, Villages; and also a good descripc[i]on of the Realm of Ireland, a note of the Noblemen and surnames English or Irish of their septs, Enraghes, Galloglasses, Kerns and followers, and if anie other plotts or maps come to his handes, let them be kept safelie."[106]

It has been repeatedly pointed out elsewhere that this advice reflects Burghley's (and probably Walsingham's) practice.[107] It does so only in part, however, and represents the ideal rather than the reality. Despite his best efforts, Burghley's filing system was clearly far from perfect, and occasionally he was unclear as to the maps he had or did not have.[108] At times, his practice deviated seriously from the principles propounded by Beale.[109] In fact, throughout his career and for reasons such as the turbulence of the times and the relative poverty of the Crown, Burghley's accumulation of maps and cartographic information had to be on an ad hoc basis. Because of this, as John Andrews has penetratingly observed, "his instinct was to preserve everything that came in (irrespective of merit), to meet each new crisis by asking for a new map of the area concerned, and in resolving the crisis to consult a wide selection of the maps already in his possession."[110]

Burghley was, however, well placed to procure maps of all sorts.[111] As Skelton has observed, he received many excellent maps, and

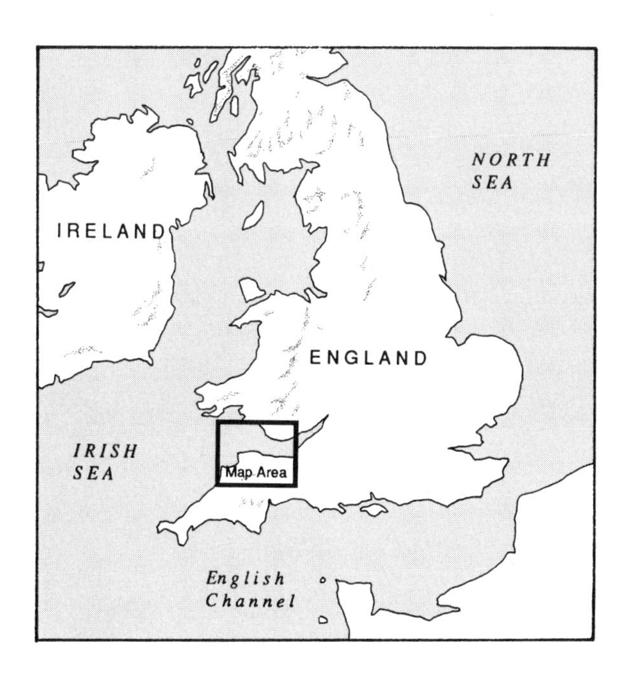

works on or containing maps, from supplicants for his favor and support. His private wealth ensured the purchase of multiple copies, in some cases, of the major printed maps and atlases published abroad. His public offices also made available to him plentiful military and civilian maps. As secretary of state, he had a hand in most domestic and foreign affairs. As master of the Court of Wards (1561–98), with responsibility for administering the estates of minors, he found himself involved with parochial matters and estate management, with which as a landowner he was already familiar.[112] As lord treasurer and, in effect, first minister, he could call on the maps produced for the Exchequer, the duchies of Cornwall and Lancaster, and other government departments. He could acquire copies of Saxton's maps as they appeared, as well as part of the archives of Francis Walsingham on the latter's death in 1590 and those of Robert Adams on his death in 1595. Furthermore, he pillaged Henry VIII's library at Whitehall for earlier maps whenever he felt the need—a practice for which he was indirectly criticized by Beale, who complained at the way in which "those thinges w[hi]ch weare publicke have bine culled out and gathered into private bokes, wherby no meanes are left to see what was donne before or to give anie light of service to yonge beginners."[113] Though this was to some extent rectified by Burgley's son Robert Cecil, earl of Salisbury, when he was lord treasurer,[114] it accounts for the presence of some plats of the 1530s and 1540s in Hatfield House.

Most of the maps from this rich harvest seem to have been kept loose by Burghley.[115] But as time went on, manuscript maps that he considered particularly informative were incorporated into specially compiled atlases, of varying sizes, which in two cases included printed and manuscript maps. These atlases he supplemented with numerous notes and annotations to increase their utility for government and administration.[116] The earliest of the surviving volumes is the so-called Nowell-Burghley atlas, containing Nowell's map of England, Wales, southern Scotland, and Ireland of about 1564 and another map by Nowell of Sicily, included because of the island's strategic importance. The atlas is actually little more than a small notebook, and according to a plausible tradition Sir William Cecil, as he then was, always carried it about with him for reference.[117]

After 1570, the Nowell-Burghley atlas was almost certainly superseded by two much larger and more informative volumes. One, and the earlier in probable date of compilation, related predominantly to foreign affairs and had at its core Ortelius's atlas of 1570.[118] The other, centered on the proof sheets for Saxton's atlas, was clearly intended mainly for consultation on British affairs.[119] These two volumes served as Burghley's main geographical references on foreign and domestic policy and administration from 1570 to the time of his death, with further maps, annotations, or notes being added as occasion offered or demanded from the 1570s until 1591 and 1595 respectively. In this period the Nowell-Burghley atlas seems to have been relegated to the office of his secretary, Sir Michael Hicks.[120]

Ireland was catered for in neither of these atlases, and at first Burghley probably relied on loose maps. He annotated John Gough's manuscript map of 1567 to show suggested locations for garrisons, perhaps as proposed by Sir Henry Sidney in that year.[121] By 1598, however, Burghley had put together no fewer than two books of maps devoted to Ireland—an indication of its importance in English policy. Described respectively as a "Book of Mappes of Ireland in large" and a "Book of Mappes of Ireland in colours" in a list of Burghley's books "at the Court" of 1598, it is not certain whether these volumes still exist, though in the view of J. H. Andrews they may now be items SP 64/1 in the Public Record office, London, and P.39 in the National Maritime Museum, Greenwich.[122]

Over and above these books, Burghley seems to have retained most if not all of the maps that ever passed through his hands, and these survive in abundance in Hatfield House, the Public Record Office, the British Library, and Trinity College, Dublin. They were often endorsed by or for him in ways that indicate how he envisaged their long-term reference use. Thus a plan of the siege of Thérouanne (1553) is endorsed "the platte of Toweryn,"[123] a plan of the siege of Rouen (1591) "Plott of Roun,"[124] and of the siege of Nijmegen (1586) "Platt of Newmegen,"[125] clearly showing that henceforth they should serve, faute de mieux, as town as well as siege plans.

Burghley's notes, annotations, and endorsements of maps give a clear idea of the manner in which and extent to which he used them and provide further evidence of his grasp of their potential as aids in government and administration. First and foremost, the maps in the made-up volumes were geared, as had been the maps of 1539, to national defense. Several of the Saxton maps showing the southwestern coasts of England, which were under particular threat of invasion from Spain, contain notes on "ye places of Descent . . . y^t are most dangerous and require greatest regard and assistance"[126] (Devon) or "dangerous places for landing of men in the county"[127] (Dorset), with more information being given in the manuscript maps showing some of the coastline in greater detail.[128] Further annotations relate to military stores and the military divisions of the counties of Westmoreland, Cumberland, and Northumberland, which bordered on Scotland.[129] And again there is an additional manuscript map showing the western marches in greater detail.[130] But the point does not need to be labored, because of the sheer number of Elizabethan maps that directly served the same defensive end, ranging from William Lambarde's rough but effective series showing the beacons system in Kent (1585–96)[131] to the finely executed survey of the coast of Sussex made by Nicholas Reynolds for the deputy lieutenants of the county in May 1587.[132]

More subtle are the plentiful annotations on the maps that passed through Burghley's hands which identify the homes of leading families inside the kingdom or name the governors of the major French ports.[133] This led Andrews perspicaciously to remark that "one of the chief uses of [Burghley's] collection was as a geographical index to the government's friends and enemies."[134] This was as true with native lords in Ireland[135] and on the Anglo-Scottish border[136] as with recusant families in Suffolk,[137]Sussex,[138] and (perhaps above all) in Lancashire,[139] for all were potential or actual rebels. Friends could, of course, also be identified in this manner, and, as Victor Morgan has pointed out,[140] maps could be an aid in ensuring an acceptable geographical spread throughout each county of justices of the peace, the agents of Tudor government at the lowest local level. Indeed, lists of justices accompany every one of the Saxton maps in the made-up atlas.[141] Maps were also much consulted when other administrative and legal decisions had to be taken. Appropriate levels of taxation or levies could, for instance, be estimated from the distribution of population, which, in a rough-and-ready manner, could be seen from the density of towns and, at the other end of the scale, sparsely occupied hills on the Saxton maps. Notes on the internal division of counties, or maps illustrating the limits of jurisdictions—for instance, bodies like the duchy of Lancaster or the Crown—could also be used to clarify disputes, and to guess at the likely yield from taxes or the number of men who could be raised in case of emergency.[142]

Side by side with men and money, communications were of vital importance from the standpoint of effective government, foreign policy, administration, and defense particularly in the case of a peripatetic court like Elizabeth's.[143] One leitmotiv of Burghley's notes in the made-up atlases is his concern about routes on land and at sea. The earliest notes in the Nowell-Burghley atlas, probably dating from the mid-1560s, are lists of the postal stages between various Continental channel ports and Augsburg in Germany. The Ortelius atlas at Burghley House contains notes on routes to the West Indies. In the manuscript map of Durham in the Saxton atlas, probably compiled by a local man at the time of the Northern Rebellion of 1569, the emphasis is on the routes taken by government agents and, potentially, by enemy foreign auxiliaries to the troubled area. The very latest of the annotations in the Saxton atlas is one of 1595 listing "stages for Postes to be laid between London and Plimouth," a point illustrated in Burghley's own manuscript map of the west country dating from a few years earlier.[144]

Maps could also help to meet the needs of colonial expansion in Ireland, albeit to a limited extent.[145] Very occasionally they were used to draw boundaries for newly created counties.[146] More commonly they could be of assistance in deciding the location of fortresses or in planning plantations, be they in Leix and Offaly in the 1560s, Munster in the 1580s, or Ulster twenty years later.[147]

Maps were also of great use simply as a means of illustration. Throughout Elizabeth's reign, although her defense program paled in significance against her father's, military engineers worked continuously, as has been said, on schemes for England's main ports and harbors, illustrating their proposals and progress with plans. Occasional bids by the corporations of minor ports, such as Great Yarmouth, for assistance from central government also left their not unattractive cartographic mark.[148]

The official and semiofficial expeditions to the Netherlands, West Indies, and France which formed an ever larger part of England's foreign policy between 1570 and 1595 equally found illustration in maps and plans. Some, such as Robert Adam's magnificent portrayals of the defeat of the Armada[149] and his surviving plan-views of Flushing,[150] Battista Boazio's manuscript picture-map of Drake's raid on Santiago in 1585 and the engraving of his map of the raid on Santo Domingo in the next year,[151] and Walter Morgan's account of the expedition of English auxiliaries in Holland in 1572,[152] were works of art in themselves. They were intended, of course, for the gratification of the recipient, but also for display to others and not infrequently for eventual publication as propaganda on broadsides or in books. The majority of the maps that arrived in ministers' hands were, however, modest sketches of even more modest cartographic pretension, penned simply to illustrate a reference in a letter.[153] Nevertheless, they sufficed to inform distant ministers of the progress of foreign sieges in which England's scarce military and financial resources were involved—even if, as in the case with the siege of Rouen late in 1590, queen and ministers seem occasionally to have suffered from a lack of adequate maps.[154]

With Burghley, as with his colleagues, public merged imperceptibly into private, and several examples exist of his semiofficial or private use of maps. Some of his annotations show his affinities with Elyot and the humanists and antiquarians of his day who believed that maps could clarify history.[155] The map of England in his Saxton atlas is accompanied by notes on the country's history since "A° mundi 2390, when Brutus came to Britain."[156] The same volume has a manuscript map of northeastern Kent, possibly supplied by William Lambarde, the first historian of that county, with notes on the village of Milton in the time of kings Alfred and Edward the Confessor.[157] The names of coastal features on Burghley's autograph map of southwestern England are rendered in their Ptolemaic forms, though the map itself, giving the location of lead and copper mines in Cornwall and Devon, was far from antiquarian in purpose.[158]

Not surprising in an age when descent was paramount for gaining a crown, a title, or an estate, Burghley took a lively interest in genealogy. Maps of Denmark in his Ortelius atlas and of Northamptonshire in the Saxton atlas contain genealogies of the kings of the one and the major gentry families of the other.[159] Burghley's particular interest in the latter is natural, since the Cecils were great landowners in the county. In this capacity, Burghley, in common with a growing number of other squires, increasingly made use of maps. The searches of Peter Eden have revealed the existence of at least one surviving plan, drawn to scale in 1593 by his steward Richard Shute, of Burghley's estate of Cliffe Park in Northamptonshire,[160] and there are several annotated maps and plans of Waltham, Cheshunt, and Enfield, northeast of London, where he also owned several estates and two great houses, Pymmes and, particularly, Theobalds.[161]

Theobalds was the house where Burghley did much of his official entertaining, because of its situation, spaciousness, and proximity to London.[162] There cartography played its part in creating a suitably imposing public image for the lord treasurer. In a manner recalling the royal court since medieval times, the great hall is recorded as having a cunning cosmographical ceiling with "the twelve signs of the zodiac so constructed that at night you can see distinctly the stars proper to each; on the same stage the sun performs its course, which is without doubt contrived by some concealed ingenious mechanism." His galleries were embellished, in more modern style, with "correct landscapes of all the most important towns in Christendom" and, appropriately for the chief minister of England, with a depiction, which may have been a version of Saxton's 1583 wall map, "of the Kingdom, with all its cities, towns and villages mountains and rivers; as also the armorial bearings and domains of every esquire, lord, knight, and noble who possesses lands and retainers to whatever extent."[163] Even when contemplating the walls of his house, Burghley did not wish to escape, or wish his visitors to escape, from the concerns that occupied his mind when consulting his atlases and other works of reference in the solitude of his study.

There can be little doubt that most of Burghley's colleagues emulated this, if only to bolster their image as well-educated governors of the queen's realm.[164] John Dee remarked in a much-quoted passage of 1570 on the growing passion of gentlemen to "beautify their Halls, Parlers, Chambers, Galeries, Studies or Libraries" with "maps, charts and geographical globes,"[165] and Victor Morgan has shown how the popularity of Saxton's maps led to the production of map tapestries, based on his work, which were meant for show as well as insulation.[166] Some of the larger early estate maps, too, such as William Gier's highly decorative 1614 map of the manor of Hammerden in Ticehurst, now in the East Sussex Record Office,[167] and Samuel Piers's map of the Cordell family's Suffolk estates in 1613, now at Melford Hall in Suffolk,[168] were also meant for public display. For, while their contents reflected the patron's wealth and power, the decoration, often including coats of arms and more rarely, portraits, emphasized his birth, rights, and taste.[169]

Queen Elizabeth, Maps, and Propaganda

Yet the use of maps and globes as symbols in the creation of a desired image, and more generally in propaganda in late sixteenth-century England, is most obviously associated with Elizabeth I herself. During her reign, cartography and astronomy continued to be used for court pageantry. The celestial ceilings that greeted the arrival of the French ambassadors in 1581, when negotiations for a marriage between the queen and the French king's younger brother, the duc d'Alençon, were at a delicate stage, seem, for instance, to have been very similar to those that had been such a feature of court life in the 1520s.[170]

Under Elizabeth, however, cartography came most characteristically to be utilized, in a symbolic manner, in the creation of a personal, imperial imagery that was particularly associated with the queen herself, though it derived from the imagery associated with Charles V and, beyond him, with rulers in medieval Burgundy and ancient Rome.[171] The frontispiece to Saxton's atlas, in which Elizabeth is shown enthroned between figures representing geography, astrology, and, at her feet, cartography and astronomy, reflects her supposed familiarity with and her role as patron of these sciences. Frances Yates has also, perhaps fancifully, argued that it is representation of "Virgo-Elizabeth as a celestial portent whose advent has been mysteriously foretold."[172] At a more down-to-earth level, the royal arms that appear prominently on each of the maps in the atlas serve as a reminder of the queen's legitimate rule over and concern with every part of her dominions, with Seckford's arms—the only other arms on the maps—expressing his patronage of Saxton and also, perhaps, his own aspiration to royal favor.[173]

John Case's striking image of Elizabeth embracing the "sphere of the state" in his book *Sphaera Civitatis* of 1588, which every Oxford undergraduate was compelled to purchase, shows how a celestial map based on Ptolemaic theories could be used in an allegory of the benefits of the queen's rule and her godlike imperial aspirations.[174]

Maps and globes as symbols for her identity with, and protection of, England and its imperial destiny also appear in several of the best-known portraits of Elizabeth. Many are connected with the festive tilts held annually on the Queen's Accession Day, 17 November, in highly elaborate, allegorical settings. These, and the sophisticated allegories found in the portraits, were the work of Sir Henry Lee, the queen's champion from 1559 to 1590. Given his position as a soldier who had close links with the Office of Ordnance, his appreciation of the potential of maps and globes might be expected to be particularly acute, even at a time when map consciousness was growing in educated society.[175]

In the paintings Elizabeth is regularly shown in close proximity to a globe or map. The "Ditchley" portrait by Marcus Gheeraerts the Younger, now in the National Portrait Gallery, London, commemorates her visit to Sir Henry Lee's home in 1592. The queen is actually shown standing on Saxton's map of England and Wales (1583), the counties on it differentiated by colors, with her feet on Ditchley, as if she were protecting England from storms, which are shown clearing from the west. The armillary spheres she wears as earrings hint at England's imperial future.[176] Similar imagery is employed in Crispin van de Passe's somewhat later engraved portrait[177] and in the contemporary "Dangers Averted" medals, though in the latter the queen is symbolized as a bay tree, which was impervious to lightning, and England as an island prospering under her protection.[178]

The latter idea is further developed in the so-called sieve portrait of 1583, whose prototype, by Quentin Matsys the Younger, is now in the Pinacoteca Nazionale in Siena. It contains, as one of many other elements, a globe placed behind Elizabeth. On it the British Isles are shown bathed in sunlight, with much (presumably English) shipping moving towards the West, while mainland Europe lies plunged into darkness.[179] The theme of imperial expansion—as advocated by John Dee and others and attempted by Sir Francis Drake in claiming "Nova Albion" for England in 1579 and in the planting of Raleigh's Roanoke Colony in the 1580s—was also picked up in the "Armada" portrait. There, to a background of scenes depicting the destruction of the Spanish fleet, the queen is shown with her hand over the representation of North America (not, it should be emphasized in passing, over the whole world) on a globe.[180] The year 1593 saw the publication of Cesare Ripa's *Iconologia,* in which globes and spheres feature as ingredients in many symbols. Its influence can be seen in the so-called Rainbow portrait of Elizabeth, in Hatfield, which abounds in spheres conveying a variety of meanings.[181]

Finally, in this context, mention should be made of Michael Mercator's silver medal commemorating Drake's circumnavigation of the globe in 1577–80. Produced a decade later, in 1589, and probably meant only for limited presentation in court circles, it contains a further visualization of English imperial claims in North America. For the west of the continent is almost entirely covered with "New Albion," and "Virginea" can be discerned in the east.[182]

Of Elizabeth and her sister Mary's actual interest in cartography, as opposed to their readiness to use maps and globes for administration and propaganda, we know very little. The interest of Mary's husband, Philip II, in maps is well known, and Mary and her courtiers may have pampered it.[183] She is said to have commissioned the splendid "Queen Mary Atlas" from Diogo Homem in 1558 as a gift for Philip, though she died before it could be completed (color plate 3).[184] At the beginning of the previous year an old and trusted Tudor servant, Sir John Mason, had presented Mary with the New Year's gift of "a mappe of Englande stayned upon cloth of silver in a frame of wodde having a drawing cover painted with the King and Queen's arms."[185] But this more likely reflected the general esteem in which such maps were then held than any particular interest in them on Mary's part.

The Mapping of the Royal Estates

At the time of Elizabeth I's birth in 1533, maps were, as we have seen, first being used for the purposes of government. She was educated by John Cheke's favorite pupil, Roger Ascham. There can be little doubt, therefore, that she was familiar with maps from her earliest childhood. We have evidence of her asking for them, for instance, during the Rouen campaign of 1591,[186] and a few months later she was presented with the celestial and terrestrial globes newly produced by Emery Molyneux at the expense of William Sanderson.[187] Maps also featured prominently in her portraits. But, judging from the surviving evidence, she does not seem to have taken any particular interest in them, in the way that her father or Lord Burghley did. Indeed, in the one particular respect where one might expect her to have shown personal interest—a cartographic survey of the Crown lands, on income from which her government was heavily dependent—there is strikingly little.

A survey could be, and usually was, written. But since the late 1570s estate plans, some of considerable cartographic and mathematical distinction, had been produced in increasing numbers to illustrate or complement the written surveys. In the same period treatises on surveying also came to discuss mapping techniques.[188] Some of the queen's younger ministers and courtiers were in the vanguard of those commissioning estate maps. As early as 1580, Ralph Treswell the Elder was preparing plans of Sir Christopher Hatton's Northamptonshire estates,[189] and by the 1590s Sir Walter Raleigh was having even his Irish estates mapped.[190]

Contemporaries recognized that the Crown lands needed to be systematically and accurately surveyed if the Crown was to receive a proper price from their sale or lease. It was felt that, in the absence of such surveys, the queen was losing out—a conclusion that, to some extent, is borne out by recent research.[191] In 1602 Sir Robert Johnson wrote to Robert Cecil that "whenever I have heard of the sale of Her Majesty's lands, I have observed that the value was seldom known." "The chief foundation of mischief," he thought, "has been the want of authentic surveys. . . . of every ten mannors there is not one perfect survey."[192] About three years later John Norden made much the same point, when he complained to James I that under his predecessor the surveying of the royal lands had been "committed (in favour not by desert) unto such (for the most part) as were very unfit to exequute the same to the great Prejudice of her late Majesties revenews."[193]

Johnson was probably thinking primarily of written surveys, but Norden presumably also had plans in mind, and the objects of this criticism may have been the traditional sort of surveyor: lawyers without any training in the geometrical measurement and depiction of land.[194] Few enough plans of royal estates survive; apart from William Hombrestone's surveys of confiscated lands and a few surveys of lands in Wales and Cornwall, there is apparently only a plan of Tottenham Court in Middlesex of 1591, by William Necton, who is otherwise unknown.[195] Norden also mentioned, in

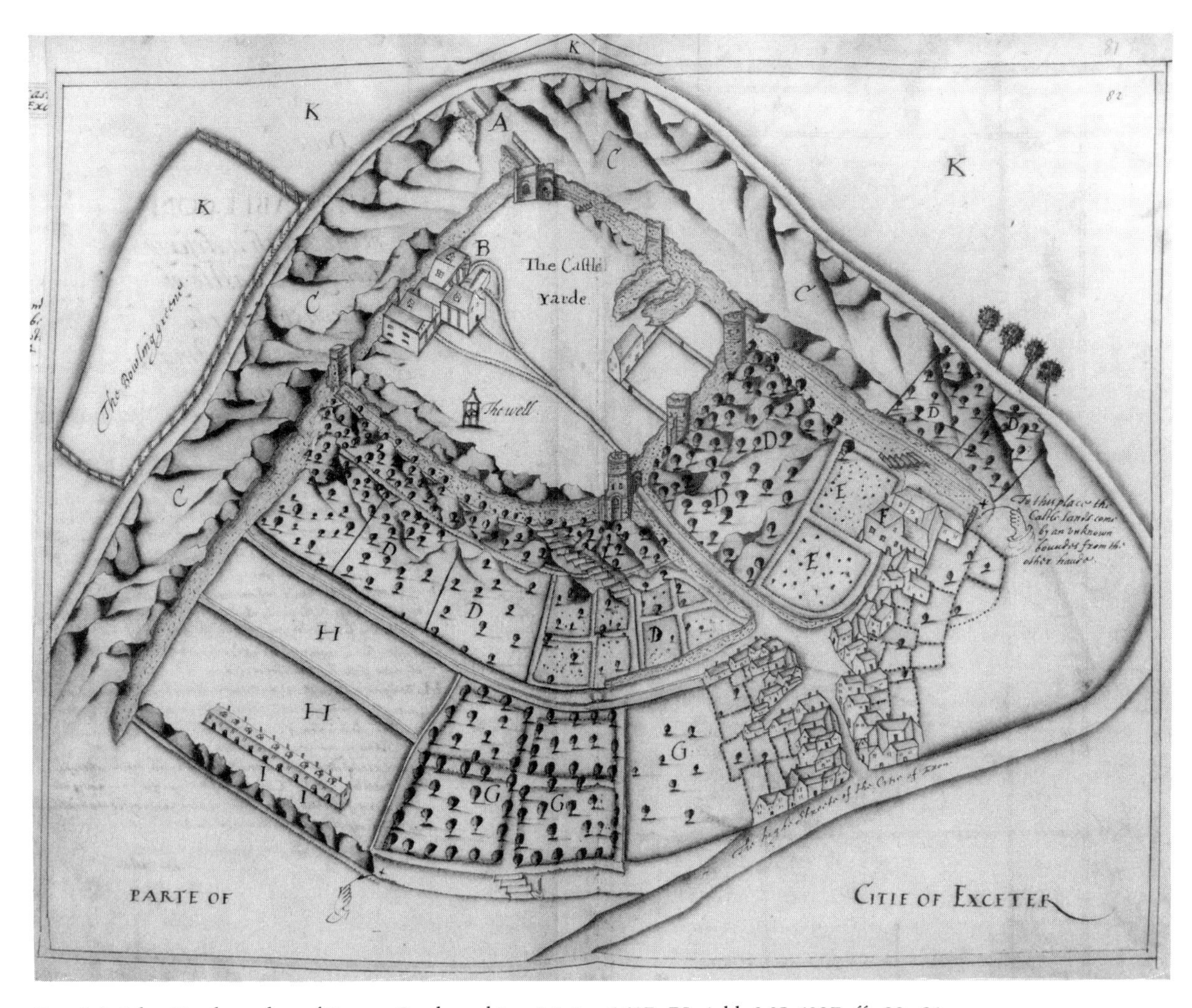

FIG. 3.8 John Norden, plan of Exeter Castle and its vicinity, 1617. BL Add. MS 6027, ff. 80–81

the petition quoted above, that he had been "formerly imployed in Survey of some of her Late Majestie's landes belonging to the Dukedome of Lancaster and of other landes,"[196] and these surveys may have been at least partially cartographic. Even assuming that the surviving plans represent only a small proportion of those originally made, the number could never have been great. Yet in the 1590s estate surveyors of the caliber of Ralph and Robert Treswell and Christopher Saxton himself were in the service of the Crown—accompanying military expeditions abroad, or employed as surveyor of Hertfordshire or as bailiff of some Duchy of Lancaster lands in Yorkshire—without apparently producing estate plans.[197]

There are several plausible explanations. The Crown was not subject to one of the main pressures that seem to have led private landowners to commission the earliest measured plans: the specter of having to defend the boundaries of their estates in courts of law—though the Crown did need to defend itself against illicit encroachments on its lands, and here plans would have been to distinct advantage (figure 3.8).[198] Again, given the Crown's chronic shortage of money, particularly in the crisis years of the 1580s and 1590s, estate mapping could well have been considered a low priority. It seems improbable that the queen's younger, cartographically minded ministers, who were potential purchasers of Crown lands, would have sought to discourage the mapping of these lands. For they were likely recipients of gifts of twenty-one year leases of Crown estates, a fairly common form of reward for loyal service to the queen, and so would have had a personal interest in knowing the lands' precise value.[199] Certainly, if Elizabeth had been enthusiastic about estate mapping she would have had it undertaken regardless.

The most probable explanation for the relative paucity of plans of the Crown lands is likely (with pressing calls on money elsewhere) to have been Elizabeth's age. However cartographically interested the queen may or may not have been in her earlier years, she was nearing sixty by the time the first measured estate plans appeared and may well have been too old to grasp their potential fully.

The contrast with James I is telling. There are various indications of an interest in cartography that may have been implanted in childhood; his tutors are known to have purchased *The Boke named the Governour* for him to study.[200] He certainly had a copy of Saxton,[201] and it is reasonable to suppose that it was through his interest that the Burghley-Saxton atlas (BL Royal MS 18.D.III) came also to form part of the Royal Library, following the partial dispersal of the Cecil papers on Robert Cecil, earl of Salisbury's, death in 1612. When Norden remarked in his dedication of the manuscript of his survey of Cornwall in 1604 that "it well befitteth a Prince to be trulie acquaynted with his own territories,"[202] he was, consciously, preaching to the converted. Barely had he arrived in England in the previous year than James ordered the stewards of Crown estates to provide details of the rent, leases, and value of different types of land held. In the next year surveyors were appointed to survey the royal lands.[203] There were about 125 in all, and many, possibly even most, being traditional surveyors, may have had no mapping experience. But the number included such well-known mapmakers as Norden, the Treswells (Ralph senior and junior and Robert; figure 3.9), Aaron Rathborne, and Saxton's critic George Owen of Henllys.

On 21 December 1607 a comprehensive survey of the Crown woods was ordered, timber being a major source of income when well managed and controlled. Finally, from 1608 onward a concerted effort, known as "the Great Survey," was undertaken to measure and value all the Crown lands. These orders were issued under the names, successively, of Lords Buckhurst and Salisbury (Robert Cecil), but it seems extremely likely that the impetus came from the king himself, for Salisbury had shown little or no inclination to order such a survey under Elizabeth. The failure of "the Great Survey" has not been investigated in detail, but antipathy on the part of tenants, and of surveyors—who seem to have been keen to minimize the actual work they did for the king, preferring written surveys wherever possible to the more time-consuming mapping—may have been responsible.[204] Its lasting legacy is probably Nor-

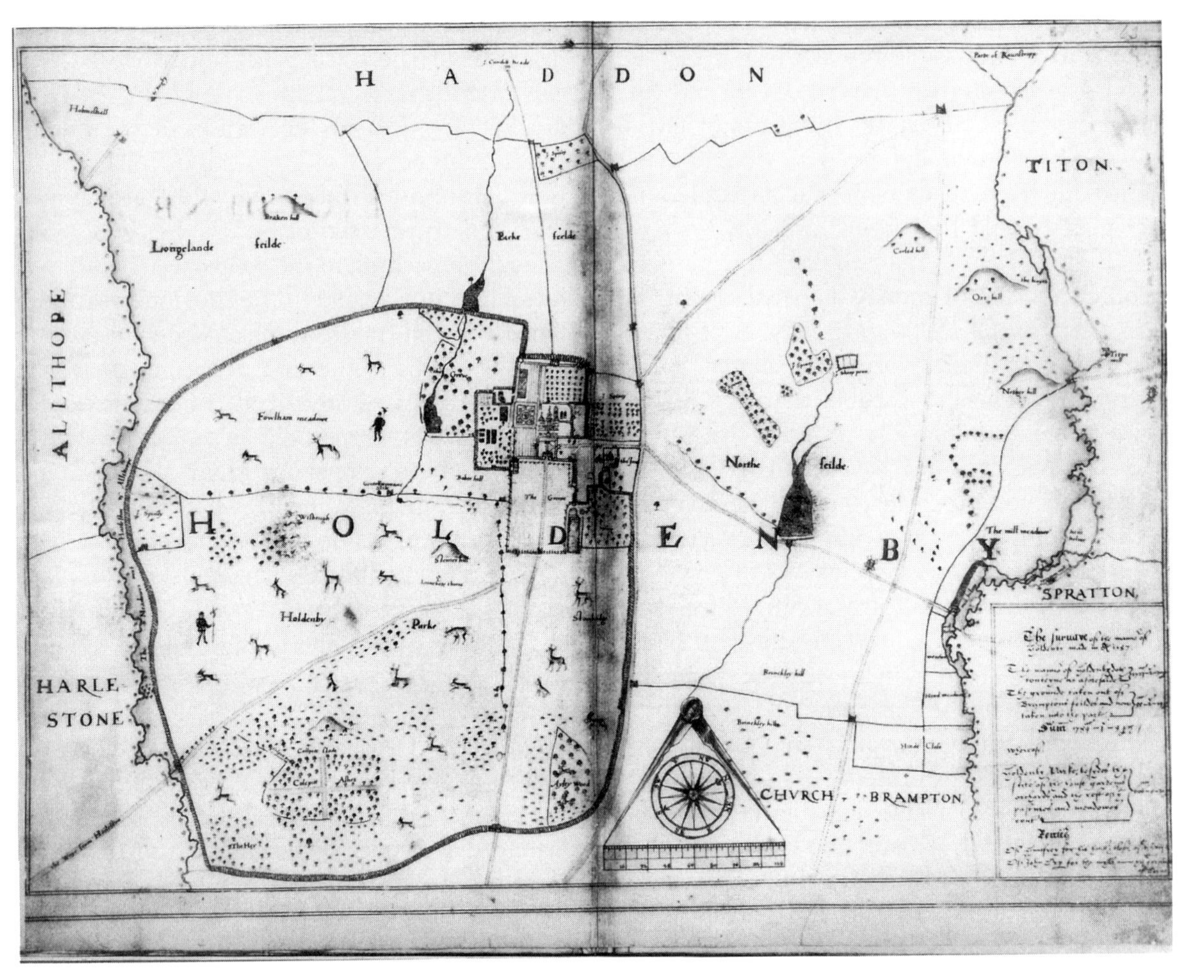

Fig. 3.9 Ralph Treswell, plan of Holdenby, 1587. Northamptonshire County Record Office, FH 272

den's magnificent "Survey of the Honour of Windsor," of 1607, which survives in two copies: one (dedicated to the prince of Wales) in the Royal Archives in Windsor, and the other (to the king) among the Harleian manuscripts in the British Library.[205] Since the survey was meant for the king himself, Norden exceeded himself as an artist. Its cartographic accuracy, however, like that of most of his royal surveys, remains to be tested.

James I: Collection, Compilation, Conservatism

If the practice of estate surveying under James I almost represents a break with the practice of his predecessor, in other respects the relationship of the king and his ministers to maps shows a natural development. John Speed had, in Fulke Greville, a patron who served in government; he received official passes, similar to Norden's, to facilitate his county mapping, and, like Saxton, he was rewarded with land and with official positions.[206] As Drake had presented a map of his world voyage to Elizabeth, so Robert Tindall probably presented his map of the Jamestown colony of Virginia to Henry, prince of Wales, the Virginia Company's patron, in 1607.[207] James's pacific policies removed the stimulus to mapping provided by the threat of foreign invasion during most of Elizabeth's reign, but his diplomatic activity had cartographic benefits. Sir John Digby, the cartographically minded English envoy in Spain, was able to transmit a map of the Chinese provinces and surrounding states from Madrid in 1609,[208] and a few years later, on his return from a mission to the Moghul court, Sir Thomas Roe supplied William Baffin with the information necessary for his map of the Moghul empire of 1619.[209]

Ministers, while continuing to call for new maps, notably from Ireland, now began to take more coherent measures for their conservation and utilization. In May 1608 Salisbury had his loose maps, including the enormous number inherited from his father, bound up into three "great books" of "mapps," "fortificacions," and architectural plans.[210] In 1610, the State Paper Office was formally established.[211] The ancestor of an important section of today's Public Record Office, it ensured the survival of large quantities of official maps and papers, including many formerly owned by Burghley and Salisbury. The process of creating the State Paper Office was not smooth, however. Large portions of the Cecil papers, for instance, remained at Hatfield House, and after its creation the office was subjected to pillaging by well-placed outsiders. One such was the antiquarian Sir Robert Bruce Cotton (1571–1631), who, under the guise of creating a national library, took maps and official papers from the State Paper Office and also, probably, Salisbury's house, as well as acquiring many other maps, manuscripts, and papers by purchase or as gifts from friends.[212] He placed the main sequence of Tudor and early Stuart official maps on shelves under a bust of Emperor Augustus in his library. Today the Cotton collection forms part of the British Library, but the maps retain their "Augustus" designation.

Another antiquarian who may have taken maps from the State Paper Office was George Carew (1555–1629), Baron Carew of Clopton (1605) and earl of Totnes (1626).[213] Altogether a more politically significant figure than Cotton, he was probably the most important later Tudor or early Stuart minister after Burghley to show a lively interest in maps. Much of his early career was spent in Ireland, where he rose to be president of Munster (1600–1603) when Lord Mountjoy, probably Elizabeth I's greatest general, was lord deputy of Ireland. He was successfully employed elsewhere. As well as serving on the expeditions to Cadiz (1596) and the Azores (1597), he acted as envoy to France (1598), as governor of Guernsey (1610–21), and as master-general of the Ordnance (1608–17). His long experience at the Offices of Ordnance in England and Ireland, with their close links to the planning of fortifications, strengthened his natural liking for maps. It was only on the eve of his departure from Munster in 1603, however, that he seems to have begun retaining maps and papers and then collecting them seriously. These all related, not surprisingly, to Ireland. Unlike Burghley's maps, however, they did not constitute a personal, working archive. They were, instead, accumulated as a collection of sources for Irish history, and a few were published in *Pacata Hibernia*, a narrative of the wars of Munster published in 1633. They came from various sources, one of the more important being the Cecil papers. Most of the manuscripts are today housed in Lambeth Palace Library (MSS 596–638), but the largest number of maps can be found in Trinity College, Dublin (TCD MSS 1209, 2379), with others in Lambeth and—in a volume probably acquired whole from Lord Burghley's papers—in the National Maritime Museum, Greenwich.[214]

Carew and Salisbury were not only distinguished from Burghley in the way in which they conserved their maps. Burghley accumulated map after map of the same area, regardless of quality and without thought of synthesizing the information (and other information available to him in written reports and the like), except in his own head, whence the result were expressed as annotations on other maps.[215] Carew was more selective, and Andrews has written that "whereas Burghley played safe with a variety of conflicting outlines of Ireland, Carew could always recognise the best."[216] Salisbury's practice also differed from his father's in that he employed draftsmen such as Norden, Francis Caundell, and Battista Boazio, the first two of whom seem never even to have visited Ireland, to prepare compilation maps of that kingdom on the basis of the earlier, Elizabethan surveys of individual provinces.[217]

An uncompleted series of estate maps and written surveys; cautiously worded official passes; diplomatic spinoffs; map conservation and compilation mapping—all of these have a sad, rather autumnal air. They stand in enormous contrast to the cartographic achievements most commonly associated with James I's reign, such as John Smith's mapping of the Jamestown, Virginia, colony,[218] Richard Norwood's sophisticated survey of the new colony of Bermuda,[219] the maps associated with the northern voyages of Henry Hudson and William Baffin,[220] and, above all, the maps in John Speed's *Theatre of the Empire of Great Britain*. All of these were not primarily made for Crown, court, or government but for the world beyond: the merchants, landowners, and noblemen who formed the backbone of the great companies and who had the money to invest in voyages of discovery and to buy atlases for their libraries. Indeed, the decoration of Speed's maps symbolizes the level of cartographic patronage under James I. There was a degree of royal patronage, and the royal arms are usually (but not invariably) there, but they are swamped by arms and illustrations designed to appeal to the antiquarian interests and local pride of the merchants, squires, and noble lords on whose purchasing power the commercial success of Speed's venture depended.[221]

There was nothing intentionally political about this. It simply reflected the circumstances of the time. The Crown, which late in Henry VIII's reign had taken the lead in patronizing mapmaking and exploiting the potential of maps, no longer had the means or direct interest to maintain this position by the time of Elizabeth I's death. It had to a great extent become a mere consumer. Its appetite for maps produced at minimum cost to itself, or at the cost of others, indeed remained voracious. But it was only one among the many consumers upon whom mapmakers—and what by then had become the map trade—depended for survival. And it was far from being the most important.

Notes

1. Joyce Youings, *Sixteenth-Century England* (Harmondsworth, 1984), 206, 214–6.

2. Howard Colvin, ed., *A History of the King's Works* (London, 1951–82), 4:398–400.

3. L. R. Shelby, *John Rogers: Tudor Military Engineer* (Oxford, 1967), 116–26.

4. Colvin, *King's Works*, 4:400, 409.

5. P. D. A. Harvey, "Estate Surveyors and the Spread of the Scale-Map in England, 1550–80" (forthcoming). Victor Morgan, "The Cartographic Image of 'the Country' in Early Modern England," *Transactions of the Royal Historical Society*, 5th ser., 29 (1979): 146–49; though the sentiments he attributes to Blundeville, Burton, and others clearly derive from Elyot, thereby pushing "map consciousness" further back and associating its development with the use of *The Boke named the Governour* in schools from the 1530s on—when leaders of the new generation were adolescents. It is indicative of the spread of map consciousness beyond the court that as early as 1556 an inventory of the possessions of a Surrey gentleman, Sir William More, who was Burghley's exact contemporary (1520–1600), should include a collection of maps (*The Treasure Houses of Britain: Five Hundred Years of Private Patronage and Art Collecting* [exhibition catalog, National Gallery of Art, Washington, D. C., 1985], 84, item 3). Compare this with the statement about the "scarcyte and price of . . . cartes within this realme of England" made in judgment of Diogo Homem's suite in the Court of Admiralty a mere nine years earlier (see chap. 2, n. 132).

6. E. G. R. Taylor, *The Mathematical Practitioners of Tudor & Stuart England* (Cambridge, 1954), 17–39, 166–73, 315–24. G. L'E. Turner, "Mathematical Instrument-Making in London in the Sixteenth Century," in *English Map-Making, 1500–1650: Historical Essays*, ed. Sarah Tyacke (London, 1983), 93–106, includes a list of the extant instruments of the earliest known native-born instrument maker, Humfrey Cole, seventeen out of twenty-two of which date from between 1564 and 1580 (100–101). R. A. Skelton and J. Summerson, *A Description of Maps and Architectural Drawings in the Collection Made by William Cecil, First Baron Burghley, Now at Hatfield House* (Oxford: Roxburghe Club, 1971), 30; henceforth referred to as Hatfield House Maps.

7. Kenneth R. Andrews, *Trade, Plunder, and Settlement: Maritime Enterprise and the Genesis of the British Empire, 1480–1630* (Cambridge, 1984), 29–30, 64ff. The earliest surviving charts produced by Englishmen seem to be those produced in 1539 (see Alwyn Ruddock, "The Earliest Original English Seaman's Rutter and Pilot's Chart," *The Journal of the Institute of Navigation* 14 [1961], but it is under Elizabeth I that more scientifically precise charts, by the likes of William Borough, Robert Norman, Richard Poulter, and others, appear (A. H. W. Robinson, *Marine Cartography in Britain: A History of the Sea Chart to 1855* (Leicester, 1962), 25–36, 92 no. 81, 152–57. The search for new markets also encouraged the creation of new terrestrial maps such as Anthony Jenkinson's map of Muscovy (1562), for which see Walker Oakeshott, "A Tudor Explorer and His Map of Russia," *Times Literary Supplement*, 22 June 1984, 703–4.

8. See Thomas R. Smith, "Manuscript and Printed Sea Charts in Seventeenth-Century London: The Case of the Thames School," in *The Compleat Plattmaker: Essays on Chart, Map, and Globe Making in England in the Seventeenth and Eighteenth Centuries*, ed. Norman J. W. Thrower (Berkeley, Calif., 1978), 45–100. Also, Tony Campbell, "The Drapers' Company and Its School of Seventeenth Century Chart-Makers," *My Head is a Map: A Festschrift for R. V. Tooley*, ed. Helen Wallis and Sarah Tyacke (London, 1973), 81–106.

9. E.g. John Rudd, Vicar of Dewsbury and prebendary of Durham Cathedral (David Marcombe, "Saxon's Apprenticeship: John Rudd, a Yorkshire Cartographer," *Yorkshire Archeological Journal* 50 [1978]: 171–75); the antiquarian Laurence Nowell (e.g. P. M. Barber, "A Tudor Mystery: Laurence Nowell's Map of England and Ireland," *The Map Collector* 22 [March 1983]: 16–21); John Dee, a fellow of Trinity College, Cambridge—for one of his maps, see BL Cotton MS Augustus I.i.1 (F. Smith, *John Dee* [London, 1909]; Peter J. French *John Dee: The World of an Elizabethan Magus* [London, 1972]). For examples from the gentry, see Thomas Digges (below) and Edmund Yorke (Robinson, *Marine Cartography*, 5).

10. Andrews, *Trade, Plunder, and Settlement*, 7–8, 29–30, 59–60, 66, 69–70. He points out (69) that after 1555 "the Muscovy Company naturally became the leading sponsor and patron of geographical science and enterprise for the next twenty or thirty years." For the drying-up of England's (and particularly London's) foreign markets and the creation of joint-stock companies, see Youings, *Sixteenth-Century England*, 235–36. The charts were of home waters such as the mouth of the Thames (e.g. BL Cotton MS Augustus I.i.17) as well as of distant shores such as the northern coasts of Russia (e.g. BL Royal MS 18.D.III f. 123).

11. See the memoir of Sir William Sanderson (1541–1631) in BL Harley MS 5208, ff. 29–30 (quoted and amplified in Helen Wallis, ed., *Raleigh and Roanoke: The British Library Exhibit Hosted by the North Carolina Museum of History* [Raleigh, N. C., 1985], 44–45, items

31, 32). For the mapping: ibid., 53, 55–57, items 48, 54–58. Some years later Sanderson was, for good financial reasons, to finance the creation, by Emeric Molyneux and Jodocus Hondius, of the first surviving globes to be produced in England (Helen Wallis, "The First English Globe: A Recent Discovery," *The Geographical Journal* 117 [September 1951], particularly 276–77; D. Waters, "English Navigation," in *Sir Francis Drake and the Famous Voyage, 1577–80: Essays Commemorating the Quadricentennial of Drake's Circumnavigation of the Earth*, ed. Norman J. Thrower [Berkeley, Calif., 1984], 20). Waters also mentions (29) how Thomas Gresham, the leading Elizabethan merchant, left money in his will (1575) for the establishment of a college of navigation in his former home. This was duly opened in 1598.

12. Andrews, *Trade, Plunder, and Settlement*, 17–18; Youings, *Sixteenth-Century England*, 249–50. However, as Andrews emphasizes, the distinction between merchants and gentry should not be drawn too sharply, since merchants were often the younger sons of gentry and tended, if successful, to buy country estates and marry into gentry families.

13. Youings, *Sixteenth-Century England*, 154–77 passim; Sarah J. Tyacke and John Huddy, *Christopher Saxton and Tudor Map-Making* (London, 1980), 46–60; Peter Eden, "Three Elizabethan Estate Surveyors," in *English Map-Making*, ed. Tyacke, 68–84; Harvey, "Estate Surveyors"; J. B. Harley, "Meaning and Ambiguity in Tudor Cartography," in *English Map-Making*, ed. Tyacke, 37–38. The work of an early estate surveyor, Israel Amyce, seems to epitomize what was required. With accurate physical measurement, he combined *social* measurement, in that the size selected for the depiction of houses seems to have been primarily dictated by the social standing of its owner (see his plan of the Melford Hall estate in Suffolk, which is still housed in Melford Hall). See A. C. Edwards and K. C. Newton, *The Walkers of Hanningfield: Surveyors and Mapmakers Extraordinary* (London, 1984), for a family of talented land surveyors whose surviving work was invariably commissioned by country gentlemen and never by the Crown. Just as groups of merchants, collectively transformed into joint-stock companies, were often patrons of chart-making, so groups of gentry and merchants, transformed into city companies or the governing bodies of colleges in Oxford or Cambridge, because frequent patrons of land surveyors, as the essays of Peter Eden and John Schofield in *English Map-Making*, 68–84, 85–91, demonstrate.

14. For this and the rest of the paragraph, Andrews, *Trade, Plunder, and Settlement*, 11–13, 15–16, 18.

15. A particularly good example is provided by Drake's "Famous Voyage" of 1577–80, which, apart from the backing provided by Drake himself and his Drake and Hawkins cousins, was covertly aided by the queen and sponsored by a string of leading ministers and courtiers led by Walsingham, Hatton, and the queen's favorite, Leicester, and by important merchants such as Sir Richard Martin, the master of the Mint. Nevertheless, as J. H. Parry has pointed out ("Drake and the World Encompassed," in *Sir Francis Drake*, ed. Thrower, 3–4), "Drake was not . . . a professional officer. . . . he was the agent of a private, profit-making syndicate." See also, in the same volume, D. B. Quinn, "Early Accounts of the Famous Voyages," 35; Kenneth Andrews, "Drake and South America," 49–59; and, for the cartography, Helen Wallis, "The Cartography of Drake's Voyage," 121–63. And see also Helen Wallis, ed., *Sir Francis Drake: An Exhibition to Commemorate Francis Drake's Voyage around the World, 1577–1580* (London, 1977), 48–49, item 37, and for the maps, 54–56, 66–67, 78–87, 94–95.

16. Colvin, *King's Works*, 4:411, 515ff. Robinson, *Marine Cartography*, 19–20, 149. Edward Lynam, "English Maps and Mapmakers of the Sixteenth Century," *Geographical Journal* 116 (1950): 8, even called him "England's most notable native mapmaker before Saxton." He was active throughout Hampshire, the Isle of Wight, and the Channel Islands. His work is well represented in the BL Cotton MS Augustus I.i&ii series. And see Donald Hodson, *Maps of Portsmouth before 1801* (Portsmouth, 1977), 140 (39), and catalog nos. 2, 32a, 35, 151, 161–63, 165, 166.

17. Colvin, *King's Works*, 3:94–96; 4:412–13.

18. Ibid., 4:413. *Dictionary of National Biography* 29:78. Robinson, *Marine Cartography*, 151. Rolf Loeber, "Biographical Dictionary of the Engineers in Ireland, 1600–1730," *Irish Sword: The Journal of the Military History Society of Ireland* 13 (1977–79): 240–41.

19. Colvin, *King's Works*, 4:410–11; Hatfield House Maps, 29, catalog nos. 57–58, 69, 70, 72–76, 78–80, pl. 4 (plans of Berwick and environs, 1561–1570).

20. Colvin, *King's Works*, 4:409–10. John Andrews, "The Irish Surveys of Robert Lythe," *Imago Mundi* 11 (1965): 22.

21. E.g. Taylor, *Mathematical Practitioners*, 322–24.

22. Colvin, *King's Works*, 4:757–62. For a possible example of his work, see BL Add. MS 11815.b. (of 1581). There seems to be no surviving plan that can with cer-

tainty be ascribed to him. BL Cotton MS Augustus I.i.46 (of 1595), attributed to him by Robinson (*Marine Cartography*, 26, 208, pl. 9), is more likely to be by John Hill (*Catalogue of Manuscript Maps, Charts, and Plans and of the Topographical Drawings in the British Museum* [London, 1844, 1861, 1962], 1:95). Edward Croft-Murray and Paul Hulton, *Catalogue of British Drawings*, vol. 1, *Sixteenth and Seventeenth Centuries* (London, 1960), xxv, suggested on stylistic grounds that a plan of Dover in 1582–83 (BL Cotton MS Augustus I.i.45) and another of 1581 in Hatfield (CPM I. 58) might also be by Digges (and see Hatfield House Maps, 46–47, nos. 37, 38). At the moment it is impossible to make firm attributions, because none of these plans is signed, no signed plats by Digges are known, and there were a multiplicy of potential platmakers at work in Dover in the early 1580s (Colvin, *King's Works*, 4:757–62). That Digges *did* make plats of Dover can nevertheless hardly be doubted. He himself made a statement to that effect in a memorial to the queen in 1582 (Taylor, *Mathematical Practitioners*, 175, n. 35).

23. Colvin, *King's Works*, 4:411–12; Hatfield House Maps, 51–52; Robinson, *Marine Cartographer*, 151. For his work see, e.g., BL Cotton MS Augustus I.ii.87, 89, 90, 91, and M. S. Rodriguez-Salgado et al., *Armada 1588–1988* (exhibition catalog, National Maritime Museum, Greenwich and London, 1988), 148–49.

24. BL Cotton MS Caligula E. ix. f. 276; Hatfield House CPM. I. Supp. 5 (Hatfield House Maps, 67, no. 11). See also Colvin, *King's Works*, 4:409, 663, for a fortification plan of Berwick (now PRO MPF 222 [1 & 2]) by the soldier Lord Willoughby de Eresby, who was adamant that he was not an engineer. The practice of rudimentary cartography by generals themselves or by their staffs was, by the 1590s, well established in western Europe: see Geoffrey Parker, *The Army of Flanders and the Spanish Road, 1567–1659: The Logistics of Spanish Victory and Defeat in the Low Countries' Wars* (Cambridge, 1972), particularly 83–86, 103ff.

25. Hatfield House CPM II.37a (Robert Norman, 1580), reproduced and discussed in Hatfield House Maps, 49, no. 44, ill. 7. For Norman, see Taylor, *Mathematical Practitioners*, 173–74, and Robinson, *Marine Cartography*, 27–28, 31, 150.

BL Cotton MS Augustus I.i.17 (Borough, 1596); Hatfield House CPM I.56 (another version), is reproduced and discussed in Robinson, *Marine Cartography*, 29–31, 208, pl. 11, Hatfield House Maps, 49–50, no. 45, and Rodriguez-Salgado, *Armada*, 209. As well as semiofficial positions as a chief pilot of the Muscovy Company and master of Trinity House (1585), Borough served as treasurer of the Queen's Ships (1582) and comptroller of the Navy (1588–98; Taylor *Mathematical Practitioners*, 173). For both men see also D. Waters, *The Art of Navigation in England in Elizabethan and Early Stuart Times* (London, 1958), passim.

26. BL Cotton MS Augustus I.ii.58. His name is contained in the compass star. Annotations make plain that the map was sent from Brittany; Treswell must have been there too and did not compile it in London. For more familiar examples of his work, see John Schofield, "Ralph Treswell's Surveys of London Houses c. 1612," in *English Map-Making*, ed. Tyacke, 85–90, John Schofield, *The Building of London from the Conquest to the Great Fire* (London, 1984), 157–62, and his *The London Surveys of Ralph Treswell* (London, 1987). See also the entry in Peter Eden, ed., *Dictionary of Land Surveyors and Local Cartographers of Great Britain and Ireland, 1550–1850* (London, 1975–79).

27. For Treswell's "private" work, see n. 26 above. For Norman and Borough, see n. 25, and also Hatfield House Maps, 36. Andrews, *Trade, Plunder and Settlements*, 72, and see n. 204 below.

28. French, John Dee. Taylor, *Mathematical Practitioners*, 176, 182–83, 196. Paul Hulton, *America 1585: The Complete Drawings of John White* (Chapel Hill, N.C.; 1985), 7–16, does, however, mention (7) that this John White *may* have been the same as a John White who was active as a military engineer in Ireland in 1567–68. It should be emphasized that Gresham College, where Gunter lectured, and the Roanoke Colony were both essentially private ventures. The private favors—even if, in many cases, they may have been no more than a vain hope—are suggested, for example, by Harriot's employment in Raleigh's household and by the readiness of Dee and other theorists and practitioners to present material to Burghley (Hatfield House Maps, passim).

29. Morgan, "Cartographic Image," 147, referring to BL Cotton MS Otho E. xi ff. 245–98, a muster roll of 1591 with miniature county maps by Pieter van den Keere, stuck in after 1617.

30. A somewhat later example of ministerial use of printed maps is provided by a group of engraved Dutch maps dating from the middle of the seventeenth century, which were annotated in the 1670s by or on behalf of Charles II's minister Thomas Osborne, earl of Danby, to illustrate the constitution and politics of the Dutch Republic, and which remain among his papers (BL Add. MS 28093 ff. 279–86).

31. Particularly Robert Dunlop, "Sixteenth-Century Maps of Ireland," *English Historical Review* 20 (1905): 309–37; John Andrews, *Ireland in Maps* (exhibition catalog, Dublin, 1961); Andrews, "Robert Lythe"; John Andrews, "Geography and Government in Elizabethan Ireland," in *Irish Geographical Studies in Honour of E. Estyn Evans*, ed. Nicholas Stephens and Robin E. Glasscock (Belfast, 1970), 178–91; and John Andrews, *Plantation Acres: An Historical Study of the Irish Land Surveyor and His Maps* (Belfast, 1985). Gerard A. Hayes-McCoy, *Ulster and Other Maps, c. 1600* (Dublin, 1964). Paul Ferguson, *Irish Map History: A Select Bibliography of Secondary Works, 1850–1983, on the History of Cartography in Ireland* (Dublin, 1983), gives a fuller listing.

32. For instance, see BL Cotton MS Augustus I.ii.21 (c. 1535); PRO MPF. 72 (c. 1560?, annotated by Burghley).

33. For instance, Trinity College, Dublin, 1209.15 (Ulster, c. 1590, by Francis Jobson); BL Cotton MS Augustus I.ii.30 (Ulster etc., c. 1602–3 by Battista Boazio); PRO MPF 36, 37 (Ulster, c. 1602–3, by Richard Bartlett); Hatfield House CPM Supp. 2 (Ulster, 1609, by John Norden [?]).

34. See, e.g., D. B. Quinn, *The Elizabethans and the Irish* (Ithaca, N.Y., 1966), particularly 106–22 and passim.

35. E.g., Andrews, "Geography and Government," 180–81. There was little alternative, given the physical, political, and financial restraints. To an extent it was true in England, too, and meant that ministers and cartographers, regardless of their personal mathematical and cartographic sophistication, often had to lower their standards and make do with any maps that could be produced, regardless of technical quality, so long as they seemed to serve the desired purpose. Edmund Yorke, surveying East Anglian forts in 1588, when a Spanish invasion seemed imminent, could excuse his sketch map of Warborne in Norfolk, "mad in hast this fyrst of May 1588," on the grounds that "reason would a scall but tyme permits not" (Hatfield House CPM II.36; Hatfield House Maps, 31, 52, no. 54; Edward Lynam, *British Maps and Mapmakers* [London, 1944], 22). As a result Lord Burghley's maps came to embrace a wide variety of styles and techniques that belie cozy assumptions of a continual, almost linear improvement in English cartography under the Tudors (Hatfield House Maps, 28–29; R. A. Skelton, "The Military Surveyor's Contribution to British Cartography in the Sixteenth Century," *Imago Mundi* 24 [1970]: 79).

36. The map, which has been attributed to John White and Francis Jobson, "the father of the Irish surveying profession" (John Andrews, "Appendix: The Beginnings of the Surveying Profession in Ireland—Abstract," in *English Map-Making*, ed. Tyacke, 21; and W. A. Wallace, *John White, Thomas Harris, and Walter Raleigh in Ireland* [Durham, 1985]) is reproduced in John Andrews, *Irish Maps* (Dublin, 1978), 10, ill. 12.

37. Taylor, *Mathematical Practitioners*, 171–72, 181–82. Turner, "Mathematical Instrument-Making," 98–101. Hatfield House Maps, 21, n. 3. Edward Wright seems to have played the same role with the young Prince Henry that John Cheke had played with Edward VI (see chap. 2 in this volume), and T. A. Birrell (*English Monarchs and Their Books*, Panizzi Lectures [London, 1987]) has found many geographical, cosmographical, and mathematical books that he acquired for the prince among the Old Royal Library books now in the British Library.

38. N. Kratzer to A. Dürer, 24 October 1524: "In meinem Namen grieß mir Herrn Pirckomair. Ich hoff, ich soll in kürz Engelland machen, das ein groß Land ist, und Ptholomeo nit bekannt ist gewesen; das wird er gern sechen. Es haben all, die darvon haben geschrieben, einen kleinen Teil Engellandt gesechen, nit mehr" (Greet Mr. Pirckheimer in my name. I hope to depict [lit. "make"] England soon. It's a big country and was not known to Ptolemy. He [Pirckheimer] will be glad to see it. All those who have written about it in the past have only seen a small part of England, not more]. H. Rupprich, ed., *Albrecht Dürers schriftlicher Nachlaß* (West Berlin, 1956–69), 1:111. Willibald Pirckheimer was then contemplating the publication of a new edition of Ptolemy's *Geographia* (eventually published in 1526). Kratzer presumably felt he had seen "all" England in part because of his travels with the peripatetic court of Henry VIII. This often-cited letter is translated in John D. North, "Nicolaus Kratzer—the King's Astronomer," *Science and History: Studies in Honor of Edward Rosen*, Studia Copernicana 16 (Wrocaw: Polish Academy of Sciences, 1971), 225–26.

39. That is, by talking of the governor's setting out his dominions "in figure."

40. Rodney W. Shirley, *Early Printed Maps of the British Isles* (London, 1980), xi, 10–11, 31, no. 41; Tyacke and Huddy, *Christopher Saxton*, 7–8; J. B. Trapp and Hubertus Schulte Herbruggen, *"The King's Good Servant": Sir Thomas More, 1477/8–1535* (exhibition catalog, National Portrait Gallery, London, 1977) 135, item 269. The political nature of the map is shown by its failure to indicate the bishoprics of Oxford, Chester, Glouces-

ter, Bristol, Peterborough, and Westminster, created by Henry VIII following the dissolution of the monasteries, though the older ones are shown. There can be no doubt that Lily was aware of the new creations. For Thomas Gemini: *Biographie nationale de Belgique* (1961), 31:386–94; C. D. O'Malley, "Thomas Geminum," in *Dictionary of Scientific Biography* (1972), 5:347–49; Taylor, *Mathematical Practitioners*, 165–66.

41. Shirley, *Early Printed Maps*, nos. 60–63, lists the derivatives from the map which continued to appear until at least 1589.

42. D. G. Moir, "A History of Scottish maps," in *The Early Maps of Scotland to 1850* (Edinburgh: Royal Scottish Geographical Society, 1973), 1:10–12; and see n. 40 above.

43. D. Marcombe, "Saxton's Apprenticeship"; Tyacke and Huddy, *Christopher Saxton*, 6–7.

44. P. M. Barber, "A Tudor Mystery: Laurence Nowell's Map of England and Ireland," *The Map Collector* 22 (March 1983): 16–21. For more recent information on Laurence Nowell's identity, see Carl D. Berkhout, "The Pedigree of Laurence Nowell the Antiquary," *English Language Notes* 23 (December 1985): 15–26. I am grateful to Carl Berkhout for having informed me of, and for supplying me with, copies of Nowell's autograph notes on his trip to Ireland in June-July 1560, contained in his commonplace book now in the Department of Special Collections, University of California at Los Angeles.

45. Almost certainly Saxton also enjoyed free access to the New Library at Whitehall. This casual and hence undocumented ministerial and royal support probably accounted for the speed with which Saxton completed his survey. For much of the time his surveying activity would have concentrated on expanding, confirming, and correcting the work of earlier cartographers, probably including that of Henry VIII's engineers, which he had consulted during the winter months (and see M. Evans and H. Lawrence, *Christopher Saxton, Elizabethan Map-Maker* [London, 1979], chaps. 2–6, pp. 9–73, 143–54, 163–70; among other things, Evans and Lawrence describe the Saxton proof maps in Burghley's atlas [now BL Royal MS 18.D.III]). As will be seen, Burghley also brought the maps of the 1530s and 1540s back into use as necessary. For Saxton see also Hatfield House Maps, 20–21, 23; R. A. Skelton, *Saxton's Survey of England and Wales, with a Facsimile of Saxton's Wall-Map of 1583* (Amsterdam, 1974); Tyacke and Huddy, *Christopher Saxton*, particularly 5–7, 25–45; W. Ravenhill, "As to Its Position in Respect to the Heaven's," *Imago Mundi* 28 (1976): 79–93; and the contributions of the same author and of Peter Eden to *English Map-Making*, ed. Tyacke, 69, 112–19; Shirley, *Early Printed Maps*, 53–54 no. 128, 55–56 no. 137; Morgan, "Cartographic Image," 129–54, and particularly 143. These works collectively largely supersede Edward Lynam's biography of Saxton (1927) and his later chapter on the atlas in his *Mapmaker's Art* (London, 1953), chap. 4.

46. For these: Robert Beale, "A Treatise of the Office of a Councillor and Principall Secretarie to her Ma[tie]," published in Conyers Read, *M[r] Secretary Walsingham and the Policy of Queen Elizabeth* (Cambridge, 1926), 1:428–29.

47. The Welsh gentleman-surveyor George Owen of Henllys complained of this at the time: Morgan, "Cartographic Image," 138.

48. W. Ravenhill, *John Norden's Manuscript Maps of Cornwall and Its Nine Hundreds* (Exeter, 1972), 11–23. Also, Hatfield House Maps, 23–25.

49. This recently discovered map, which is in private hands, is illustrated in *The Map Collector* 19 (June 1982): 18. The countess of Warwick nevertheless had close links with court as the widowed sister-in-law of Elizabeth's favorite "Sweet Robin" (Robert Dudley, earl of Leicester), as a close personal friend of the queen herself, as aunt of her current favorite, the earl of Essex, and as a member of the influential Russell family. The map was presumably prepared late in 1598, following the death that year of Burghley, to whom Norden had presented another manuscript version (now Lambeth Palace Library Codex 521), and would seem to illustrate Norden's growing and ultimately unfortunate ties with the group surrounding Essex.

50. Ravenhill, *John Norden's*, 15.

51. Northampton (the first to be completed, in 1591; Paris, Bibliothèque Nationale, Collection Gaignières, Manuscrits Anglais [series 58], no. 706), Essex (1594; Hatfield House, Cecil Papers, 326 [1]), and Hertfordshire (1598; Lambeth Palace Library Codex 521), are all dedicated to Burghley.

52. Ravenhill, *John Norden's*, 14–15. Heather Lawrence, "Permission to Survey," *The Map Collector* 19 (June 1982): 18–19, points out the significant difference in tone between the passes of January and July 1594.

53. R. A. Skelton, comp., *County Atlases of the British Isles, 1579–1850* (London 1964–), 119–222. The British Library has recently completed the purchase of all twelve printed maps (Sarah Tyacke, "Useful Maps: Themes in European Cartography," *The Map Collector* 28

[September 1984]: 39). A further example of the printed map of Northamptonshire has recently come to light among the Althorp Papers in the British Library, illustrating a seventeenth-century manuscript volume of statistics relating to the county.

54. W. H. Godfrey and A. R. Wagner, *The College of Arms: The Sixteenth Monograph of the Survey of London Committee* (London, 1963), 220–21. As a herald, Smith could expect to supplement his income through fees, in a manner akin to the offices that came to be granted to Saxton and Norden. It may be significant that he had to wait two years from the time of his original petition in 1595 (apparently without ministerial support) before his appointment as *dragon rouge poursuivant*. Another herald who also worked as a cartographer and was employed as a surveyor by the Crown was Smith's contemporary Robert Treswell. I am grateful to John Schofield for informing me of this.

55. Ravenhill, *John Norden's*, 19–20.

56. E.g. through his "Particular Description of England" of 1588 (BL Sloane MS 2596). He had developed his skills during years of residence in Nuremberg. Relics from this are a manuscript translation of the German text version of the 1572–73 edition of Ortelius's *Theatrum Orbis Terrarum* (described in *Sotheby*, catalog of the map sale, London, 11 July 1986, lot 359), which was clearly meant for presentation to a notable in England, and a manuscript description of Nuremberg with maps, for which see W. Roach, "William Smith: A Description of the Cittie of Noremberg 1574," *Mitteilungen des Vereine für die Geschichte der Stadt Nürnberg* 48 (1958): 194–245.

57. For this and the rest of the paragraph, David Waters, *The Art of Navigation in England in Elizabethan and Early Stuart Times* (London, 1958), 103–14, and, more generally, Andrews, *Trade, Plunder, and Settlement*, 29–30.

58. Waters, *Art of Navigation*, 39–74. Smith, "Manuscript and Printed Sea Charts," 189–90, has stated that even a century later Eden's book was the "most complete set of instructions on chart making . . . available . . . to . . . practitioners."

59. BL Cotton Roll XIII. 48 is a copy, prepared for John Dee in 1580, of a chart of the North American coastline by Fernandes (Wallis, *Raleigh and Roanoke*). For Fernandes, who was to be the pilot for the Roanoke voyages of 1584–88, see David B. Quinn, *England and the Discovery of America, 1481–1620* (New York, 1974), 246–63. See also n. 131, chap. 2 in this volume, for other Portuguese navigators who served in English ships during Elizabeth I's reign.

60. Better known as Lord Howard of Effingham and first earl of Nottingham. Howard's interest in promoting chart-making as an aid to navigation was not purely patriotic: as a large-scale investor in the Roanoke venture and in privateering, he had a vested interest in skilled navigation (Andrews, *Trade, Plunder, and Settlement*, 205, 244).

61. For the whole of this paragraph: Waters, *Art of Navigation*, 168–75; and the introduction by R. A. Skelton to the facsimile of L. J. Waghenaer, *The Mariner's Mirrour* (Amsterdam, 1966). Taylor and Skelton, however, distinguished two stages: first the Privy Council's decision to translate the *Spieghel der Zeevaerdt* into Latin ("a language familiar to all nations") at Howard's suggestion (the Latin translation appeared in 1586, and the first volume was dedicated to Elizabeth I), and *then* Hatton's sponsorship of the English translation (Taylor, *Mathematical Practitioners* 327, no. 68; Hatfield House Maps, 6).

62. Ruddock, "Earliest Original," 415.

63. See Ashley's introduction to *The Mariner's Mirrour*, which was dedicated to Hatton and bears his arms. Like Howard, Hatton had a personal interest in skilled navigation, being a major investor in Drake's and Frobisher's ventures (Andrews, *Trade, Plunder, and Settlement*, 144, 148, 150, 175). George Best's account of the three Frobisher voyages, *A True Discourse* (1578), had been published by "Henry Bynnyman, servaunt to the right Honourable Sir Christopher Hatton" (Waters, *Art of Navigation*, 146, n. 1). Hatton's interest in the practical application of the latest cartographic techniques is also shown in the scale plans of his Northamptonshire and Dorset estates that he commissioned from Ralph Treswell, senior, from 1580. Some of the very earliest of the genre, they are now among the Finch-Hatton papers on loan to the Northamptonshire Record Office in Delapré Abbey, and in the Dorset Record Office. See also note 189, below.

64. Hatfield House Maps, 6. The English version, still in Hatfield House, was kept as a working copy and annotated.

65. For France and Spain, see others chapters in this book.

66. E.g. the sketch map of Roanoke Island (now PRO MPG 584) sent to Walsingham by Ralph Lane, probably with his letter of 6 September 1585 (PRO C.O. 1/1/6) (Wallis, *Raleigh and Roanoke*, 53, nos. 47, 48).

67. I.e., the manuscript charts by William Borough

of the Northeast and Northwest passages, respectively to be found in the "Burghley-Saxton Atlas" (BL Royal MS 18.D.III f. 123) and in the "Ortelius" atlas now at Burghley and owned by his eldest son's descendants, the marquesses of Exeter. If it is arguable whether, by the time he acquired it, the Northeast passage chart was still an "explorer's map," there can be no doubt that the now-lost charts of the coasts of Guinea by Harriot, presented to Burghley by Raleigh in 1595–96, fell into this category. (Hatfield House Maps, 6). Edward Fenton was instructed, prior to his ill-fated voyage to South America of 1582, to present the Lords of the Council with all the charts from the voyage on his return, together with any charts or maps acquired from third parties during its course (Wallis, "Cartography of Drake's Voyage," 136). Thus was the Crown able to acquire the latest geographical information on the cheap—and prevent it from slipping into dangerous hands.

68. BL Cotton MS Augustus I.i.1, described in Wallis, *Raleigh and Roanoke*, 34, item 18.

69. Kenneth Andrews, "The Aims of Drake's Expedition of 1577–80," *American Historical Review* 73 (1968): 724–41, and his *Trade, Plunder, and Settlement*, chap. 7 (pp. 135–66 passim). Judging from Purchas's description of 1625, Drake went out of his way in the world map to give the impression that the Pacific coast of South America was accessible to English navigators and that—via the royal arms over the "Elizabethides Islands" at the southern tip of the continent—the region was already associated with England (and see Wallis, "Cartography of Drake's Voyage," 122, 129).

70. Wallis, ("Cartography of Drake's Voyage," 133–37) shows how Elizabeth and her ministers tried, for political reasons, and with a large degree of success, to suppress all public knowledge of Drake's map and details of his achievements for a decade after his return from the "Famous Voyage" in 1580.

71. Wallis, "Cartography of Drake's Voyage," 122–23.

72. Colvin, *King's Works*, 4:394, 637. L. R. Shelby, *John Rogers: Tudor Military Engineer*, (Oxford, 1967), 76–77.

73. Moir, "History of Scottish Maps," 20.

74. Taylor, *Mathematical Practitioners*, 18, 168; for Dee see French's biography, *John Dee*.

75. Sir Thomas Wyatt (d. 1554), the son of the poet, was himself noted by Sir William Paget when serving in Boulogne in 1545 as "a great forseer in fortifications" who could "make his plattes artificially" (Colvin, *Kng's Works*, 4:392).

76. Colvin, *King's Works*, 3:384–85, 388; Taylor, *Mathematical Practitioners*, 22–23, 165–67, 175, 315–17, 320–21; Waters, *Art of Navigation*, 96–97. For Gemini see n. 40 above.

77. Waters, *Art of Navigation*, 84, goes so far as to say that Northumberland "was the power behind Cabot" from 1548 onward.

78. Andrews, *Trade, Plunder, and Settlement*, 65; and Krystyna Szykula, "The Newly-Found Jenkinson Map of 1562," Twenty-third International Congress on the History of Cartography (Amsterdam, 1969), Abstracts, 109–11. See also Oakeshott, "Tudor Explorer," 703–4. Oakeshott mentions that "Sidney's interest in cartography became a passion." Sidney had also been responsible for the appointment of Clement Adams as tutor to the royal pages in 1551 (Taylor, *Mathematical Practitioners*, 169), and supported his work in editing Jenkinson's map of 1562. For the basic facts on Henry Sidney see the DNB article; for his activity in Ireland, see Edward Lynam, "Sir Henry Sidney," *Studies* 2 (September 1913): 185–203. Northumberland's sons, Ambrose, earl of Warwick, and (to a lesser extent) Robert, earl of Leicester, were very interested in exploration and overseas expansion; Ambrose invested particularly heavily in Frobisher's search for the Northwest Passage in the mid-1570s (Andrews, *Trade, Plunder, and Settlement*, 18, 168, 169, 175, 180), but they do not seem to have had Henry Sidney's feel for maps.

79. Andrews, "Robert Lythe" Dunlop, "Sixteenth-Century Maps," 331–35.

80. E.g. in the Public Record Office alone, MPF 148 (Ludlow, 1577?); MTB 17 (2) (Quatford, 1582): MPC 36 (Grossmont, White, and Skenfrith castles); MPC 69 (Skenfrith); MPC 93 (White Castle); G.178/4428 (Forest of Morfe), with many more to be discovered. The Burghley-Saxton atlas (BL Royal MS 18.D.III) has more maps devoted to Shropshire, the heart of the Welsh marches, than to any other English or Welsh county (in addition to the printed Saxton map [f. 85], a sketch-map of the county [f. 75*], the Forest of Morfe [f. 28], Shrewsbury [ff. 89–90], Clun, Oswestry, and Purslow hundreds [ff. 42–43]). Another relevant plan is of the Marcher lordship of Caus (Landsowne MS 104, ff. 106–7). Collectively they may reflect Sidney's particular encouragement of mapping for administrative purposes, though the destruction of the records of the Council of Wales, late in the seventeenth century, probably entailed the loss of most of the work commissioned by him.

81. Quinn, "Early Accounts," 35. Read, *Mr Secretary Walsingham*, 1:431. And see n. 86 below.

82. Read, *Mr Secretary Walsingham*, 3:370–410. Andrews, *Trade, Plunder, and Settlement*, 15 and passim. Anna Carleill née Barne was his first wife.

83. Hulton, *America 1585*, 20. This is a surmise, since no dedication has survived and there are several possible alternatives. The surviving reconnaissance map of Roanoke Island (PRO MPG 584) was almost certainly intended for him; see n. 66 above.

84. D. B. Quinn, ed., *The Roanoke Voyages, 1584–1590* (London, 1955), 847–84; and see Hulton, *America 1585*, 32–34, pl. 60.

85. Clara Egli Le Gear, comp., *A List of Geographical Atlases in the Library of Congress* (Washington, D.C., 1963), 6:113–14. *The Rosenwald Collection: A Catalogue of Illustrated Books and Manuscripts* (Washington, D.C., 1954), 134. Evans and Lawrence, *Christopher Saxton*, 153, no. 44.

86. The map (Hatfield House CPM II. 43) bears Walsingham's arms and is dated 1585. It is described and illustrated in Hatfield House Maps, 65, no. 104 and pl. 17. For a later map of Flushing by Adams (1588): BL Cotton MS Augustus I.ii.105. For the "Cautionary Towns": R. B. Wernham, *Before the Armada: The Emergence of the English Nation, 1485–1588* (New York, 1972), 318, 321, 334–35, 371, 386, 391. Another surviving map that was sent to Walsingham on business is one of Roscommon in Ireland (PRO MPF 95), sent by Sir Nicholas Malby to Walsingham on 18 July 1581 (Dunlop, "Sixteenth-Century Maps," 324–25, no. 76).

87. BL Harley MS 6035, ff. 2, 4, 24, 31. Richard Hakluyt the younger was particularly associated with Walsingham and was, in 1589, to dedicate his best known work, the *Principall Navigations*, to him.

88. E.g. BL Additional MS 11815.b. And see n. 22 above.

89. L. St. Lomas and A. B. Hinds, eds., *Calendar of State Papers: Foreign, April-December 1587* (London, 1929), 21, part 3, pp. 270, 320 (letters of Charles Francx to Walsingham, 23 August and 22 September 1587). I am grateful to my colleague Anne Payne for drawing my attention to this.

90. Hatfield House Maps.

91. For this and the following paragraphs: B. W. Beckingsale, *Burghley: Tudor Statesman* (London, 1967), 3–25.

92. Ibid., 17, 195.

93. Letter to Sir William Cecil, June 1563 (BL Lansdowne MS 6, f. 135). And see the opening quotations to this paper. The letter is published in Dunlop, "Sixteenth-Century Maps," 330–31.

94. Beckingsale, *Burghley*, 18.

95. Taylor, *Mathematical Practitioners*, 170.

96. Hatfield House Maps, 69–70, no. 124. It forms part of the Burghley-Ortelius atlas.

97. Hatfield House Maps, 27, 34; Eden, "Three Elizabethan Estate Surveyors," 69–70.

98. Hatfield House Maps, 6, 26.

99. Hatfield House Maps, 21, 26; Taylor, *Mathematical Practitioners*, 321–22 (nos. 43, 44, 46), 326 (no. 61), 327 (no. 65), 332 (no. 87). Bourne's treatise (BL Sloane MS 3651, later published as *The Arte of Shooting in Great Ordnance*), Blagrave's *The Mathematical Jewell*, and Norden's *Northamptonshire* (1591), *Preparative* (1596), and *Hertfordshire* (1598) were also dedicated to Burghley.

100. Hatfield House Maps, 36; Andrews, *Trade, Plunder, and Settlment*, 15.

101. Andrews, "Geography and Government," 180.

102. To the seven listed by Skelton in 1971 (Hatfield House Maps, 27)—all in the Public Record Office or Hatfield House—can now be added another two in the British Library; Lansdowne MS 104, ff. 101, 104. (The volume itself contains miscellaneous, partly private papers from the Burghley Papers, which form Lansdowne MSS 1–122.) The sketch plan on folio 101 (actually two plans, both of the Bristol Channel from west of Cowbridge to east of Chepstow on the Welsh side and from west of Portishead to east of Aust on the English side) perhaps dates from the 1580s and locates the homes of the leading gentlemen of Gloucestershire (listed on folio 100) and south Wales. The other plan is a sketch of the boundaries of the Honour of Windsor from Wisley to Woking in Surrey, apparently illustrating a dispute over jurisdiction between keepers of the various walks in the Honour, with Burghley's proposals for a settlement. On comparison of the names of keepers with those given in Norden's *Survey* of 1607 (BL Harley MS 3749), this plan can be dated to the 1590s and may be the latest surviving example of Burghley's sketch-maps. Also in the same volume (f. 106–7) is a "Plott of the Lordship of Caws" (Caus, in western Shropshire), possibly dating, on stylistic and paleographic grounds, from as early as the 1530s, but so thoroughly annotated by Burghley as to make it virtually another of his plans. Caus was a decayed marcher lordship (Trevor Rowley, *The Shropshire Landscape* [London,

1972], 110, 178). The plan is annotated to show land ownership, values, and capacity (e.g., Minsterly Park "will kepe lx kyne"), but its precise purpose is unclear.

103. Hatfield House Maps, 22, 30. But see also Eden, "Three Elizabethan Estate Surveyors," 70.

104. Ibid., 69. Hatfield House Maps, 7, 19, 25–26, and passim. The "Yelverton Papers" (so named after a later owner, and now BL Add. MSS 48000–48196), contain the largely intact working reference library of Robert Beale, which gives a better idea of what Burghley's own reference library must have been than do the scattered remnants of the latter. It is to be the subject of a detailed, forthcoming British Library catalog. Unfortunately, the maps in Beale's library have suffered more than its other parts, and there is now only one surviving volume, containing charts of the coasts of France and Spain (BL Add. MS 48021). Nevertheless, the totality of the library, like the totality of Beale's "Treatise of the Office . . . of a Principall Secretarie," makes it quite clear that for Tudor ministers maps were never more than one source of information and not necessarily among the most important.

105. Read, *Mr Secretary Walsingham*, 1:423–43.

106. Ibid., 1:428–29. Note, in view of the number of times this passage has recently been quoted in works on the history of cartography, that this—a few words in a single paragraph—is all Beale says of maps or plans in a treatise on government that occupies twenty closely printed pages.

107. E.g., Skelton in Hatfield House Maps, 3, 22; Morgan, "Cartographic Image," 138, n. 24.

108. Hatfield House Maps, 26, for two examples.

109. E.g., his methods of acquisition.

110. Andrews, "Geography and Government," 181. Hatfield House Maps, 26.

111. For this paragraph, except where otherwise stated, ibid., 5–9, 21, 25, 65, no. 104.

112. Eden, "Three Elizabethan Estate Surveyors," 69. Examples of this interest are the maps of Wainfleet in Lincolnshire (BL Cotton MS Augustus I.i.82; Hatfield House CPM I. 48) described in Hatfield House Maps, 54–55, no. 60.

113. Read, *Mr Secretary Walsingham*, 1:431. In context it is clear that Beale was talking of the "New Library" in Whitehall (see n. 158, chap. 2 in this volume) and criticizing Burghley and Walsingham, who occupied the secretaryship successively for all but eight of the years after 1547. The point is emphasized by Beale's recommendation that a secretary should henceforth "[keep] such things [the public papers that have been 'culled out'] aparte in a chest or place and n[o]t . . . confound them w[i]th his owne"—as Burghley had, accounting for the survival of official maps and papers of Henry VIII in Hatfield House. Skelton (Hatfield House Maps, 14–15, 25) is vague about the provenance, simply implying that they stemmed from the papers of previous secretaries (in contrast to the Exchequer material, which had long been housed in The Tower).

114. See the section "James I: Collection, Compilation, Conservatism," in this chapter, and following note.

115. They were, however, grouped together (and divorced from their related documents) in his archives (Hatfield House Maps), 8–9.

116. Hatfield House Maps, 19.

117. Ibid., 19, 36, 38–39, 64, no. 95; Barber, "Tudor Mystery," 19–21; P. M. Barber, "The Minister Puts His Mind on the Map," *British Museum Society Bulletin* 433 (July 1983): 18–19.

118. Now in the possession of the marquess of Exeter, Burghley House, Stamford, Northamptonshire.

119. BL Royal MS 18.D.III.

120. Hatfield House Maps, 19–20, 36–37. Barber, "Minister Puts His Mind."

121. Andrews, "Robert Lythe," 22, n. 5. The map is now PRO MPF 68. For his consultation of another loose map of Ireland, see n. 101 above.

122. Hatfield House Maps, 7, 15–16, 18–19. The list is now PRO S.P. 12/268 f. 65. Many of Burghley's Irish maps were not bound for him and survive loose in the Public Record Office, Hatfield House, the British Library, and (in a later binding) Trinity College, Dublin.

123. BL Cotton MS Augustus I.ii.72.

124. BL Cotton MS Augustus I.ii.90.

125. BL Cotton MS Augustus I.ii.92.

126. BL Royal MS 18.D.III f. 11.

127. BL Royal MS 18.D.III f. 13v.

128. BL Royal MS 18.D.III, ff. 9v, 15, 17. Hatfield House Maps, 41 no. 18, 43 no. 26, 44 no. 29.

129. BL Royal MS 18.D.III, ff. 71v-75.

130. BL Royal MS 18.D.III, f. 76. Hatfield House Maps, 60 no. 83; and note the other maps of the area in Hatfield House (ibid., 57–50, nos. 67–85). Burghley even made a sketch-map of the area for his own purposes (PRO S.P. 59/5, f. 403, reproduced in Tyacke and Huddy, *Christopher Saxton*, 26).

131. The map in the second (1596) edition of his *Perambulation of Kent* has recently been revealed to be the latest in a series that also embraces a manuscript plan

of 1585 by Lambarde himself, which came to light in 1983 (BL Add. MS 62935) and a somewhat later printed variant in the Pepys Library in Magdalen College, Cambridge. I am grateful to Sarah Tyacke for information about the latter, which she has described for her forthcoming catalog of the Pepys maps. Lambarde's 1585 map of the beacons in Kent was reproduced in Rodriguez-Salgado, *Armada*, 148.

132. BL Add. MS 57494. The British Library's Map Library has long possessed an eighteenth-century copy (K. Top 42.10.a). And see Wallis, *Raleigh and Roanoke*, 93, no. 112, and more generally Morgan, "Cartographic Image," 39.

133. For the latter see Hatfield House Maps, 36 (Burghley-Ortelius atlas map 9 verso).

134. Andrews, "Geography and Government," 180.

135. Andrews, "Robert Lythe," passim and particularly 27: "what interested Lord Burghley most about Irish maps . . . was the location of chiefs and lords." This is a notable feature of Nowell's depictions of Ireland (BL Add. MS 62540; BL Cotton MS Domitian xviii, ff. 101, 103).

136. Morgan, "Cartographic Image," 137, n. 21, and see Burghley's own map of Liddesdale (PRO S. P. 59/5, f. 403).

137. Annotation showing the Howard family's seat of Framlingham on Nowell's map of England and Ireland (BL Add. MS 62540).

138. The home of the then-Catholic Gage family, at Firle in Sussex, is specially marked on Reynold's survey of the Sussex coast of 1587 (BL Add. MS 57494).

139. BL Royal MS 18.D.III ff. 81v-82 (a copy of PRO MPF 123; Hatfield House Maps, 56, no. 65); and see J. Gillow, "Lord Burghley's Map of Lancashire in 1590," *Miscellania* 4 (1907): 167–222. Morgan, "Cartographic Image," 136–37.

140. Ibid., 138.

141. BL Royal MS 18.D.III. BL Lansdowne MS 104 ff. 102–3 contains a schematic plan of the "gentlemen of Gloucs. and their habitations," listed in geographical sequence along the principal highways and the River Severn. See also n. 102 above.

142. Morgan, "Cartographic Image," 137–38, 147, quotes the observation of Burghley's nephew Sir Francis Bacon, "The numbers of Cities & Townes [may appear] by Cartes and Mappes" ("Of the Greatness of Kingdoms and Estates," in *Works*, ed. J. Spedding et al. [London, 1878], 6:445). For further examples: BL Royal MS 18.D.III ff. 31v-32 (Honour of Windsor in Berkshire), f. 87 (Forest of Morfe, Shropshire), ff. 12, 74 (military divisons of Devon and Westmoreland); and see also BL Lansdowne MS 104 f. 104 (Honour of Windsor in Surrey).

143. Morgan, "Cartographic Image," 139.

144. Hatfield House Maps, 36–37, 56 no. 63 (Durham), 40 no. 12 (pl. 22; coasts of the West Country). Skelton's suggestion (19) that the itineraries of the routes between Antwerp and Dunkirk to Augsburg and from London to Edinburgh in the "Nowell-Burghley Atlas" may date from before 1560 is mistaken: they occur on the verso of Nowell's map of England and Ireland and must have been added after the presentation of the map in 1564.

145. Andrews, "Geography and Government," 178–91 passim. This is of value not least because it gives evidence for the limits of the government's use of maps.

146. Andrews, "Robert Lythe," 23; Andrews, "Geography and Government," 183.

147. Ibid., 185–88 (where Andrews shows the negative consequences that flowed from the English failure to utilize Lythe's survey when planning the Munster plantation); Andrews, "Robert Lythe," 22 n. 5, 23. John Andrews, "The Maps of the Escheated Counties of Ulster, 1609–10," *Royal Irish Academy Journal* 100 (1974): 133–70. Hatfield House Maps, 62–63, no. 91. The relevant maps are Trinity College, Dublin (TCD), 1209.9 and BL Cotton MS Augustus I.ii.40 (Dunlop, "Sixteenth-Century Maps," 317–18, nos. 27–28; Leix and Offaly); PRO MPF 305 (Munster); PRO MPF 35–65 (S.P. 64/2), Hatfield House CPM Supp. 2, BL Cotton MS Augustus I.ii.44 (Ulster).

148. BL Cotton MS Augustus I.i.74 (partly reproduced in Tyacke and Huddy, *Christopher Saxton*, opposite p. 16). It can be dated to about 1588 (see Colvin, *King's Works*, 4:406, 408). For Irish examples, see Hayes-McCoy, *Ulster and Other Maps*, and his "Contemporary Maps as an Aid to Irish History, 1593–1603," *Imago Mundi* 11 (1965): 32–37.

149. Engraved by Augustine Ryther to accompany Petruccio Ubaldini's *Expeditionis Hispaniorum in Angliam vera descriptio* (1589). Reproduced in Rodriguez-Salgado, *Armada*, 243–48. They provided the basis for a set of tapestries commissioned in 1595 by Lord Howard of Effingham, sold by him to James I, and displayed in the House of Lords from 1650 to 1834 (ibid., with reproductions, 248–51).

150. Hatfield House CPM II 43; BL Cotton MS

Augustus I.ii.105, discussed in Hatfield House Maps, 65 no. 104 and pl. 17 (though there are greater differences between the maps than Skelton suggests). Also Rodriguez-Salgado, *Armada*, 118.

151. BL Egerton MS 2579 (Santiago) discussed in Wallis, *Raleigh and Roanoke*, 85, no. 100. The plan of Santo Domingo was probably published, with three other town-plans, to illustrate Walter Bigge's *A summarie and true discourse of Sir Francis Drake's West Indian Voyage*, in 1588. It is discussed and illustrated in Wallis, *Sir Francis Drake*, 108–111, no. 117.

152. Duncan Caldecott-Baird, *The Expedition In Holland 1572–4: The Revolt of the Netherlands: The Early Struggle for Independence from the Manuscript by Walter Morgan* [All Souls' College, Oxford, MS 129] (London, 1976). On pp. 41–45 Caldecott-Baird emphasized Morgan's cartographic plagiarism, and it has since been demonstrated that all the maps in MS 129 are copied from Dutch printed examples (S. Groenwald, "Het Engelse Kroniekje van Walter Morgan en een onbekende reeks historieprenten (1572–4)," *Bijdragen en Mededelingen voor de Geschiedenis van Nederland* 98 [1983]: 19–74). I am grateful to Geoffrey Parker for this reference. Morgan's manuscript, which is dedicated to Burghley, is no less attractive for this plagiarism. The rediscovery of manuscript plans of Flushing and Bergen-op-Zoom of 1588, which are, on stylistic grounds, undeniably by Walter Morgan, though he signs himself "Walter Morgan Woulphe, Professor at Armes" (BL Cotton MSS Augustus I.ii.107, 115) establish that Caldecott-Baird's suggestion (4) that he died in about 1574 is incorrect.

153. E.g., Norris's map of Brest of 1591 (BL Cotton MS Caligula E.ix, f. 276).

154. H. A. Lloyd, *The Rouen Campaign, 1590–2* (Oxford, 1973), 27, 64–65. Later that year, regional (e.g., BL Cotton MS Augustus I.ii.59) and local Rouen maps (BL Cotton MS Augustus I.ii.87, 89, 90, 91) were being created in relative abundance, judging by the number that survive, while Henrician maps, such as the map of the English Channel possibly by Rotz (BL Cotton MS Augustus F.ii.65, 66), were annotated to follow the course of the campaign.

155. See examples cited in Morgan, "Cartographic Image," 147–48.

156. BL Royal MS 18.D.III f. 2.

157. BL Royal MS 18.D.III f. 21. Lambarde was a noted Anglo-Saxon scholar, but the map may have been among those left to him by his friend the still more notable Anglo-Saxon scholar and cartographer Laurence Nowell (see Barber, "Tudor Mystery").

158. Hatfield House CPM Supp. 1, illustrated and discussed in Hatfield House Maps, 28, 40 no. 12, pl. 22.

159. Hatfield House Maps, 36; BL Royal MS 18.D.III f. 42. A similar juxtaposition of cartography and genealogy occurs in the context of Burghley's sketch-plan of the Bristol Channel and seems to illustrate the interrelationships of the Gloucestershire gentry, who are listed on one page and cartographically located on another (BL Lansdowne MS 104 ff. 100–101).

160. Tyacke, *English Map-Making*, 69–70, pl. 32. The plan is at Burghley House (MS 57/7).

161. Hatfield House Maps, 72, nos. 132, 133, 134, and essay by Sir John Summerson, 77–79, and 81, nos. 147–51, 170, 171. Tyacke, *English Map-Making*, pl. 31.

162. J. Summerson, "The Building of Theobalds, 1564–1585," *Archaeologia* 97 (1959): 107–26, pl. 23–32; and Hatfield House Maps, 78.

163. Frederick, duke of Württemberg, in *England as Seen by Foreigners*, ed. William Brenchley Rye (London, 1865), 44. Beckingsale, *Burghley*, 40. Morgan, "Cartographic Image," 142.

164. Ibid., 141–42.

165. John Dee, "Mathematicall Preface" to Henry Billingsley's English translation of Euclid's *The Elements of Geometrie* (1570). Taylor, *Mathematical Practitioners*, 320, no. 41.

166. Morgan, "Cartographic Image," 152–53, and see 149 for a related example.

167. SAS/CO/d3.

168. This map is still owned by the Cordells' successors, the Parker family. Recently Israel Amyce's large, but less decorative scale map of the same area dating from 1580 has been put on display in the Elizabethan Great Hall, giving visitors an impression of the impact that might originally have been hoped for—though the map (one of the very earliest estate maps drawn to an explicit scale) is damaged and faded, perhaps from overexposure in the past.

169. J. B. Harley, "Meaning and Ambiguity in Tudor Cartography," in *English Map-Making*, ed. Tyacke, 22–45, and particularly 31, 36–38. P. M. Barber, "Wall Maps, 'Some to Beautify Their Walls,'" *Country: The Magazine of the Country Gentlemen's Association* 89 (July 1989): 41–42; and T. Campbell et al., *What Use Is a Map?* (exhibition catalog, London, 1989).

170. Sydney Anglo, *Spectacle, Pageantry, and Early Tudor Policy* (Oxford, 1969), 163.

171. Frances A. Yates, *Astrea: The Imperial Theme*

in the Sixteenth Century (London, 1985), 1–120. For the following paragraphs I am indebted to Frances Yates, Roy Strong's many works on the cult of Elizabeth I, and his *Art and Power: Renaissance Festivals, 1450–1650* (London, 1985), and particularly to J. B. Harley, not only through his essay, "Meaning and Ambiguity" but also through many stimulating conversations.

172. Yates, *Astrea*, 63; Harley, "Meaning and Ambiguity," 38–39.

173. The association of royal arms with legitimacy of object and of the monarch portrayed on the obverse is also to be found in the Tudor coinage. As Harley has noted (ibid., 36–37), the royal arms on Saxton's 1583 wall map underline England's political unity. Elizabeth Danbury's paper "Patronage and Propaganda in Medieval English Art: The Decoration of Royal and Private Charters," delivered at the 1985 Conference of Anglo-American Historians in London, included a discussion of related examples where decoration sometimes rendered the item a visual petition for favor and an expression of loyalty, by analogy with contemporary and earlier paintings portraying donors being presented by their patron saint to, e.g., the Virgin.

174. Yates, *Astrea*, 64–65. Wallis, *Raleigh and Roanoke*, 100, 105 no. 143.

Contemporaries may also have realized that the pose of the queen, embracing the universe, was similar to that of God on several medieval world maps such as the verso of the "Psalter" world map as well as the map itself (BL Add. MS 18601 f. 9).

175. Yates, *Astrea*, 88–111, 115–17; E. K. Chambers, *Sir Henry Lee* (Oxford, 1936). It is significant in this connection that in the portrait of 1568 by Antonis Mor in the National Portrait Gallery, London, Sir Henry Lee wears sleeves decorated with armillary spheres.

176. Yates, *Astrea*, 106, 219. Harley, "Meaning and Ambiguity," 33–34. Morgan, "Cartographic Image," 133–34. Roy Strong, *The Elizabethan Image: Painting in England, 1540–1620* (exhibition catalog, Tate Gallery, London, 1969), 45, no. 78; Strong has repeatedly discussed it in his other publications.

177. Illustrated in Yates, *Astrea*, pl. 6b. Wallis, *Raleigh and Roanoke*, frontispiece.

178. E. Hawkins, A. W. Franks, and H. A. Grueber, *Medallic Illustrations of the History of Great Britain and Ireland to the death of George II* (London, 1885), 1:154–56, nos. 129–33. For an echo of this approach, but employed by the Austrian Habsburgs a century later, see the engraving "Austriaca Olea Contra Oleastrum Turcicum" (illustrated and discussed in K. Gutkas, ed., *Prinz Eugen und das barocke Österreich* (exhibition catalog, Schlosshof, Austria, 1986), 227–28, no. 9.3).

179. Yates, *Astrea*, 114–20, particularly 115–16; C. Beltramo Ceppi and N. Confuorto, eds., *Firenze e la Toscana dei Medici nell'Europa del Cinquecento* (exhibition catalog, Consiglio d'Europa Sedicesima Esposizione Europea di Arte, Scienze, e Cultura, Florence, 1980), 1:67, 85 no. 435; R. Strong, *Gloriana: The Portraits of Queen Elizabeth I* (London, 1987), 101–7; Rodriguez-Salgado, *Armada*, 86–87. It has also been argued that Sir Christopher Hatton is portrayed in the background.

180. Strong, *Elizabethan Image*, 44, n. 76. Harley, "Meaning and Ambiguity," 33. Neither Strong nor Harley appears to notice that Elizabeth, while expressing imperial aspirations through the globe/orb, is placing her hand over *North America* and is pointing at the Spanish Main and South America, possible areas of English expansion. John Dee and other theorists had particularly emphasized England's special claims "partly Jure Gentium, partly Jure Civile, and partly Jure Divino" to "a great parte of the Sea Coastes of Atlantis (otherwise called America)! . . . and of all the Iles nere unto the same from Florida, Northerly, and chiefly of all the Islands Septentrionall (great and small)" (Dee, BL Cotton MS Augustus I.i.1. verso).

181. Yates, *Astrea*, 216–19.

182. Wallis, "Cartography of Drake's Voyage," 148–51. Wallis, *Sir Francis Drake*, 94, no. 96. Hawkins, Franks, and Grueber, *Medallic Illustrations*, 1:131, no. 83. Sir George Hill, *Medals of the Renaissance*, rev. and enl. by Graham Pollard (London, 1978), 150, 187, 387.

183. Geoffrey Parker, *Philip II* (London, 1979), 48; and see Parker's essay in this volume.

184. BL Additional MS 5415 A. For a full discussion and reproduction of all the charts in the atlas (which never seems to have had a frontispiece), see A. Cortesao and A. Texeira da Mota, *Portugaliae Monumenta Cartographica* (Lisbon, 1960), 2:13–15, pl. 100–108.

185. BL Additional MS 62525. It may be significant, given the link between military engineering and cartography, that Mason prided himself on his expert knowledge of fortification (Colvin, *King's Works*, 4:401).

186. Lloyd, *Rouen Campaign*, 27, n. 50; and for an earlier example, Andrews, "Robert Lythe," 26.

187. Helen Wallis, "The First English Globe: A Recent Discovery," *The Geographical Journal* 117 (September 1951): 276. See also Helen Wallis, "Further Light on the Molyneux Globes," *The Geographical Journal* 121

(September 1955): 304–11; and the same author's "Cartography of Drake's Voyages," 153.

188. See particularly Harvey, "Estate Surveyors," and his *The History of Topographical Maps* (London, 1980), 158, 168. Tyacke and Huddy, *Christopher Saxton*, 52–60. Eden, "Three Elizabethan Estate Surveyors," 68–69. As late as 1649, however, detailed instructions on estate surveying that did not mention mapping could be compiled for the surveyors of the Duchy of Cornwall (S. J. Madge, *Domesday of Crown Lands* [London, 1938], 334–39).

189. Northamptonshire Record Office, Finch Hatton MS 272. I am most grateful to John Schofield for drawing this to my attention.

190. See n. 36 above.

191. Youings, *Sixteenth-Century England*, 160–62, 166–67.

192. Heather Lawrence, "John Norden and His Colleagues: Surveyors of Crown Lands," *The Cartographic Journal* 22 (June 1985): 54.

193. Quoted by Ravenhill, *John Norden's* 21, citing BL Royal MS 18 A xxiii ff. 38–39.

194. Eden, "Three Elizabethan Estate Surveyors," 69. For the traditional survey: P. D. A. Harvey, *Manorial Records*, Archives and the User 5 (London, 1984), 15–24.

195. Lawrence, "John Norden," 54; Hatfield House Maps, 50, no. 48 (CPM II.19).

196. Ravenhill, *John Norden's* 21. In 1593 he had surveyed some Crown lands in Wales, but it is unclear whether these included mapping (W. B. Gerrish, *John Norden, 1548–1626(?): A Biography* (Ware, 1903), iii). I am grateful to Graham Haslam for this reference.

197. See Hatfield House Maps, 22 (Robert Treswell); Tyacke and Huddy, *Christopher Saxton*, 52.

198. See, for example, John Norden's later (1617) plan illustrating the townspeople's encroachments on Exeter Castle, which was owned by the Duchy of Cornwall (BL Additional MS 6027 ff. 80v-81), and compare Eden, "Three Elizabethan Estate Surveyors," 76–78.

199. Youings, *Sixteenth-Century England*, 175.

200. Thomas Elyot, *The Book Named the Governor*, ed. S. E. Lehmberg (London, 1962), vii. Scotland's own cartographic expertise and its mapmaking tradition (see n. 81, chap. 2 in this volume) must have been influential too.

201. Note by Howard Nixon in the copy that is now owned by the British Library (BL MAPS C.7.c.1).

202. BL Harley MS 6252 f. 4, quoted by Morgan, "Cartographic Image," 141.

203. For this and the following paragraph, Lawrence, "John Norden".

204. Norden's *Surveyors' Dialogue* (London, 1607) gives space to the tenants' viewpoint, though only to refute it. For Norden's own efforts to wriggle out of having to do numerous "tedious" plans of whole manors: Lawrence, "John Norden," 54–55. For a written survey by Norden, see Graham Haslam, "John Norden's Survey of the Manor of Kennington," *London Topographical Record* 25 (1985): 59–62. It could be that private practice was generally more lucrative, but published evidence is lacking. A comparison between the figures cited by Eden ("Three Elizabethan Estate Surveyors," 73–75) and by Lawrence, "John Norden," 55, suggests that the Crown paid better than Oxford colleges, but like is not compared with like, and the impression could well be misleading. At all events, a distinction should be made between the surveys paid for by the Crown and those paid for by the Duchy of Cornwall. The latter was far wealthier, because it could increase rents whenever a new duke (heir to the throne) was installed or when the duchy went into commission, whereas the income from Crown estates remained fixed for the duration of leases (I am grateful to Graham Haslam, the archivist to the Duchy of Cornwall, for this information).

205. BL Harley MS 3749.

206. Information from Heather Lawrence, who was completing a biography and full bibliography of John Speed at the time of her untimely death in October 1985. For further details on Speed's life, see the article in the DNB, and, for his county maps, E. Lynam, *British Maps and Map Makers* (London, 1944), 25. For Fulke Greville, later First Baron Brooke, see DNB and Lawrence, "Permission to Survey," 19–20.

207. BL Cotton MS Augustus I.ii.46; R. Tindall to Henry, prince of Wales, 22 June 1607 (BL Harley MS 7007 f. 139); Wallis, *Raleigh and Roanoke*, 96, no. 115; W. P. Cumming, R. A. Skelton, and D. B. Quinn, *The Discovery of North America* (London, 1971), 236–37. Roy Strong, *Henry Prince of Wales and England's Lost Renaissance* (London, 1986), 61, and T. E. Birrell, *English Monarchs*, have shown Henry to have been an apt pupil of Edward Wright (see n. 37). He took a great interest in the sciences and colonial expansion, and many up-to-date geographical books, and probably maps, were acquired for him. Almost certainly his view of the world would have been partly derived from maps and have been influenced by them; but having died, aged only eighteen, before acceding to the throne, he is beyond the scope of

this study.

208. BL Cotton MS Augustus I.ii.45.

209. "A description of East India conteyning th' Empire of the Great Mogull," 1619 (copy in BL King's Topographical Collection 115.22).

210. Hatfield House Maps, 10–11.

211. Ibid., 14–15.

212. Ibid., 14–17. There is no good study of Cotton as a collector. The latest and most reliable life is by Kevin Sharpe, *Sir Robert Cotton, 1586–1631: History and Politics in Early Modern England* (Oxford, 1979). On analyzing the printed books in James I's library, part of which still survives in the British Library, Birrell (*English Monarchs*, 26–27) has deduced that it, too, was assembled in part as an official, printed archive. As such, it is likely to have contained many maps, but it would appear that the printed maps and all but one of the atlases owned by James I can no longer be firmly identified.

213. For Carew and his maps see John Andrews, *The Irish Maps of Lord Carew: An Exhibition in the Library of Trinity College, Dublin* (Dublin, 1983). Also Hatfield House Maps, 9, 10, 17–18, 20.

214. J. S. Brewer, *Calendar of the Carew Manuscripts Preserved in the Archiepiscopal Library at Lambeth*, 6 vols. (London, 1867–73). National Maritime MS P. 39.

215. Hatfield House Maps, 25.

216. Andrews, *The Irish Maps of Lord Carew*, 6–7.

217. Hatfield House Maps, 62–63, nos. 62–63. Carew tried to do likewise, but while he had several maps revised in the process of having them copied (unlike Cotton and Burghley, who had maps copied but without revision), he never managed to secure the services of a cartographer who could synthesize information from a variety of maps (Andrews, *The Irish Maps of Lord Carew*, 4–6).

218. "Virginia. Discovered and Discribed by Captain John Smith" (engraved by William Hole). Usually found in John Smith, *A Map of Virginia* (Oxford, 1612), and see Coolie Verner, *Smith's Virginia and Its Derivatives*, Map Collectors' Circle 45 (London, 1968).

219. Jeannette Black, *The Blathwayt Atlas* (Providence, R.I., 1975), 2:149–53.

220. Cumming, Skelton, and Quinn, *The Discovery of North America*, 236–37, 242–47.

221. R. A. Skelton, *County Atlases of the British Isles, 1579–1850: A Bibliography*, Map Collectors' Circle (London, 1964), 1:30–44. Lawrence ("Permission to Survey," 18–19) detected the change in tone between the wording of Saxton's passes and the first set of Norden's passes, on the one hand, and Norden's later passes and those of Speed, on the other. This reflected the growing inability of the Crown to sponsor a national survey alone. A similar inability can be detected in the wording of the privilege granted to the land surveyor Aaron Rathborne by James I, in connection with his abortive plan to undertake a detailed survey of London and other English cities (PRO c.66.2152). I am most grateful to Ralph Hyde for bringing this document, which he discovered, to my attention. The other national survey undertaken under James VI and I, Pont's survey of Scotland (1583–1601), received even less royal support than Speed's, despite the king's declared intention of giving "some moneyis" toward Pont's expenses (Moir, "History of Scottish Maps," 37–53, particularly 42).

FOUR

Monarchs, Ministers, and Maps in France before the Accession of Louis XIV

David Buisseret

Louis XIV began his personal government of France in March 1661. He was resolved from the start to put an end to the disturbances associated with the Fronde, and to set the administration of the kingdom on a new footing. So in September 1663 his chief minister, Colbert, sent out a major *instruction* to all the provincial officers, setting up a general investigation of the state of the realm. The first clause of this document required the provincial officers to send in all the maps they could find of their areas, so that the royal cartographer, Nicolas Sanson, could correct them if they were defective.[1] Louis and Colbert envisaged using maps in four main areas of their activity:

1. For military and naval purposes. They wanted maps not only for use in the field but also to regulate the system of fortifications, and to define the responsibilities for defense of each sector of France's frontier. They also needed maps of coasts and naval installations, both at home and overseas, as well as maps drawn by exploring expeditions.

2. For making political and judicial decisions. Here it was chiefly a question of defining areas of responsibility, both inside and outside the realm. Cartographic material was to be used to establish the boundaries of the realm, particularly on the disputed frontiers of the northeast and southwest; here it was a long time before maps could do justice to the intricacies of boundaries established by wordbound treaties. Within the kingdom, Louis and Colbert commissioned maps setting out the areas for which different courts were responsible. This may not seem to us a very remarkable step, but it marked a decisive phase in the attempt to make sense of the many conflicting jurisdictions into which France was divided under the *ancien régime*.

3. For economic and financial planning. Maps were commissioned to show where the various fiscal divisions, or *généralités*, ran, and where specific taxes like the *gabelle* (salt tax) were to be paid. Figure 4.1. for instance, shows the *Carte Generalle des Gabelles de France* published in 1665; the keyed symbols indicate not only the *greniers à sel* ("Gr") but also the means of enforcing this unpopular tax ("Cont" indicates *Controlle*). Other maps were ordered when great public works like the *canal du Midi* were being planned; this canal had a very rich cartography associated with it. Others, again, were commissioned to show the sites of the mines in France, or the nature and extent of its forests.

4. For setting out the ecclesiastical structure. The church was still very much the business of the monarchy, which therefore needed to know the boundaries of each diocese. Often

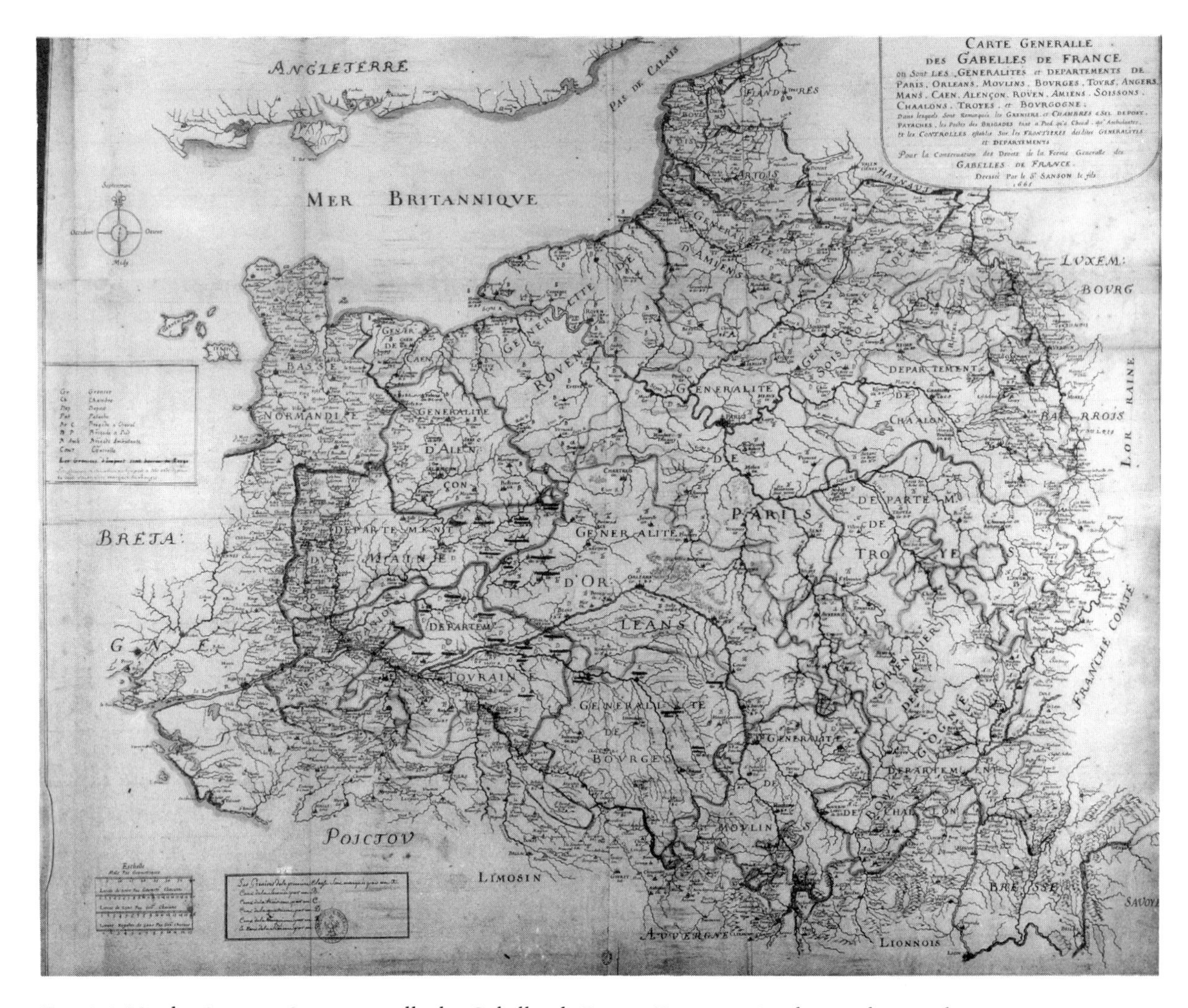

FIG. 4.1 Nicolas Sanson, *Carte generalle des Gabelles de France* (Paris, 1665). The Newberry Library (NL)

diocesan maps had been commissioned by individual bishops, and there was under Colbert an attempt to draw these maps together, and to consolidate them into a general map of ecclesiastical boundaries in France.

At the time of Louis XIV's accession, then, French governing circles possessed a well-developed sense of the usefulness of maps, and there were cartographers capable of responding to their needs. There was also—in contrast, for instance, to Spain—an abundance of presses, mostly concentrated in Paris, capable of printing and diffusing large maps in considerable quantities. But these characteristics, so well developed in the middle of the seventeenth century, had taken about 150 years to emerge. About 1500, where we begin our analysis, map consciousness at all levels of French society was

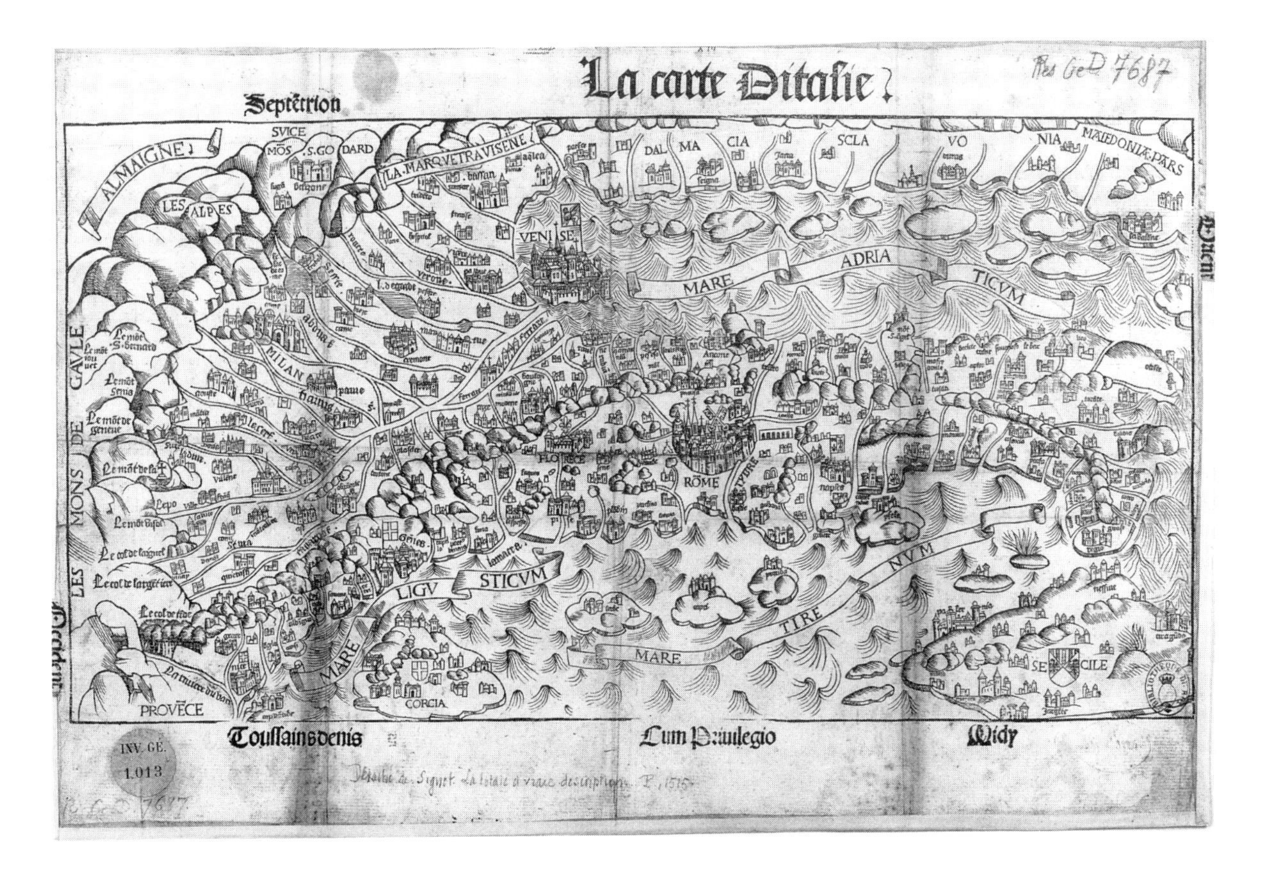

FIG. 4.2 Jacques Signot, *La Carte d'Italie* (Paris, 1516). Bibliothèque Nationale (BN), Paris

rudimentary, particularly when compared with the ways in which people were already using maps in some societies of northern Italy and southern Germany.

FRANCIS I AND HIS PREDECESSORS

The first French king whom we know to have commissioned a map was Charles VIII (1483–98), who about 1495 asked Jacques Signot to reconnoiter the Alpine passes through which the French army might march in order to invade Italy. Signot drew a map, known as the *Code Signot* (1495–98), and in 1515 this map was printed as *La Carte d'Italie* (figure 4.2).[2] It

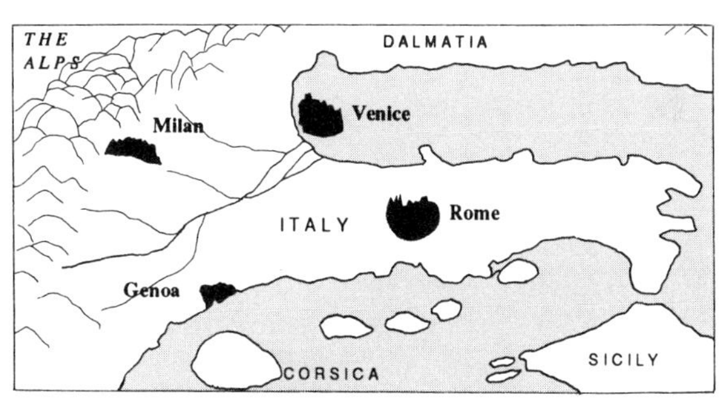

may seem rather crude at first sight, with its eastern orientation and excessively large views of Venice and Rome, but this little map did indicate the seven passes from France into Italy (on the left, leading from "Les Mons de Gaule"), and the accompanying text had helpful notes like the one pointing out that the Col du Mont-Genèvre "is the best and easiest passage for artillery."

This was the first of an enormous number of maps generated by the kings of France for military purposes. If the Italian wars, which lasted from 1494 to 1559, were thus instrumental in their earliest phase in generating a precocious map, they also affected French cartography in another, less quantifiable way. For just as many French soldiers going down into Italy were dazzled by the architecture of Humanism and returned to France resolved to build in that style, so others must have been influenced by the way in which the northern Italians were already using maps for military purposes (see chapter 1 in this volume).

During the fifteenth century, the kings of France had been acquiring in their libraries copies of various geographical manuals, and in particular of Ptolemy's *Geography*, the Renaissance text that transmitted to early modern Europe a good part of the cartographic knowledge of classical antiquity. Louis XII (1498–1515), for instance, acquired a manuscript Ptolemy copied at Ghent in 1485.[3] But it would be hard to maintain that Louis XII regarded this work other than as a fashionable ornament for his library; he may have enjoyed leafing through it and forming some idea of the shape of his world, but he surely had no idea of using Ptolemy's map of Gaul to help him govern France.

Under his successor, Francis I (1515–47), royal map use became more systematic. The mathematician, engineer, and cartographer Oronce Fine (1494–1555) had worked with the king as a military engineer during the Italian campaigns of the 1520s and, if a claim on Fine's world map of 1534 is true, had in about 1520 designed a cordiform map of the world "for the celebrated *savant* King Francis I."[4] The king was notoriously eager to encourage and promote the new learning, largely deriving from Italy, of which Fine was a leading practitioner. In 1530 Francis appointed him one of the newly established *lecteurs royaux* (royal professors) at Paris, where he continued to produce maps, as well as teaching mathematics.[5]

Francis I was involved in constant wars against the emperor, whose armies often threatened to overrun France. These armies were equipped with powerful artillery, so that Francis had to fortify a large number of French towns in the new artillery-resistant style.[6] This involved laying out a bastioned trace around the town, so conceived that each part of the wall was covered by the guns on the adjacent bastions. These traces had to be laid out in accordance with mathematical principles, bearing in mind the range of defensive weapons at any given time. They could hardly be undertaken without a detailed plan's being drawn. At first this skill was beyond the capacity of any Frenchman, and Italian engineers were called in.

We know many of their names, and we also know that many of them drew maps; in 1524–26, for instance, Anchisio de Bologna drew a plan of the Boulevard Saint-Sébastien at Lyon, and in 1541 Girolamo Bellarmato made a map of the new works at Le Havre.[7] Alas, none of these maps seem to have survived—in sharp contrast to the situation in England described in chapter 2—so that we can only guess that these "French" Italians probably had as important an influence on cartography in France as the Italian engineers hired by Henry VIII had in England.[8]

Francis was also aware of the value of maps for planning overseas expeditions, which indeed were in the sixteenth century a sort of extension of European warfare. Robert Knecht, the king's most recent biographer, describes a most interesting planning session held before the third voyage of Jacques Cartier, in 1541.[9] The French mariners of Dieppe were in close contact with the cartographers of Portugal, and it was probably through these relations that a Portuguese pilot, Joao Lagarto, was brought to see the king. He brought with him two marine charts and an astrolabe; Francis spoke to him for more than an hour "with understanding and intelligence," as the pilot reported. The following evening the king showed Lagarto two charts of his own, which the pilot described as "well painted and illuminated, but not very accurate." This account clearly shows Francis' interest in using charts to plan overseas ventures and suggests that he was aware of the work being done at Dieppe, where a great school of chartmakers was just beginning to emerge.

Henry II (1547–59)

Francis's successor, Henry II, had this same interest. He retained Oronce Fine on a royal pension, and Fine in 1549 dedicated to the king "La Sphere du Monde," a general cosmography. Henry consulted him about the possibility of finding a northwest passage around the American continent, for such a passage would enable the French to bypass Spanish possessions in the New World. Like his predecessor, Henry was in touch with the Dieppe chartmakers, who at this time were beginning to produce the maps of what is now called the "Dieppe School."[10] In 1550 Pierre Desceliers dedicated to him one of the most characteristic products of the school: a huge world map, designed to be laid flat on a table and examined from each side in turn.[11] Another marine cartographer who received royal encouragement was Guillaume Le Testu, whom Henry appointed pilot royal at Le Havre in 1556. Le Testu was the author of a superb fifty-eight-map atlas, now preserved at the Château de Vincennes. All in all, we have to see Henry as a king keenly interested in overseas ventures and well aware of the use of maps in promoting them.

He was also well aware of the importance of cartography for military operations. During his reign, the Italian engineers continued to work on the fortifications of France and to generate plans of cities, even if these all seem to have perished. But Henry carried military map use a stage further, for during the French expedition to the Rhine in 1552, he had a map made that showed not only roads—a capital innovation—but also the villages in which an army might lodge for the night. We have a passage in the memoirs of Marshal Vieilleville that describes the king on one occasion as looking at the map in order to decide the best route for the army, and Vielleville himself observed that a military leader should no more think of marching without a map than a mariner should think of setting to sea without a compass.[12] Clearly, during the reign of Henry II the military use of maps had become more extensive and sophisticated.

The king, who has often been portrayed as something of a blockhead, also used civilian maps in an innovative way. It seems to have been following his orders that the great map of Paris by Olivier Truschet and Germain Hoyau was drawn in 1550,[13] and there is another example of the same year that most tellingly reveals how he saw the use of maps in solving a specific problem. By the late 1540s the uncontrolled expansion of Paris had become a problem, as people settled outside the walls in unruly little communities. So in September 1550, the king wrote to the city hall (*hôtel de ville*) that the mayor and council should meet with "Jhérome Bellarmato, l'un de nos ingenieulx"

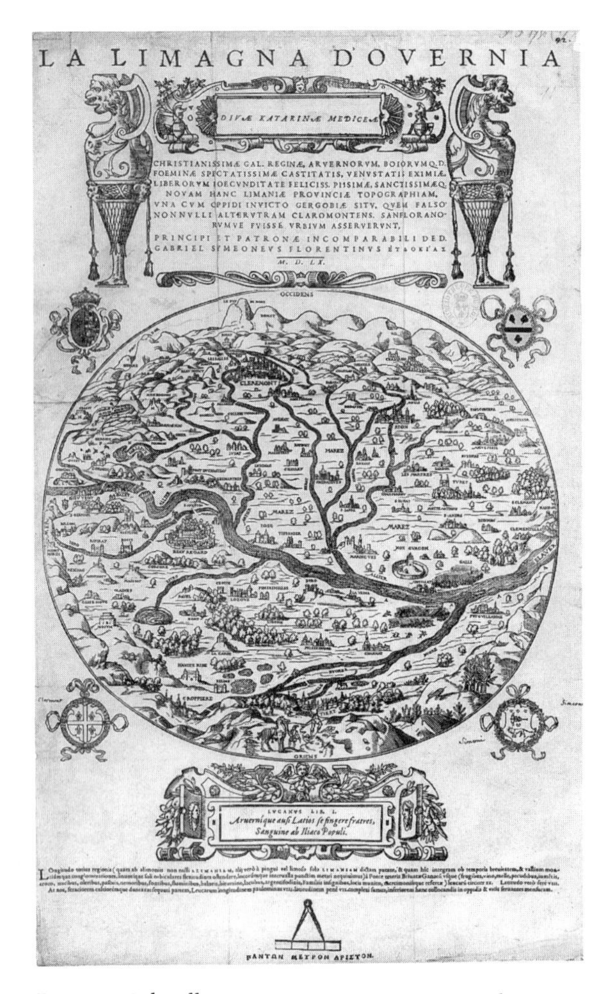

FIG. 4.3 Gabrielle Symeone, *La Limania d'Overnia* (Paris, 1560). BN, Département des Cartes et Plans

(Girolamo Bellarmato, the engineer mentioned above for his work at Le Havre), who was normally stationed at Châlons-sur-Marne, working on fortifications. Bellarmato would then make a sketch of the areas outside the walls which needed attention, and the mayor and councillors could meet on site with this sketch. Having decided what the new town limits should be, Bellarmato would then insert them on his plan.[14] It is not clear how well this procedure worked, but it is perfectly plain from Henry's instructions that he knew what he wanted done and how he meant to use Bellarmato's mapping skills.

Catherine de Medici and Her Children

Henry died in an accident in 1559, and the government of France fell into the hands of his widow, Catherine de Medici, and her sons: Francis II (1559–60), Charles IX (1560–74), and Henry III (1574–89). The forty years that followed Henry's death saw the outbreak of a series of civil wars, which greatly weakened the monarchy, diverted French rulers from considering major overseas ventures, and enfeebled those social groups that in other countries were commissioning maps.

Catherine had grown up in Florence and well knew how maps were being used in Italy. She had numerous Florentines in her suite, one of whom was Gabriello Symeone. He was primarily an astrologer, who advised her, for instance, on "the best day for the king's coronation, taking into account the position of the moon, Saturn and Mars" (1 July turned out to be the best date).[15] He was also a genealogist and a military theorist, but his role that interests us most was as a cartographer and author of a map of the province of Auvergne (figure 4.3; it is oriented westward, with "Cleremont" at the top).[16] We do not know if Catherine commissioned the map, but it was dedicated to her, and she must have known and approved of Symeone's mapping. The map is not very accurate, but it was the first of its kind, and found its way in modified form into many editions of the *Theatrum Orbis Terrarum* of Ortelius, where it was reoriented northward as *Limaniae Topographia*.

Another member of the royal entourage interested in cartography was Pierre Hamon,

FIG. 4.4 Pierre Hamon, Map of France, 1568. The Pierpont Morgan Library, New York, M 980, single leaf

calligrapher and typographer, who became writing-master to the dauphin.[17] About 1568 Hamon drew and dedicated to Catherine a superb (though not particularly accurate) map of France, which has recently been acquired by the Pierpont Morgan Library of New York. It has never been closely studied, but presumably contains a self-portrait of Hamon himself with a huge pair of dividers at the bottom left (figure 4.4). Hamon also made a copy of one of the world maps of the Dieppe School cartographer Nicolas Desliens; this is now in the National Maritime Museum in Greenwich.[18]

These involvements of Catherine with

cartographers were more or less incidental, but the queen mother had a much more systematic arrangement with Nicolas de Nicolay, Sr. d'Arfeuille (1517–83).[19] This enigmatic character, whom we have already encountered in English service in chapter 2, was a soldier, diplomatic representative, and traveler, and had in about 1552 been named by Henry II *géographe du roi*. This title made him eligible for a royal stipend and was a mark of favor; similar appointments were made in the sixteenth century for doctors, painters, mathematicians, and other skillful or learned persons.

Toward 1560, Nicolay received a commission from Catherine to draw up a general account of France, accompanied by maps of each province. He seems to have begun at once, for the next year we find Catherine writing to the duchess of Ferrara about him. "Fearing," she writes, "that Nicolay will receive some impediment on the road between Montargis and Briare," she enjoins the duchess to have him escorted as far as Briare and then as far beyond as necessary, "so that the business which he has begun does not remain incomplete."[20]

Nicolay had already, in 1558, completed a map of the area round Calais, which was adopted by Ortelius in the 1580 edition of the *Theatrum* under the title *Caletensium et Bononiensium ditionis accurata delineatio*, and after him by Mercator (1585) and by Maurice Bouguereau (1594). Now he completed maps of Berry (1567), of Bourbonnais (probably 1576), and perhaps of the Lyonnais (1573). These maps remained manuscript,[21] but the set for Berry is particularly impressive, showing how Nicolay could systematically map the component parts of a province and then draw the parts together into a striking general map (color plate 4). It was evidently the intention of Catherine to make Nicolay a sort of French precursor of Saxton, but the religious and civil disturbances that broke out in the early 1560s prevented him from completing more than a small part of his task. Moreover, it is by no means certain that an atlas taking in all the French provinces could have been printed in France at this time, and this would have added a further impediment.

As well as initiating surveys of the French provinces, Catherine continued to use maps for planning overseas ventures. In 1561, for instance, she engaged the French ambassador in Lisbon, Jean Nicot, to bribe the Portuguese cartographer Andreas Homem into making for the French a map of southern Africa.[22] At the end of the next decade, she was concerned with French designs on Brazil, and two charts by Jacques de Vau de Claye are interestingly connected with this venture.[23] The first shows Brazil about 1579. It is dominated by the arms of Philippe Strozzi, Catherine's cousin, who was to lead an expedition into Portuguese-held territory. The military purpose of the map is demonstrated by an inscription that reads, "[Here are] savages to make war on the Portuguese." The second map is also dated 1579 and shows the area round Rio de Janeiro. There are several inscriptions of a military nature, in particular the one to the north of the town that reads, "Here is the side from which to take Janeiro."

Catherine, then, used maps in planning overseas ventures and encouraged mapping in France itself. She also collected many atlases in her private library, though we have no means of knowing how central these were to her intellectual concerns. If numbers are anything to go by, she was more interested in history and theology, though the atlases certainly would have enabled her to give her historical reading a sound geographical background.[24]

Of her three sons who successively became kings of France, we know most, cartographically speaking, about Charles IX (1560–74). He inherited from Henry II as royal cosmographer the celebrated André Thevet, prolific and inaccurate author of works about French overseas ventures. According to Thevet, Charles IX often sent for him "to clear up the problems he had with respect to maps and foreign countries."[25] As far as maps of France went, Charles was sensitive, for when in 1564 he was brought a map of France by a Portuguese cartographer, he detained both of them, on the grounds that this was a "map for war, useful to an enemy, who with a compass and quadrant could lead an army through the whole country."[26]

He could also envisage peaceful uses for maps, and in 1565 and again in 1571 he called for maps of the royal forests.[27] These forests were important not only for hunting but also for their economic resources, with wood for kindling, food for swine, and so forth. The maps do not survive, which is a pity, as they might show that Charles had the precocious idea of identifying economic resources on a map.

Henry IV

Catherine died in 1589, and Henry III was assassinated in the same year. The successor to the throne was Henry IV (1589–1610), under whom the cartography of France would advance considerably. Henry was a great enthusiast for maps. We do not know whether he had any formal cartographic training in his youth. Of course, by the 1560s many young nobles would have been introduced to maps as a matter of course; this is probably why in one portrait Henry is shown as a child with a furled map and compass at his feet.[28] In any case, he came to appreciate the value of maps during the many campaigns he waged to reach the throne, and he could make topographical sketches. For instance, when in 1587 he decided to make a little fort at the abbey of Maillezais, he "made the drawing himself, calling on Sully to give him advice, knowing that he had studied mathematics, and liked making maps, drawing plans and designing fortifications."[29]

Henry was indeed, as the Père de Dainville once observed, a "visuel," or as Raymond Ritter put it, a man who "thought everything out in images."[30] Thus after he had left his château at Pau, with its gardens upon which he had lavished so much care, he twice sent artists back there to sketch them for him in their present state, and eventually he sent Agrippa d'Aubigné on a special mission to make plans of the sites of his early battles, so that he could better remember them.[31]

Five years after his accession, and partly in response to his known taste for maps, a collection of maps of France was dedicated to him and published by Maurice Bouguereau, of Tours.[32] The *Théâtre François*, as it is called, made use of a wide variety of maps of the French provinces. Bouguereau appealed to learned men to send him maps of their regions and included as well maps from the *Theatrum Orbis Terrarum* of Ortelius, like the ones drawn by Nicolas de Nicolay and Gabriello Symeone. Even so, there were whole regions without a map, and others whose coverage was very sketchy or inaccurate.

During the reign of Henry IV, the cartography of France was largely advanced by the work of two branches of the army: the corps of royal engineers (*ingénieurs du roi*) and the service of lodging-masters (*maréchaux des logis*). The royal engineers as a service had emerged from the Italian engineers, employed more or less informally from the early sixteenth century onward; they numbered four in 1597 and

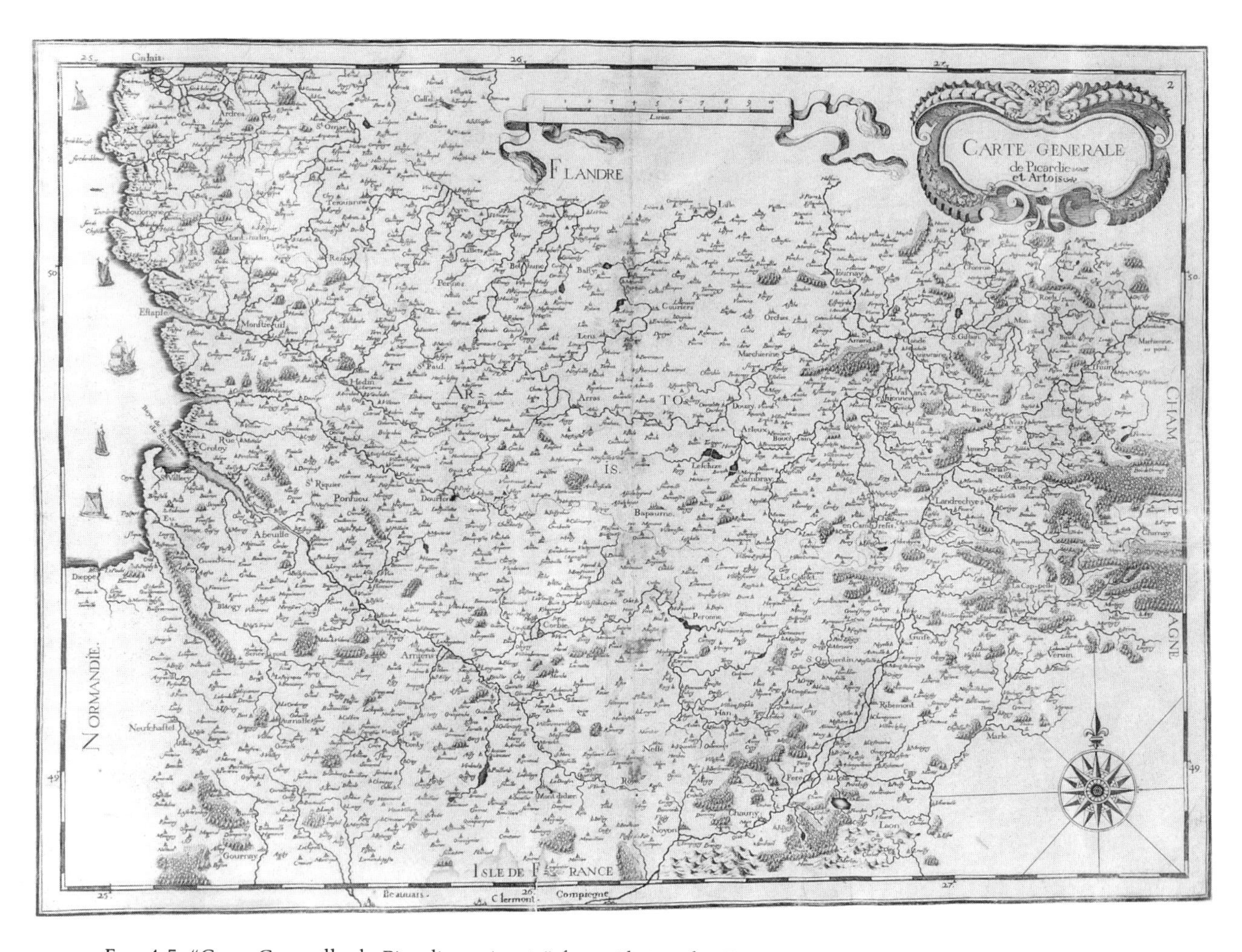

Fig. 4.5 "Carte Generalle de Picardie et Artois" from Christophe Tassin, *Les Cartes generales de toutes les provinces de France* (Paris, 1634). NL

six in 1611, and they were now nearly all French.[33] Each was responsible for the fortification of one of the great frontier provinces and had under his command a *conducteur des desseins* (roughly, a drawing-master), who was in effect the cartographer, though sometimes the engineer himself drew the maps. In Picardy, for instance, the engineer was Jean Errard, and the British Library has preserved fourteen of his plans of Picardy towns, together with thirty-six maps and plans of the province by his *conducteur des desseins*, Jean Martellier.[34]

The provinces largely covered by the maps of the royal engineers were Picardy, Champagne, Dauphiné, and Provence. Their work remained at first in manuscript, but much of it eventually found its way into print through the activities of such enterprising cartographer-publishers as Christophe Tassin. Himself an *ingénieur-géographe du roi*, during the 1630s he published a widely disseminated series of atlases, often drawing on the work of the royal engineers. Figure 4.5, for instance, shows a map published by Tassin but deriving directly from the manuscript map of the province by Jean Martellier.

The work of the engineers and their *conducteurs* naturally varied in its quality, for there was as yet no question of a central *école* in which standards might be set. In Picardy, Jean Martellier drew maps of considerable accuracy, which covered almost the whole province, as did Claude de Chastillon in Champagne. The engineer for Dauphiné was Jean de Beins, whose extraordinary skill in depicting the Alpine passes has been closely analyzed by the late Père de Dainville.[35] This author quotes a most interesting letter from the duc de Lesdiguières, governor of the province. Writing to the king in 1602, Lesdiguières tries to summarize the tactical situation in the mountain passes, explaining that "Your Majesty will understand much better than I can set it out in writing, if [you] will look at the map of Dauphiné with the Piedmont border."

In fact, the maps drawn by Jean de Beins laid the foundation for French Alpine cartography. Figure 4.6 shows his view of the country around Grenoble, roughly in the center. The River Isère comes down out of the mountains to the upper left, joining the River Drac, which flows down from the top right; together they leave the map at bottom center. This delineation, in its mastery of the shape of the mountains and the course of the hydrography, is typical of Beins's work, which was very extensive.

In Provence the engineers were Raymond and Jean de Bonnefons, who specialized in town plans and topographical views rather than in true maps; the delineation of the coast of Provence had to await the time of Richelieu. Sometimes the engineers were called upon to execute maps outside the provinces for which they were responsible. Claude de Chastillon thus drew maps and plans for sites all over France and was in addition responsible for a very early attempt to define a frontier in precise cartographic terms, in this case the border between northeastern France and the Spanish Netherlands.[36] The printed maps of Ortelius and of Mercator had offered rough delineations of such borders, but now for the first time an actual investigation on the ground was formulated in cartographic terms.

The engineer in Guyenne was Benedit de Vassallieu, "dit Nicolay," who did not carry out much work in his province but seems to have been active in the Paris Arsenal, where he composed a remarkably precocious treatise on the manufacture and operation of artillery. He also drew a plan of Paris, engraved in 1609, and probably compiled maps of the coasts of Normandy, Brittany, and Poitou, which seem to have been lost. They were no doubt in the style of his map of the site of Henrichement (1609), now preserved at the Archives Nationales.[37]

The other branch of the army that generated maps was the service of lodgings-masters, or *maréchaux des logis*. Here the key figure was Jacques Fougeu, who from 1590 onward drew a huge number of maps—four or five hundred—covering almost the whole of France, and indeed part of the territories of its neighbors to the east.[38] These maps are not very elegant, for they were basically field notes, designed to tell lodging-officers how many hearths they could find in the villages along a given line of march. But these scrappy-looking little maps together form an astonishing compendium of information about France's towns

FIG. 4.6 Jean de Beins, "La Baillage de Greyzivaudan et Trieves," 1619. BN, Département des Cartes et Plans

and villages around 1600. Take, for instance, Fougeu's map of the mouth of the River Somme (figure 4.7). It has a roughly correct outline of the river and coast, and also marks, in about their correct position, over one hundred towns and villages. The only readily available contemporary printed map, in the *Theatrum* of Ortelius, marks just twelve towns in the same area.

Very few of the maps of Fougeu have been studied in detail. But his rendering of Savoy has recently been studied by Bernard Savary, in the course of an analysis of a series of maps of this region.[39] Savary draws our attention to the astonishing profusion of place-names—799, to be exact, of which 673 are still identifiable—and to the remarkable knowledge of the topography displayed by Fougeu. For Savary, the remarkable extent of this knowledge has to be the result of a long tradition of cartography of which all evidence has been lost. But it seems more likely that Fougeu was just a very remarkable topographer who used his work in the army, and no doubt a host of military collaborators, to produce maps that amaze us by their accuracy and amplitude.

FIG. 4.7 Jacques Fougeu, map of the mouth of the River Somme, c. 1600. BN, Département des Manuscrits, ms. fr. 11224

The king thus possessed a remarkable source of information. But it remained secret and was used only on rare occasions like the Juliers campaign of 1610. None of the Fougeu maps was published, and none seems even to have influenced the publications of Tassin, as did the maps of the royal engineers of the same period. The Fougeu maps ended up at the château of Bontin, in Burgundy, where Henry's minister, Sully, seems to have maintained a regular cartography workshop, about which we unfortunately know very little.[40]

Like his predecessors, Henry was interested in overseas expansion, supporting the cartographic work of the de Vaulx dynasty at Le Havre[41] and showing a keen interest in the map showing Drake's journey round the world in 1577–80.[42] But his greatest claim to fame in this respect lies in his appointment of Samuel de Champlain as geographer to the Canadian expeditions of 1603 and 1604–7. During these expeditions, Champlain produced maps of remarkable accuracy, portraying the rivers and coastline of New France in a style that owed nothing to previous marine cartographers.[43] We know nothing about Champlain's education, but these maps are convincing evidence that he was fully aware of the latest styles and techniques practiced in early seventeenth-century France, and it seems likely that he must have had at least an apprenticeship, if not formal training, perhaps at some Jesuit academy.

The king and Sully encouraged the production of maps in the other services they commanded, like those of the artillery and of the roads and bridges, though no coherent corpus has come down to us. For instance, when Sully wanted in 1604 to know what resources existed for the artillery in the Lyonnais, he instructed his *lieutenant* there to make an inventory of all firearms and ammunitions, to list the timber available for gun-carriages, to check the supplies of various metals, and then "to make a very exact map" upon which would be marked the best river crossings and the number of pioneers available from each parish. The *lieutenant* for the Lyonnais replied with information that included a map, and Sully probably received similar answers from other provinces.[44]

Besides directly soliciting maps in this way, the king and Sully established a scientific milieu in which cartography and similar skills could flourish. At the royal palace of the Louvre they set aside a whole gallery (the *grande gallerie*) for craftsmen. Here could be found painters, gunsmiths, engravers, and other specialized workers; here also was Philippe Danfrie, who was an engraver but also an

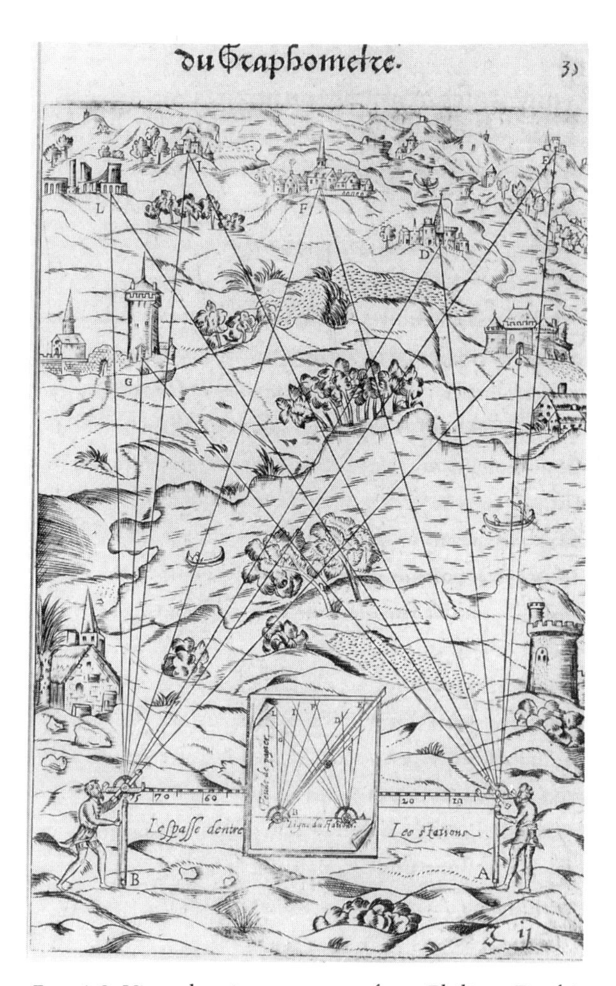

FIG. 4.8 View showing surveyors from Philippe Danfrie, *Declaration de l'Usage du Graphometre* (Paris, 1597). NL

inventor of surveying instruments, some of which no doubt served the military cartographers.[45] Figure 4.8 shows two surveyors at linked "stations" using the *graphometre* to fix the positions of various distant objects. Henry intended to set up in the *grande gallerie* a regular museum of machines, in which inventors and technicians could find inspiration. In imitation of the princes of Italy,[46] he also wished to establish at the Louvre a geographical and hydrographical museum, in which there would have been six huge maps showing France, the continents, and the oceans. There was a faint foreshadowing of this at Fontainebleau, where one gallery was given over to views of the royal châteaux; Sully also had one room at his château of Villebon decorated in this way.[47]

Another of their building projects, the Place Royale (now Place des Vosges), contained town houses for technicians and administrators of all kinds. The Fougeu family had a house here, and so did the engineer-cartographer Claude de Chastillon.[48] In short, we have to envisage Henry IV and Sully as encouraging a wide range of talent in mapping as in the other arts; whereas before them cartographic ventures had been mainly episodic or else, like Catherine de Medici's work with Nicolay, frustrated by unfavorable conditions, now the whole of France had been covered in more or less satisfactory fashion, and some parts had been mapped with great detail and precision.

Richelieu and Louis XIII

After Henry's assassination in 1610, much of this activity came to an end, for Marie de Medici was much diverted by political problems and in any case did not have the technological tastes of Henry IV. But then, in the later 1620s and 1630s, cartographic activity revived strongly under the inspiration of Richelieu. The cardinal does not seem to have been quite as obsessed by maps as was his contemporary and great rival Olivares, chief minister in Spain,[49] but Richelieu had taken instruction from the notable cartographer Nicolas Sanson, and one of his first acts after coming to power in 1624 was to commission from Sanson a map of France in thirty sheets.[50] This commission was never executed—indeed, it would probably have been impossible to execute it in 1624 or soon thereafter—but it does show the way Richelieu's mind was working.

The first surviving evidence from Richelieu's early mapping ventures also dates from 1624 and consists of an atlas of fifty-eight maps and charts, "apparently relating to a planned attack by France on Spain and on Spanish-held possessions in the Mediterranean."[51] These maps cover most of the Spanish possessions in the western Mediterranean and are anonymous, except for a superbly colored map of Italy, Corsica, and Sardinia by the celebrated royal engineer Salamon de Caus (c. 1576–1626; figure 4.9). Clearly, Richelieu was making cartographic preparations for the war with Spain that eventually broke out in 1629 over the Mantuan succession.

Many of the cartographic activities initiated during previous reigns carried over into the time of Richelieu. In Canada, for instance, the latest discoveries continued to be charted by Champlain, in a series of maps culminating in the brilliant *Carte de la Nouvelle France* of 1632. At Dieppe, the sixteenth-century tradition of cartography was revived by Jean Guérard, who, after drawing maps of the French and European coasts during the 1620s, in 1634 drew and dedicated to Richelieu a magnificently colored world map (figure 4.10).[52] This masterpiece, now preserved at the Bibliothèque Nationale in Paris, gave Richelieu a good idea of the latest discoveries by the European powers, including not only the French in the St. Lawrence Valley but also the British around Hudson Bay. Where there was uncertainty Guérard pointed it out, as in his inscription, "Great Ocean discovered in 1612 by the Englishman, Henry Hudson; it is thought that this leads to Japan." With this map Richelieu had a capital tool for planning overseas ventures.

FIG. 4.9 Salamon de Caus, Map of Italy, 1624. Private collection

At home, the work of the military engineers went on, with hitherto unparalleled quantities of maps and plans of the frontier regions being produced by an enlarged engineering service.[53] Many of these maps and plans soon found their way into the widely distributed printed atlases of Christophe Tassin, who was extremely active during the mid-1630s. One of his most original publications was the thirty-map atlas *Cartes . . . de toutes les costes de France*; this work was dedicated to Richelieu and, according to Nicolas de Fer, had been commissioned by him. Be that as it may, we may be sure that Richelieu, as grand master of navigation, greatly welcomed this first atlas of France's coasts, most of whose maps derived from the work of the royal engineers. In fact, the whole work of Tassin, which ran to nine widely diffused atlases, was in effect a popularization of the engineers' work, and one that deeply affected the cartographic image seventeenth-century French people had of their country.[54]

During the 1630s, Richelieu seems to have

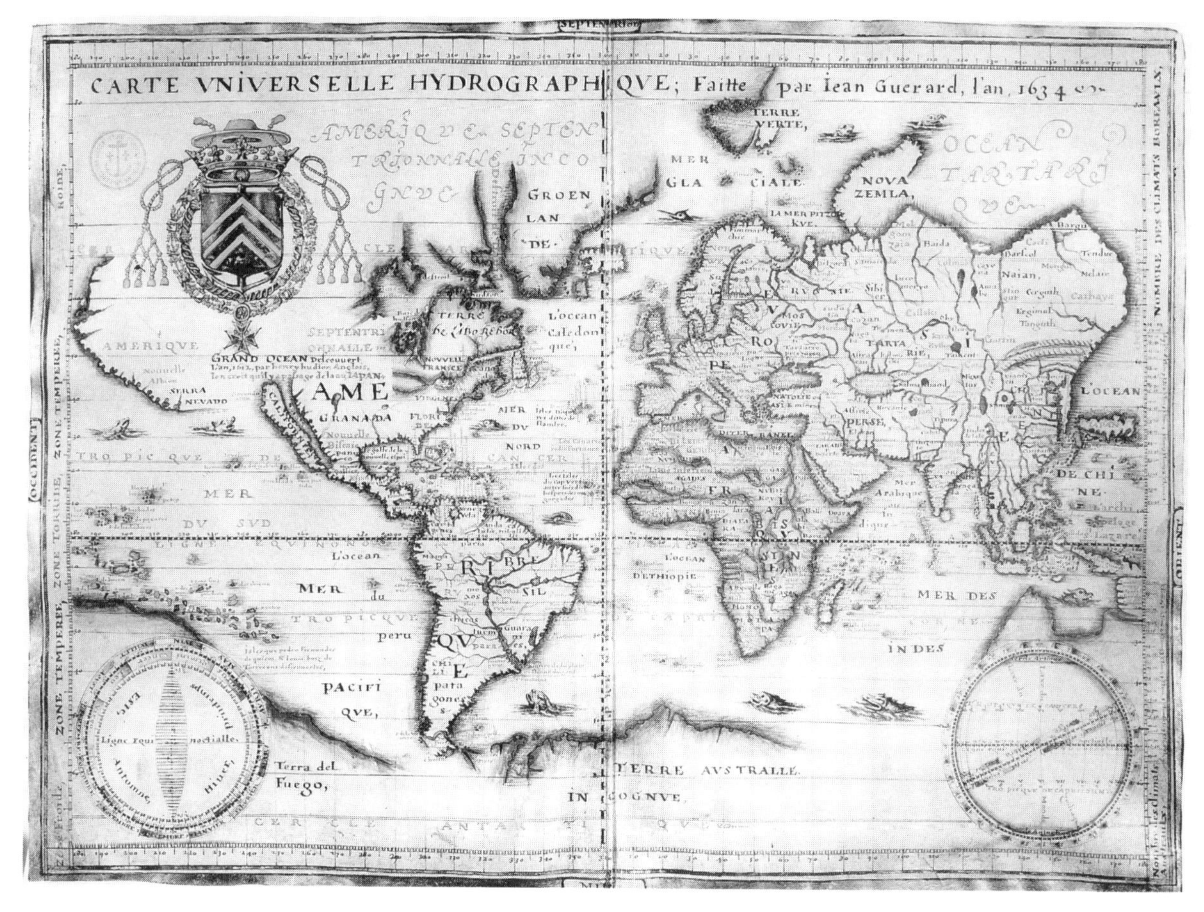

FIG. 4.10 Jean Guérard, "Carte universelle hydrographique," 1634 BN, Département des Cartes et Plans

tried more or less systematically to fill the gaps in the cartography of France left by Henry IV. In 1632, for instance, he requested his *lieutenant* as grand master of navigation in Provence, Henri de Séguiran, to map the coast of that province.[55] Séguiran enlisted the aid of Jacques Maretz, professor of mathematics at Aix and a skillful engraver; together they set out from the mouth of the Rhône eastward in January 1633. In an account of the journey for Richelieu, Séguiran describes how he "had the plans of the islands and fortresses drawn by the Sr. Maretz, professor of mathematics and His Majesty's engineer, whom we took with us, assisted by the Srs. Augier and Flour, also other engineers and painters, to draw a map of the whole coastline, and all of the towns, ports and castles which are there."

The party kept to the coast, which meant that they could not observe the interior in any detail. But the resulting map was a great improvement over the existing map of Provence, by Bompars; it was used by Tassin in his *Cartes générales de toutes les provinces de France* and

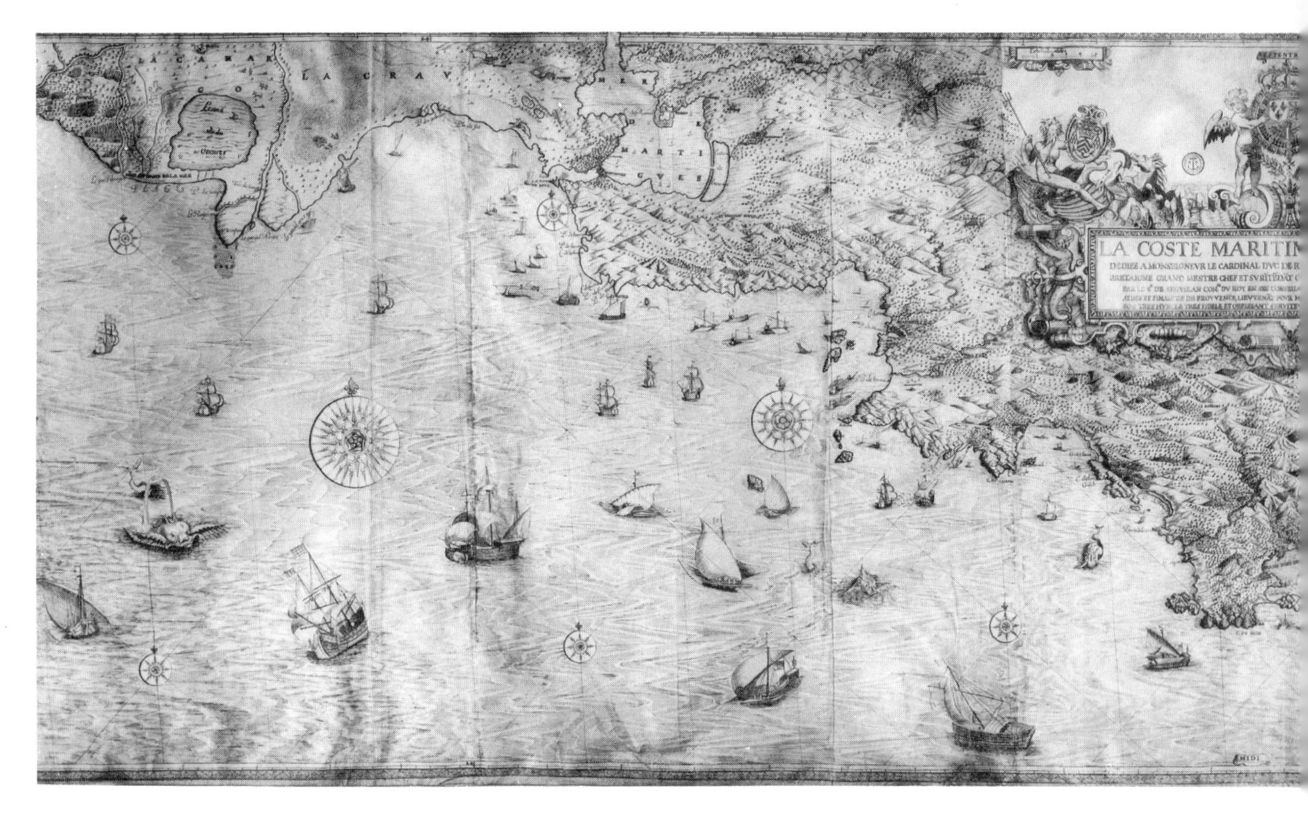

FIG. 4.11 Jacques Maretz's map of Provence, 1633. BN, Département des Cartes et Plans

remained the best map of the coast until the time of Delisle, being used both by Sanson and by Jaillot (figure 4.11). Curiously enough, Richelieu seems in that same year, 1633, to have commissioned another survey of the Provençal coast, by the Marseillais Augustin Roussin. This map survives at the Bibliothèque Nationale and is in a rather crude late portolan-chart style.[56] Perhaps Richelieu called for two maps of the same area so that he could choose the better one.

Two years later, his attention shifted to the west coast, when letters-patent were issued reading thus: "Having been assured of the capacity and experience of the Sr. J. Guimbert in geography, and in drawing up maps of countries and provinces, and not having at present any true and exact map of the provinces of Saintonge, Angoumois and Aunis, we allow him to make and print one."[57] One or two years later, Richelieu commissioned an atlas of the strongholds in Burgundy, and a splendid volume was produced, which still survives at the Bibliothèque Nationale.[58]

In 1639, he ordered a visitation of the Normandy coast, about which we are particularly well informed, since we have both the account of the journey and the resulting maps.[59] Louis le Roux, Sr. d'Infréville, *commissaire général de la marine*, was ordered to take "the

Sr. Regnier Jenssen the younger, royal engineer, or his father," and proceed southward from Calais as far as Cherbourg, "to seek out what places would be best for building a port." D'Infréville arrived in Calais in December 1639, to find that the elder Jenssen was laid up with gout and his son was working on the fortifications of Hesdin. So the younger Jenssen was called in and, early in January, set out southward with d'Infréville. They called at all the major ports and many minor ones, taking the advice of local engineers and drawing plans where necessary. At Honfleur, for instance, Jenssen and the local engineer, a certain Saint-Clair, "drew a plan [of the port], because the site seemed fairly suitable." Eventually they reached the end of the line at Cherbourg, where they signed a summary of their recommendations and sent the report to Richelieu, together with thirty maps.

Figure 4.12 shows one of them, setting out improvements to be made in the "Vallée de Clacquedan"; the River Cany flows northward into the English Channel at the right, and on the left are the proposed improvements, which include two jetties, two locks, and a dock for vessels. With a map like this, the cardinal had a detailed and fairly accurate image of a particular site on the coast; the entire set of maps could enable him to come to a decision about

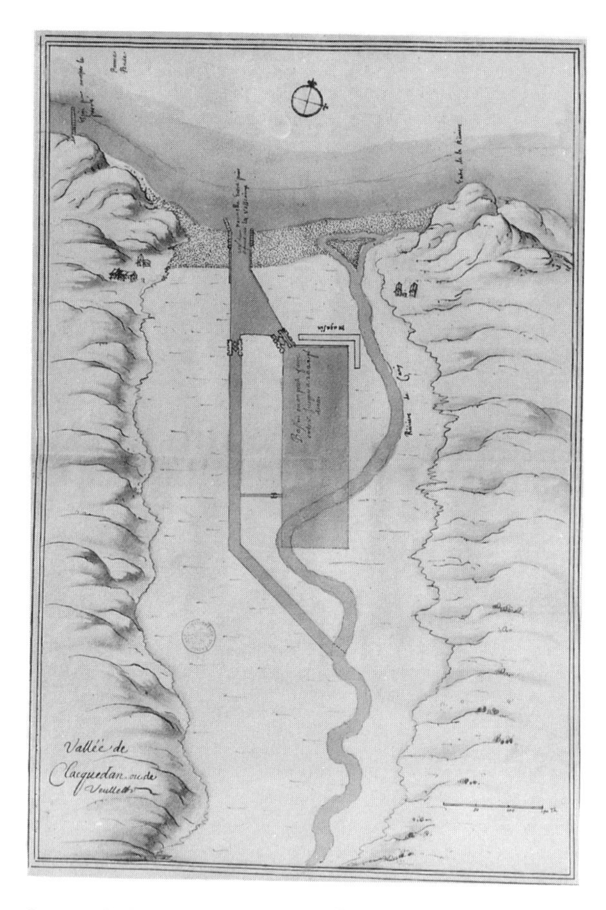

FIG. 4.12 Regnier Janssen, "Vallée de Clacquedan," 1639 BN, Département des Manuscrits, ms. fr. 8024

where to build a port. Nothing of the kind had existed before this survey, and without it Richelieu would not have been able to bring any informed judgment to bear on the problem. The report and the maps were probably copied and distributed to other ministers, though the information they contain never seems to have gone into print.

By the late 1630s, Richelieu had filled many of the gaps in France's cartography left by Henry IV. He probably continued to collect maps of particular areas, as is illustrated by a letter he wrote in July 1630 to the marshal de La Meilleraye, who was leading a French army in operations against the Spaniards near Calais. With this letter came a "very special map, which M. de Rasle gave me, from his own hand, of all the reconquered land around Ardres and in this neighborhood." This map, Richelieu goes on to say, shows rivers and dikes in particular detail and will be very useful in case the French need to fall back on Calais; he then asks La Meilleraye to return it after use.[60]

Richelieu, then, was well aware of the use of maps for a wide variety of military and colonial purposes. If we have no examples of his use of maps in the civil administration, that is no doubt because so much of his time was taken up with the war against Spain. Richelieu was not singular in this taste for maps. He belonged to a generation that had largely been schooled by the Jesuits, whose academic syllabus laid heavy emphasis on geographic and cartographic knowledge—an emphasis that paid handsome dividends when Jesuits had to construct maps of the parts of the world to which they were sent, like North America or China.[61]

Another important element was the work of Nicolas Sanson. He had instructed Richelieu in map use and also tutored both Louis XIII and the Condé princes.[62] Sanson eventually drew so many maps of France that he in effect invented the idea of the base-map with thematic variations. Beginning with a general map, this system uses it to set out a variety of phenomena, whether they have to do with political boundaries, economic resources, or military posts. This was a very important innovation, exemplified in color plate 5 and figures 4.13 and 4.14. Color Plate 5 shows French postal routes, following the map produced by Melchior Tavernier in 1632. Very few features are shown except those relevant to the theme: the towns

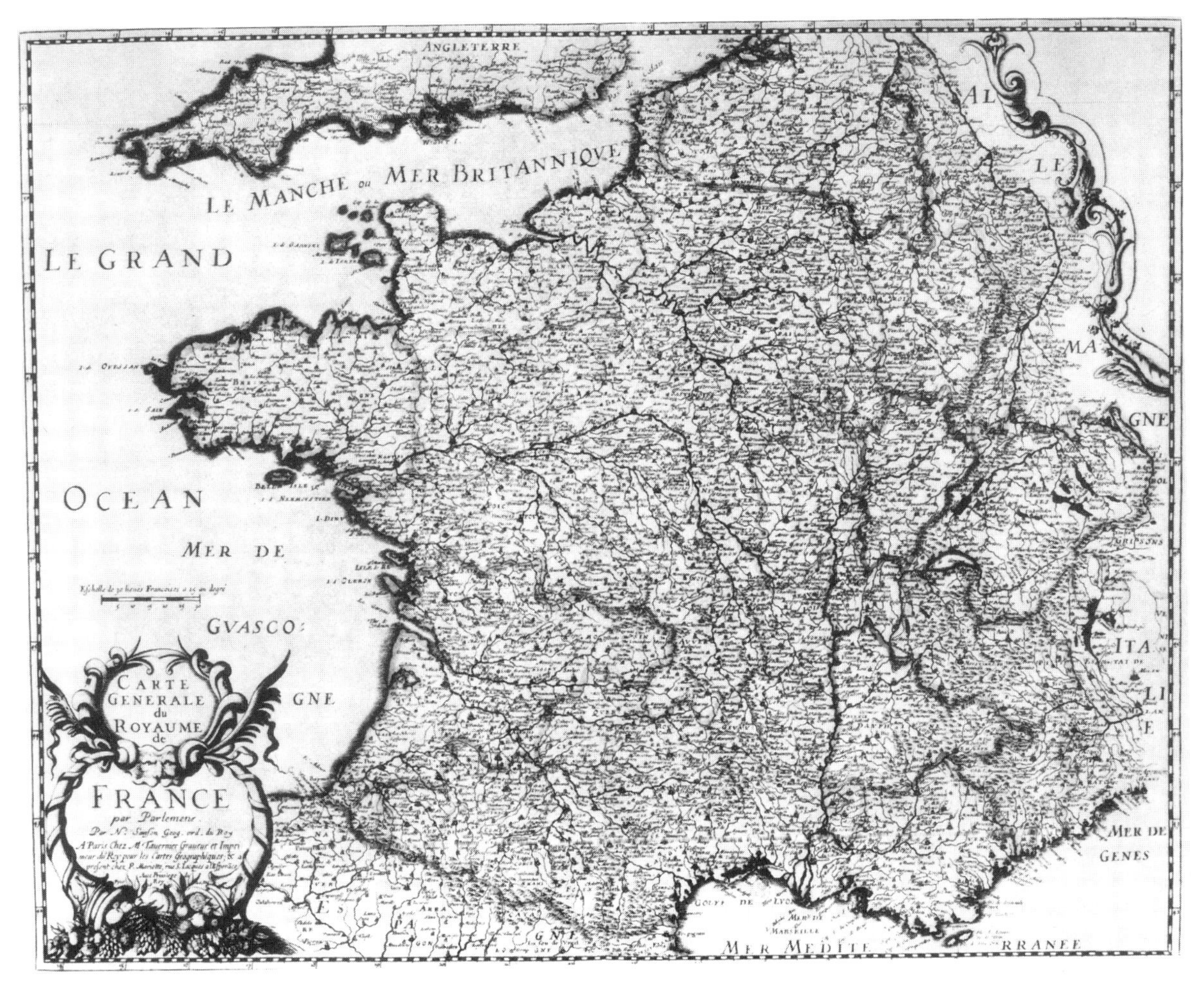

FIG. 4.13 France divided into *parlements*, from Nicolas Sanson, *Les Isles Britanniques, l'Espagne, la France . . .* (Paris, 1644). NL

where the mail service stopped, the rivers, and some major towns outside the network. Color is cunningly used to emphasize the line of the postal route and also to define some frontiers; such specialized information laid on a simple base-map is sometimes called a "thematic map."

Figures 4.13 and 4.14 show a different kind of thematic map. Here the base-map is more complicated, and upon it are laid, first, the outline of the various *parlements*, or judicial divisions, and then the outline of the ecclesiastical jurisdictions. Using a single map of France, Sanson could thus present to the king—and to the public—a wide variety of phenomena; as time went by, ever more complicated and transient developments could be plotted in this way. It was, in effect, a sort of

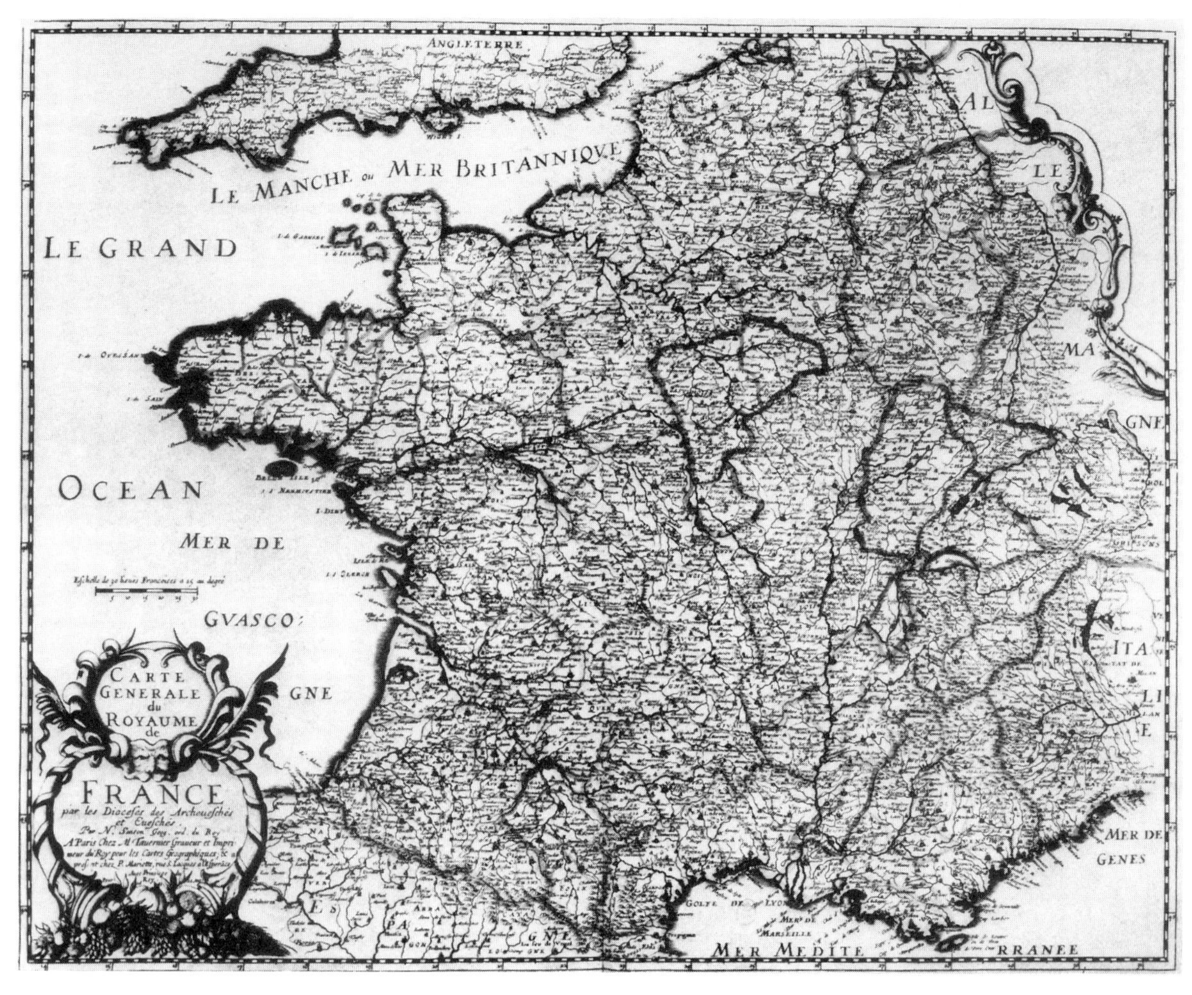

FIG. 4.14 France divided into archbishoprics and bishoprics, from Nicolas Sanson, *Les Isles Britanniques, l'Espagne, la France . . .* (Paris, 1644). NL

seventeenth-century geographic information system, to use the jargon of our day.

Conclusion

When Louis XIV came to the throne, he was able to benefit from a long period of map use by French monarchs and their ministers. If we were to attempt a typology, we could maintain that maps first came to be used for military ventures in the early sixteenth century. Henry II had a more extensive view of their possibilities, as did his widow Catherine de Medici, and under Henry IV the country came to be more or less systematically mapped for the first time. With Richelieu we come to a period of consolidation, when no new principles were put forward, but old ones were used to fill in the gaps in the system. An excellent base was therefore laid for Louis XIV, under whom France came to be very thoroughly mapped, eventually by the topographical survey that would be initiated by the first of the Cassinis.[63]

Notes

1. For what follows on map use by Louis XIV, see James E. King, *Science and Rationalism in the Government of Louis XIV, 1661–1683* (Baltimore, 1949), especially p. 130 onward.

2. See the commentary in two recent catalogs from the Bibliothèque Nationale: *A la Découverte de la terre* (Paris, 1979), 102, and *Images de la montagne* (Paris, 1984), 15.

3. On these royal collections, see Léopold Delisle, *Le Cabinet des Manuscrits de la Bibliothèque Nationale*, 3 vols. (Paris, 1868–81).

4. On this map, see Rodney Shirley, *The Mapping of the World* (London, 1983), 77.

5. See the long article by P. Hamon in *Dictionnaire de biographie française*, vol. 13 (Paris, 1975). Note also that the Houghton Library, Harvard University, holds Fine's "La Sphere du Monde," MS Typ 57.

6. For an idea of the great extent of this work, see the numerous entries in the *Catalogue des actes de François I*, and also Gaston Zeller, *L'Organisation défensive des frontières du nord et de l'est au XVIIe siècle* (Paris, 1928).

7. For these references, see Emile Picot, *Les Italiens en France au XVIe siècle* (Bordeaux, 1852), 226, and the communication by L. Préteux in *Bulletin Philologique et Historique*, 1926–27; lxv-lxvii, as well as René Herval, "Un Ingénieur siennois en France au XVIe siècle: Girolamo Bellarmato et la création du Havre," *Études Normandes* 40 (1961): 33–43.

8. On this presumed influence, see the article by Marcus Merriman, "Italian Military Engineers in Britain in the 1540s," in *English Map-Making, 1500–1650*, ed. Sarah Tyacke (London, 1983).

9. Robert Knecht, *Francis I* (Cambridge, 1982), 338–39.

10. The most recent summary of the work of this school is in the introduction by Helen Wallis to *The Maps and Text of the Boke of Idrography Presented by Jean Rotz to Henry VIII 1542* (Oxford, 1982).

11. For a good recent reproduction of this map, see Michel Mollat du Jourdin and Monique de La Roncière, eds., *Sea Charts of the Early Explorers* (New York, 1984), pl. 47. This book also has fine reproductions of the work of Guillaume Le Testu.

12. For these references, see *Mémoires de la vie de François de Scepeaux, Sr. de Vieilleville*, Petitot ed. (Paris, 1823), 440–43; I owe this citation to the kindness of Frederick Baumgartner.

13. See Jean Dérens, ed., *Le Plan de Paris par Truschet et Hoyau, 1550* (Zürich, 1980).

14. For a full account of this affair, see Gaston Bardet, *Naissance et méconnaissance de l'urbanisme* (Paris, 1951), 120–30.

15. On Symeone's astrology, see Bibliothèque Nationale, Manuscrits Français 3159, f. 16.

16. See Bibliothèque Nationale, Cartes et Plans, Rés GeD 7664 and Ge DD 627 (47).

17. On Hamon's life, see the article by T. de Morembert in *Dictionnaire de biographie française*.

18. Mollat and La Roncière, *Sea Charts*, 239.

19. See M. R. Hervé, "L'Oeuvre cartographique de Nicolas de Nicolay et d'Antoine de Laval," *Bulletin de la Section de Géographie*, 1955, 223–63.

20. *Lettres de Catherine de Médicis*, ed. Hector de La Ferrière, 10 vols. (Paris, 1880–1909), 1:238.

21. "Description generale du pays et duché de Berry" is at the Bibliothèque Nationale, Manuscrits Français 2790.

22. This incident is described in Armando Cortesao and Avelino Teixeira da Mota, eds., *Portugaliae Monumenta Cartographica*, 5 vols. (Lisbon, 1960) 2:85.

23. The charts are reproduced in Mollat and La Roncière, *Sea Charts*, nos. 60, 61.

24. See "Inventaire de la Bibliothèque de la Reine," of 1589, Bibliothèque Nationale, Manuscrits Français 5585, f. 130, and also the anonymous "Notice sur la bibliothèque de Catherine de Médicis," *Bulletin du Bibliophile*, 1858, 1–34.

25. Roger Schlesinger and Arthur Stabler, eds., *André Thevet's North America* (Kingston, 1986), 22.

26. Incident cited by Luis de Matos, *Les Portuguais en France au XVIe siècle* (Coimbra, 1952), 17.

27. Communication from Mme Mireille Pastoureau, from an article due to appear in volume 3 of *History of Cartography*.

28. See *Henry IV et la reconstruction du royaume*, Ministère de la Culture (Paris, 1989), 30, "Henri IV enfant"; and also Jacques Perot, *Musée National du Château de Pau: Quinze années d'acquisitions, 1979–1984* (Paris, 1985), 30.

29. Sully, *Oeconomies Royales*, ed. Bernard Barbiche and David Buisseret, 2 vols. (Paris, 1970, 1988), 1:177–78.

30. Raymond Ritter, *Le Château de Pau* (Paris, 1919), 248.

31. Jean Plattard, *Une Figure de premier plan dans nos lettres de la Renaissance: Agrippa d'Aubigné* (Paris, 1931), 94.

32. On this atlas, see Père de Dainville, *Le Premier Atlas de France: Le Théatre Francoys de M. Bouguereau, 1594*, in *Actes du 85e Congrès National des Sociétés Savantes* (*Section de Géographie*) (Paris, 1960).

33. On the development of this engineering service, see David Buisseret, "Les Ingénieurs du roi au temps de Henri IV," *Bulletin de la Section de Géographie*, 77 (1964): 13–84.

34. British Library, Additional Manuscripts 21, 117.

35. In *Le Dauphiné et ses confins vus par l'ingénieur d'Henri IV, Jean de Beins* (Geneva, 1968).

36. See David Buisseret, "The Cartographic Definition of France's Eastern Boundary in the Early Seventeenth Century," *Imago Mundi* 36 (1984): 72–80.

37. Reproduced in *Henri IV et la reconstruction du royaume*, 317; see also Buisseret, *Sully*, pl. 10, for a detail of Vassallieu's plan of Paris.

38. These maps are summarily described in the article by Myriem Foncin, "La Collection de cartes d'un château bourguignon," in *Actes du 93e Congrès National des Sociétés Savantes (Section de Géographie)* (Paris, 1970), 43–75. See also B. R. Kroener, "Le Développement d'un réseau routier militaire," in *Actes du 104e Congrès des Sociétés Savantes (Bordeaux, 1979)*.

39. *Images de la montagne*, 85–86.

40. See David Buisseret, "L'Atelier cartographique de Sully à Bontin: l'oeuvre de Jacques Fougeu," in a volume edited by Bernard Barbiche (in press).

41. For an example of their work, see Mollat and La Roncière, *Sea Charts*, 71.

42. David Quinn, "Henri Quatre and New France," *Terrae Incognitae*, 22 (1990):13–28.

43. See the full and excellent analysis of his work by C. E. Heidenreich, *Explorations and Mapping of Samuel de Champlain, 1603–1632* (Toronto, 1976).

44. For the Lyonnais map, see the Bibliothèque Nationale catalog *Espace français* (Paris, 1987), 89.

45. See, for instance, the instrument described by Danfrie in his *Declaration de l'usage du graphometre* (Paris, 1597), briefly commentated in a Newberry Library catalog, James Akerman and David Buisseret, eds., *Monarchs, Ministers, and Maps* (Chicago, 1985).

46. On these Italian map chambers, see the chapter by Juergen Schulz in *Art and Cartography*, ed. David Woodward (Chicago, 1987).

47. On Henry's proposed museum, see Jean-Pierre Babelon, *Henri IV* (Paris, 1982), 801–2.

48. See Sully, *Actes*, ed. F. de Mallevoüe (Paris, 1911), which has a plan of the tenants in the Place Royale.

49. On Olivares's special map room, see J. H. Elliott, *Richelieu and Olivares* (Cambridge, 1984), 28.

50. See the dense and suggestive article by Mireille Pastoureau, "Les Atlas imprimés en France avant 1700," *Imago Mundi* 32 (1980):45–72.

51. The atlas is described in Sotheby's *Catalogue of Atlases, Maps, and Printed Books*, no. 9223, June-July 1980.

52. Superbly reproduced in Mollat and La Roncière, *Sea Charts*, map 84.

53. There is very little recent work on the engineers of Louis XIII, since the thesis of Jean-François Pernoud has not been published. See his "Les Chevauchées des ingénieurs militaires en France" in the Centre National de la Recherche Scientifique publication *La Découverte de la France XVIIe siècle* (Paris, 1980).

54. On these atlases, see Mireille Pastoureau, *Les Atlas français, XVI-XVIIe siècles* (Paris, 1984).

55. See the article by Myriem Foncin and Monique de La Roncière, "Jacques Maretz et la cartographie des côtes de Provence au XVIe siècle," in *Actes du 90e Congrès . . . Nice 1965* (Paris, 1966), 11–16.

56. Reproduced in Mollat and La Roncière, *Sea Charts*, pl. 83.

57. See G. d'Avenel, *Richelieu et la monarchie absolue*, 4 vols. (Paris, 1884–95), 3:118, no. 1.

58. Cartes et Plans, Rés GeDD 2662; it bears Richelieu's arms on its cover.

59. Bibliothèque Nationale, Manuscrits Français 8024 contains both the account and the map; see also the Service Historique de la Marine at Vincennes, Ms. 81 (979).

60. See Richelieu, *Lettres et papiers d'Etat*, ed. D.-

L.-M. Avenel, 8 vols. (Paris, 1853–77), 6:449.

61. See the sections on history and geography in Père de Dainville, *L'Education des Jésuites*, rev. ed. (Paris, 1978).

62. Pastoureau, "Les Atlas imprimés," 54.

63. This chapter leaves the story of state cartography in France at just the point where Josef W. Konvitz takes it up in his *Cartography in France, 1660–1848* (Chicago, 1987).

FIVE

Maps and Ministers: The Spanish Habsburgs

Geoffrey Parker

In 1642, eighteen months after the outbreak of the Portuguese revolt against the Habsburgs, the royal cosmographer of Spain, Jean-Charles della Faille, wrote to ask a colleague in the Spanish Netherlands: "Is it possible to find in Flanders some map of Portugal or Catalonia? I would be awfully grateful if you would tell me, so that I can have it brought here, because here we know very little about this country [Spain]. I can see that the maps of Ortelius are highly erroneous on Portugal and its frontiers so I am not surprised that our enemies, with smaller forces, are getting the better of us." This despondent letter apparently sentences my subject to death before it is born, for it suggests that cartography never *did* emerge as a tool of government in Habsburg Spain. The ministers of Philip IV, it seems, had nothing more recent to assist them in planning the defeat of rebellious Portugal than the *Theatrum Orbis Terrarum* of Abraham Ortelius, whose maps of the Iberian peninsula (made in the 1560s) were indeed, as the lugubrious della Faille remarked, "highly erroneous."[1] This view is apparently confirmed by the extremely detailed posthumous inventory of Philip II's possessions, drawn up in 1607, which listed everything from toothbrushes to bone rings for the treatment of the king's hemorrhoids, but scarcely mentioned cartographic items: a few charts and maps (all of either America or the Atlantic), four globes, and three copies of Ortelius's *Theatrum*.[2]

But all this is highly misleading. We know, for example, that Philip II requested and received maps from many of his agents abroad; and indeed many cartographic items belonging to him have survived, although they are not listed in the inventory of his worldly goods.[3] There is, in fact, so much material available that this paper deals only with maps covering Europe; charts of the Ocean Sea and maps of islands and other continents have reluctantly been excluded.[4]

And even for Europe, though the surviving cartographic materials are copious, a far larger quantity has clearly been lost. There are three reasons for this. In the first place, we are searching for evidence of a *temporary* interest: maps are most useful then they are new, and outdated ones tend to be discarded, if not destroyed, in favor of more current work. Second, maps share the normal risks of destruction that threaten any ancient document: to be reused as

I am greatly indebted for invaluable references and orientation to several scholars in the field, particularly to Peter Barber, David Buisseret, Ulla Ehrensvärd, John Elliott, Richard Kagan, and above all the late Richard Boulind. I am also much obliged to the staff of the map rooms at the National Library of Scotland in Edinburgh, the Kungliga Bibliothek in Stockholm, the Newberry Library in Chicago, and the British Library in London for their assistance.

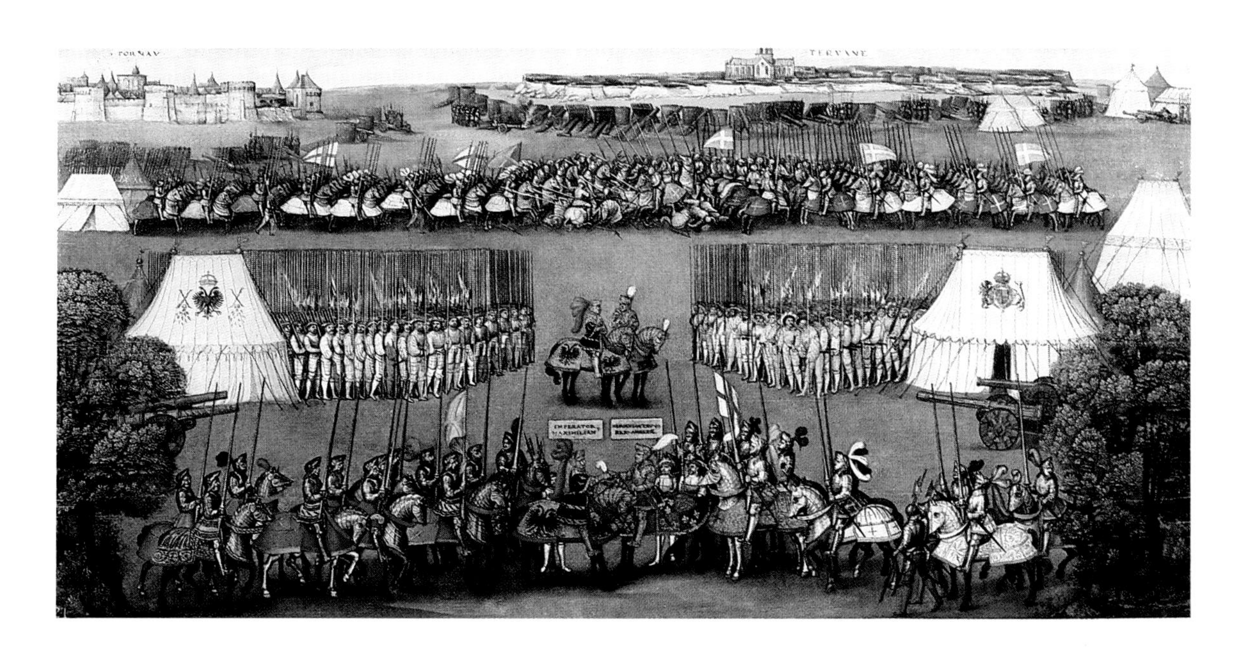

Pl. 1 Painting of Henry VIII of England meeting the Emperor Maximilian at Thérouanne, 1513. The Royal Collection, St. James's Palace, London
This painting is probably based on the earliest recorded plat associated with Henry VIII, and it seems to be related to subsequent imagery such as Holbein's painting for Henry VIII's triumphal arch at Greenwich in 1527. Paintings like these demonstrate one important use of map-related material at court.

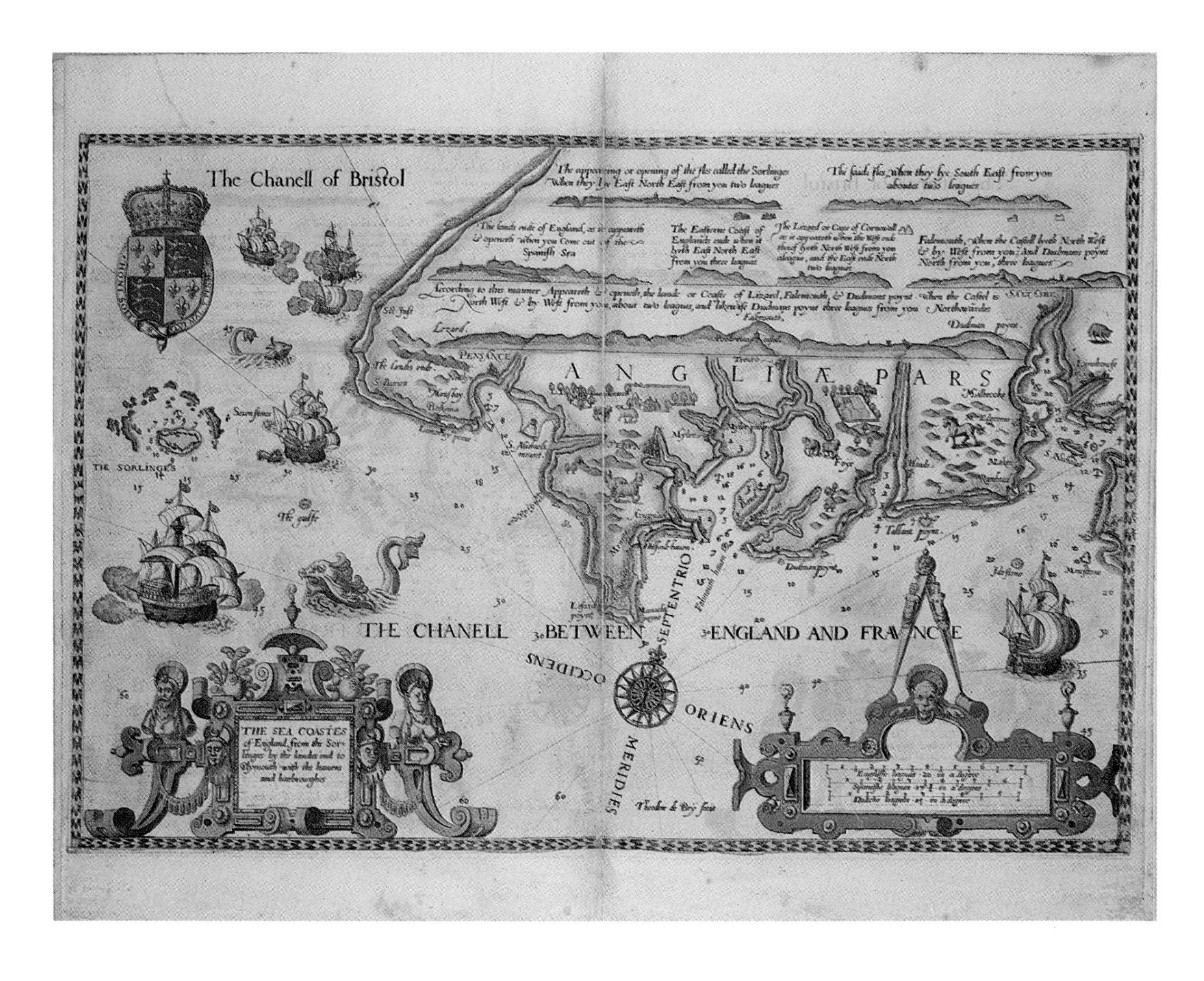

Pl. 2 "The Sea Coastes of England, from the Sorlinges by the Lande's End to Plymouth," from Lucas Waghenaer, *The Mariner's Mirrour* (London, 1588). The Newberry Library

This map comes from the English translation of Lucas Waghenaer's *Spieghel der Zeevaerdt* (Leiden, 1584–85). The translation received indirect government support for reasons of national defense.

(Facing page)

Pl. 3 Chart of the Indian Ocean from the atlas prepared by Diogo Homem in 1558. British Library, Add. MS 5415A, ff. 15–16

Homem was a Portuguese chartmaker working in England from the late 1540s. This chart was probably commissioned by Mary Tudor as a gift for her map-loving husband, Philip II, but was left incomplete on her death in November 1558.

Siria.
Mediteranū
Arabia petrea.
Terrestre mare.
Persarū Imperium.
persia
India prima caisar.
EGI
TVS
Arabia sterilis
Arabia felix
sinus persicū
Mare rubrū
Barnacais.r.
Aethiopia sub egitu.
Abeschi.r.
India maior ethiopi.
sinus arabic.
Media thiopia.
Mare Indicum.
EQVINO TIALIS.
AFRICA
Aethiopia Interior
Quiloa
TROPICVS CAP RICORNI
Bona spes.
Mare bone spei.
MERI DI ES
OCC I DE N S
O R I E N S

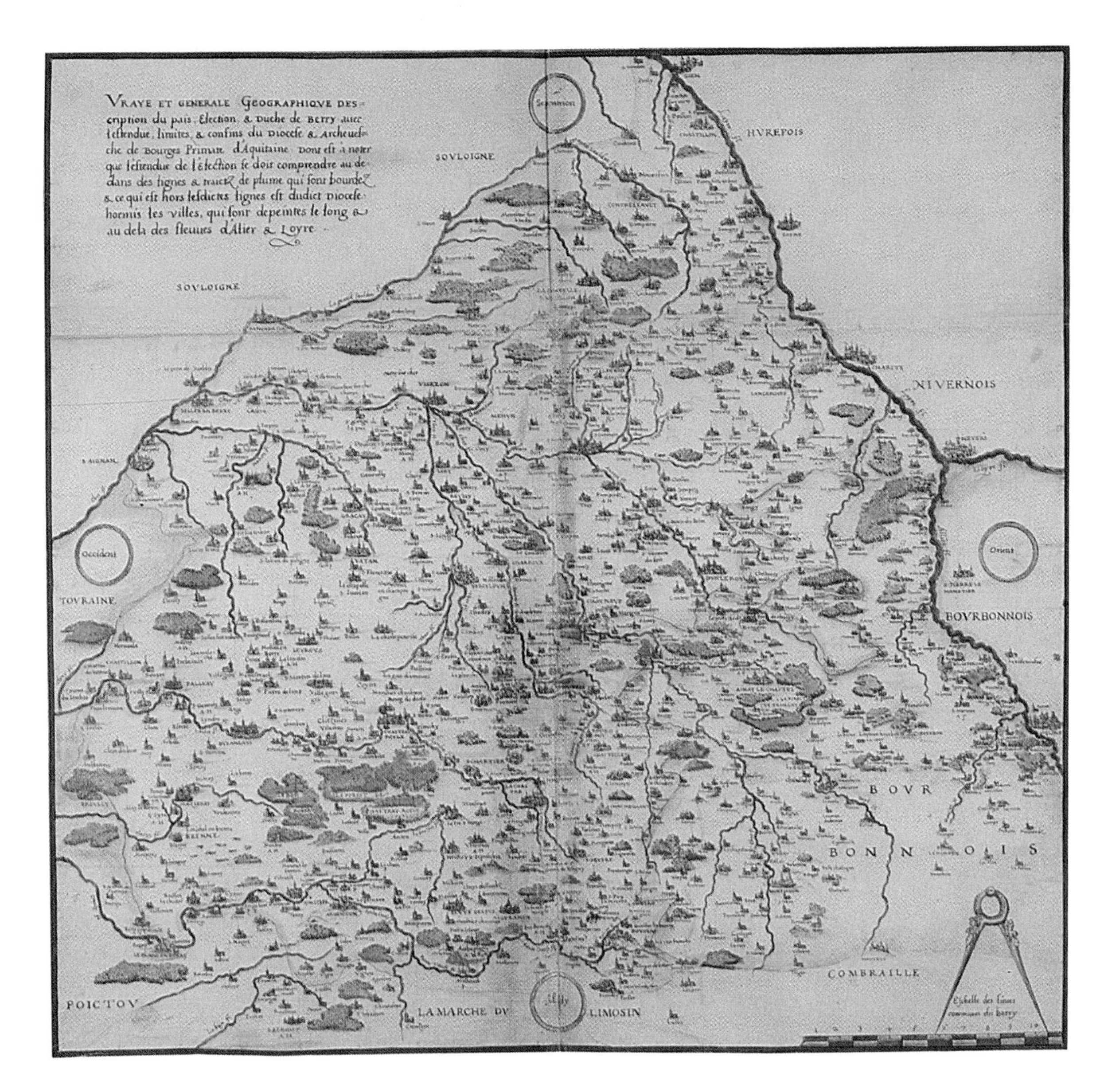

Pl. 4 Map of the Duchy of Berry by Nicolas de Nicolay, 1567. Bibliothèque Nationale, Département des Manuscrits, ms. fr. 2790
This map is the first of seven in a superb vellum atlas dedicated to Catherine de Medici by Nicolas de Nicolay in 1567. Here we see the Duchy of Berry as a whole on a scale of about 1:400,000; the six following maps show details on a larger scale. In style and degree of detail, Nicolay's work is very similar to that of his contemporary Christopher Saxton.

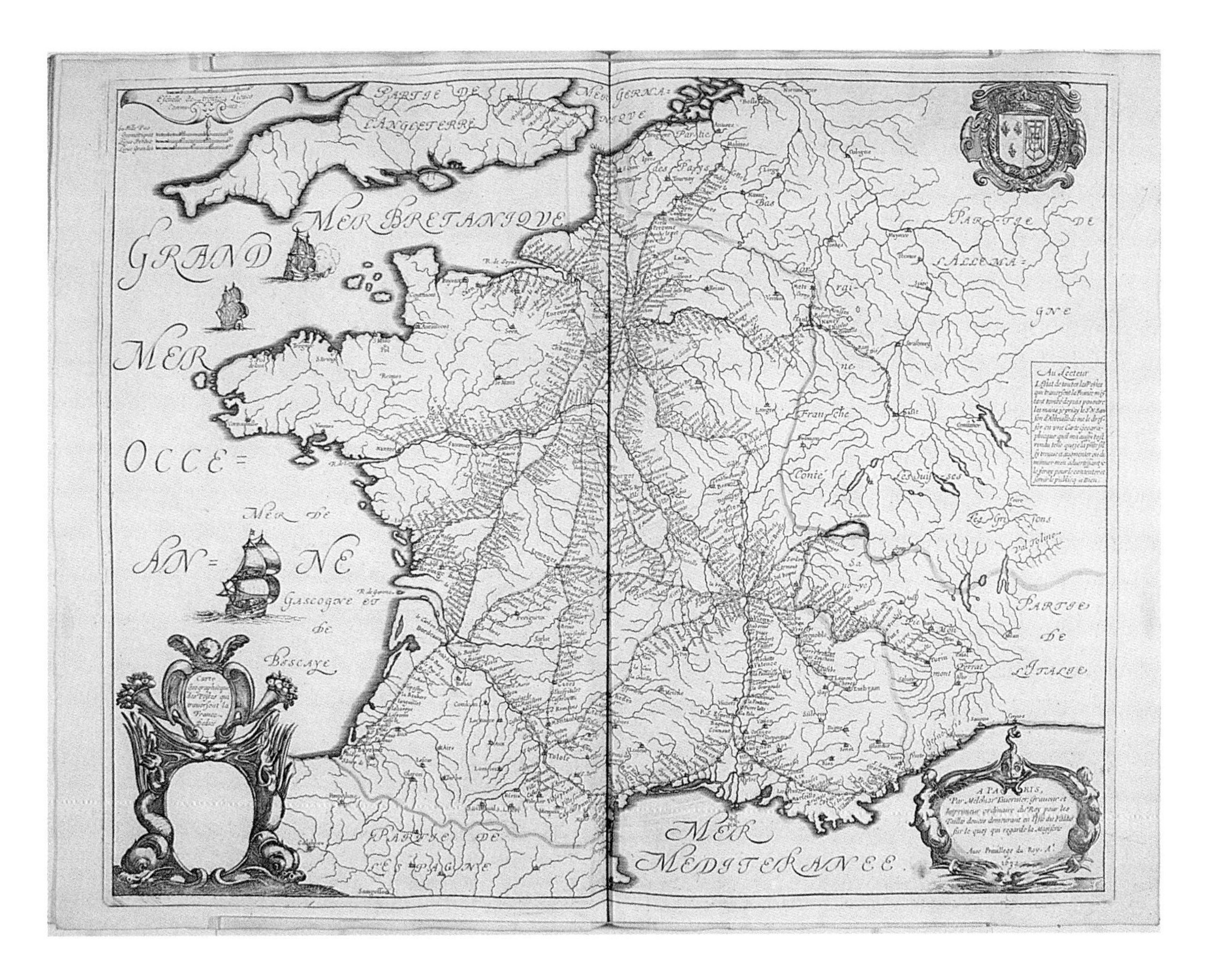

Pl. 5 "Carte Géographique des Postes qui traversent la France" from *Cartes Generales de toutes les parties du Monde* by Nicolas Sanson (Paris, 1658). The Newberry Library

This map is interesting not only for its thematic nature, but also for its use of color. France's boundaries are shown in green, and the chief mail routes in yellow; note that the latter extend outside the frontiers to such capitals as London, Brussels, and Turin.

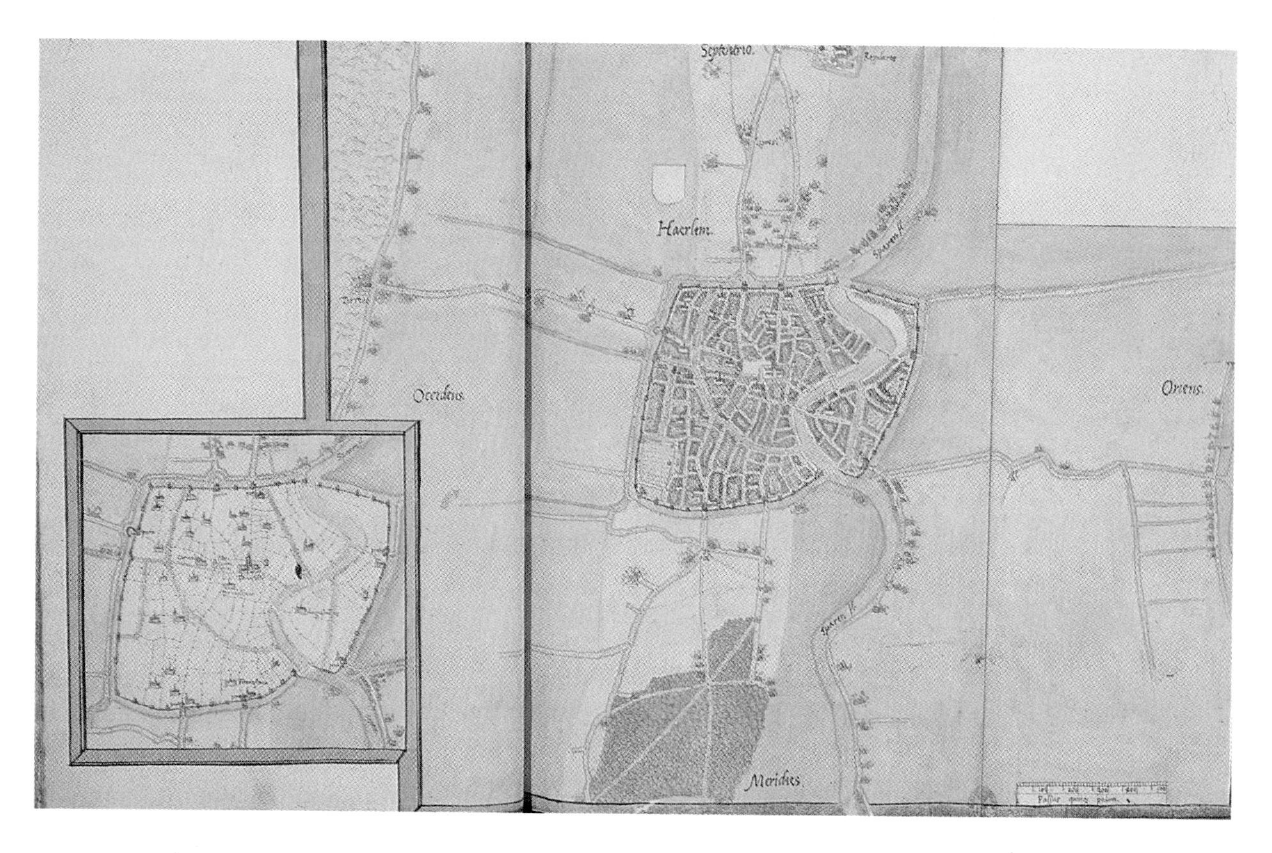

Pl. 6 Jacob van Deventer, Plan of Haarlem, c. 1560. Biblioteca Nacional, Madrid

Jacob van Deventer drew more than 200 town plans for Philip II during the 1560s and 1570s. Here we see Haarlem both in its general location amid the surrounding countryside, and also in detail, with the streets and prominent buildings emphasized. The success of this mapping project in the Netherlands encouraged Philip II to undertake a similar venture in Spain.

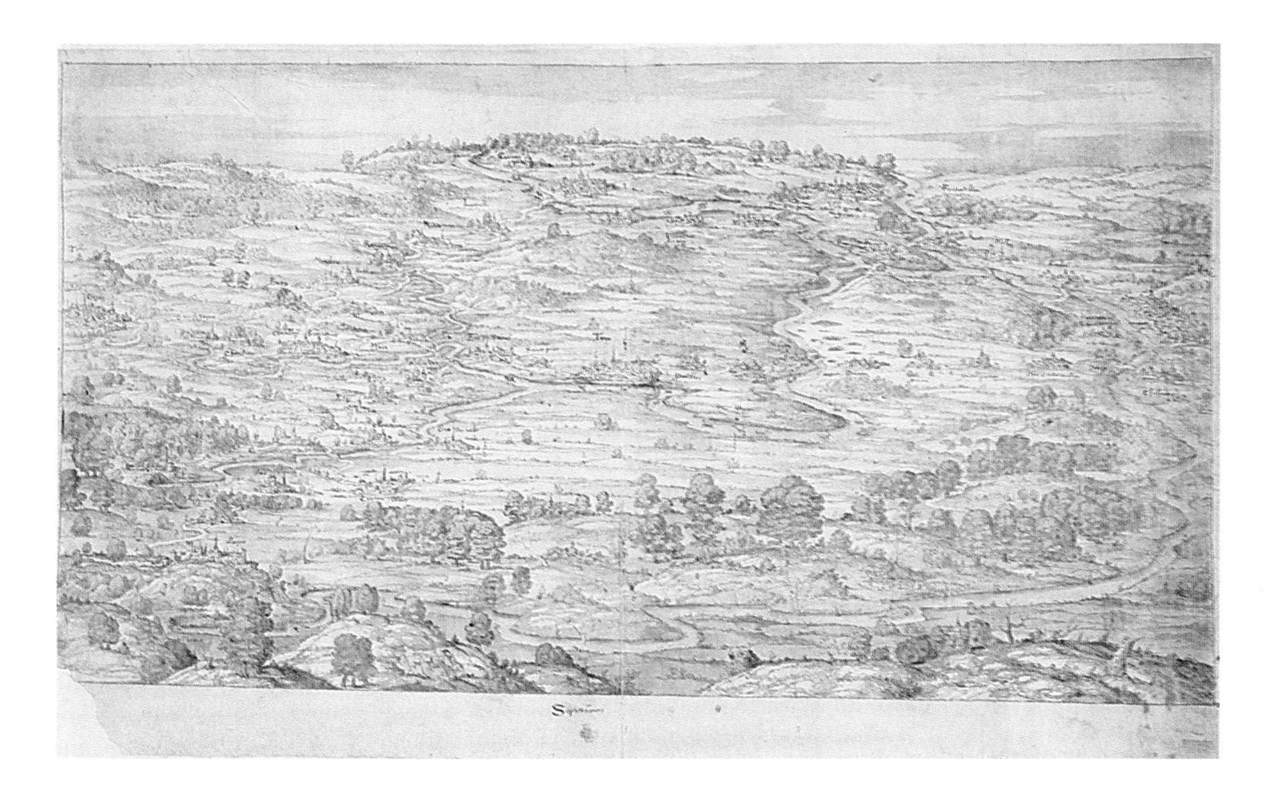

Pl. 7 Anonymous, "Discrizion de la Franzia por donde entró el Perador," c. 1544. Biblioteca Nacional, Madrid
The Emperor Charles V may have used this map, looking into France from an aerial position high to the northeast, when he invaded Champagne in 1544. It shows Troyes in the center middle ground, Auxerre in the center background, and Fontainebleau—the principal residence of his enemy Francis I—at the top.

Pl. 8 Anonymous, "Armada Chart." Prins Hendrik Maritime Museum, Rotterdam
This chart of "the Channel between England and Normandy" clearly derives from the version in Lucas Waghenaer's *Spieghel der Zeevaerdt* (fig. 5.5), with Portuguese captions (but not place-names). It was probably prepared for the use of the Armada of 1588 and may be the work of Luis Teixeira, whose workshop was in Lisbon.

binding for a book; to be scraped off to serve as a palimpsest for some later information deemed to be more important; or to be destroyed by fire or some other disaster.[5] The third reason for the high losses among early modern maps is the policy of deliberate destruction followed by many governments. The best-known example of this tendency concerns Portugal. The compilers of the comprehensive *Portugaliae Monumenta Cartographica* regarded the total disappearance of all but two pre-1500 Portuguese charts, when so many more are known to have existed, as "the most amazing mystery in the history of Portuguese cartography."[6] But the mystery was no accident; the government in Lisbon more than once ordered the total destruction of certain categories of maps, and at all times prohibited the export of cartographic instruments or materials outside the kingdom. Thus on 13 November 1504 a royal edict forbade anyone to make a nautical chart that included information on navigation south of the Congo River and ordered all existing maps containing such data to be surrendered in order to have it deleted. The edict also prohibited anyone from possessing a globe, and all globes in private hands were to be destroyed.[7] Terrestrial maps might likewise be destroyed or restricted on the grounds that the information they contained might compromise national security and should therefore be concealed from hostile eyes. Thus when King Charles IX of France (1560–74) was presented with a map of France by a Portuguese cartographer, he immediately ordered it to be kept under lock and key, "because maps are useful in war, enabling a foreign enemy to lead an army without the aid of a guide who knows the country across the terrain shown on the said maps, utilizing only a quadrant and compass."[8]

Sometimes, indeed, maps were considered so sensitive that even friendly eyes could not locate them when need arose. Thus in the summer of 1566 Philip II wished to evaluate a report from the council of the Indies about the voyage of Miguel López de Legazpi to the Philippines, but he could not find any maps to show him where the islands were. "Tell the councillors," he commanded his secretary, Francisco de Eraso,

> that they are to make every effort to find all the papers and charts which exist on this, and to keep them safe in the council offices; indeed the originals should be put in the archives at Simancas, and authenticated copies taken to the Council. I think that I have some [maps of the area] myself, and I tried to find them when I was in Madrid the other day—because if I do have them, that is where they will be. When I get back there, if I remember and if I have time (which I don't just now), I shall look again. Do you think you could find something on this, Eraso? I would like you to search, and make sure that anything you find is looked after as I have just said, with the Council always in possession of the copies.[9]

So the government of Habsburg Spain, in spite of the later lament of Jean-Charles della Faille, did in fact possess maps: of Europe, of America, of the Pacific. The principal problem facing the historian is therefore to repair the ravages of time, secrecy, and Philip II's filing system in order to establish the true extent and the nature of the cartographic resources to hand, and the uses to which the government put them.

Two main categories of maps concerning Europe were available. On the one hand stood the general surveys, some covering the dynasty's own states, other representing those of their neighbors. All were useful because they permitted the government to locate and visualize geographically places that were in the news. They helped ministers to cope with the unforeseen. On the other hand were maps drawn up in connection with a particular policy or problem (especially when it involved the use of armed force). Since the variety within each category was considerable, they will be examined separately, and in detail.

General Maps

It used to be thought that the earliest "modern" map of Spain was the one included in the 1482 printed edition of Ptolemy's *Geography.*[10] However, it has been shown that, from the 1450s onward, earlier manuscript versions of this work also contained maps—probably prepared by the noted cartographic school of Ancona in central Italy—and that the 1482 printing merely reproduced them.[11] But what, then, were the sources of the manuscript Ancona maps? Clearly, portolan charts served as the basis for the Iberian coastline, which was depicted in great detail and had begun to lose the overall appearance of "an ox-hide pegged out on the ground," to which Strabo had compared the shape of the peninsula many centuries before. But the sources for the mountains, rivers, and inland cities are less certain. It is possible that earlier maps, now lost, once existed; but it seems more likely that the compilers relied instead upon the "descriptions" written by pilgrims and other travelers in the peninsula. Whatever its derivations, however, this prototype remained the model for all representations of the Iberian kingdoms for almost a century—until the appearance of a map printed in 1544 by Giacomo Gastaldi, entitled *La Spana.* This time there was a wealth of new detail on the interior, much of it supplied (as the cartouche of the map itself tells us) by the Spanish ambassador in Venice, Don Diego Hurtado de Mendoza.

La Spana marked a distinct improvement on anything previously available, but it was soon overtaken by two other maps. First, in 1550, came *Spagna con le distantie de li loci,* compiled, after personal reconnaissance, by Vincenzo Palentino in Venice; it was followed in 1555 by the *Nova descriptio Hispaniae,* published in London by Thomas Geminus, which offered, for the benefit of Philip II's new English subjects, the first reasonably accurate delineation of the "shape" of Spain.[12] Since this map was published with official approval, it probably drew upon government sources and may even have used the detailed map made a few years before by the royal cosmographer Alonso de Santa Cruz. According to a letter written by Santa Cruz to the emperor in November 1551, "I have completed a [map of] Spain, of more or less the size of a large tablecloth [*repostero*], showing all the cities, towns and villages, the rivers and the mountains, together with the frontiers of the kingdoms and many other details." But nothing further is known of this map.[13] Whatever its sources, however, Geminus's work was used heavily in four other maps published shortly afterward: Pirro Ligorio's *Nova totius hispaniae* (1559), Vincenzo Lucini's *Hispaniae descriptio* (1559), Paolo Forlani's *Spagna* (1560), and Domenico Zenoi's *Hispaniae descriptio* (1560). In all of them, however, the interior of the peninsula was still little more than an idiosyncratic selection of half-known place-names and features.[14]

The next watershed in mapping Spain itself came with the *Theatrum Orbis Terrarum* of Abraham Ortelius, published at Antwerp in 1570. Although some thirty of the thirty-eight maps in the *Theatrum* were of Italian manufacture, "Spain" was one of the exceptions: Ortelius chose the *Hispaniae nova descriptio* of the French botanist Charles de l'Ecluse (or Carolus Clusius), recently printed by Christophe Plantin in a large format as a separate wall map. This work was the fruit of careful study and some personal reconnaissance, and it was updated in subsequent editions. For example, in 1571, Ortelius received a communication from Philip II's chief minister in Spain, Cardinal Diego de Espinosa, expressing general satisfaction with the Clusius map but regretting that his own birthplace, Martín Muñoz de las Posadas, was not shown. Since it was small, not to say obscure, the cardinal graciously described exactly where it was and even suggested a place-name that could be deleted in order to make way for the new entry. Ortelius took the point: the 1573 edition of the *Theatrum* included "Martimuñoz" about halfway between Ávila and Medina del Campo (figure 5.1).[15] Rather more usefully, throughout the 1570s Benito Arias Montano, librarian of the Escorial and a noted humanist who had resided in Antwerp for several years, sent Ortelius various maps of Spain and Portugal, and of their overseas colonies, for future editions. Best of all, permission was given to agents of the Antwerp cartographer to visit Spain and make their own maps. Not for nothing had Ortelius dedicated his *Theatrum Orbis Terrarum* to Philip II.[16]

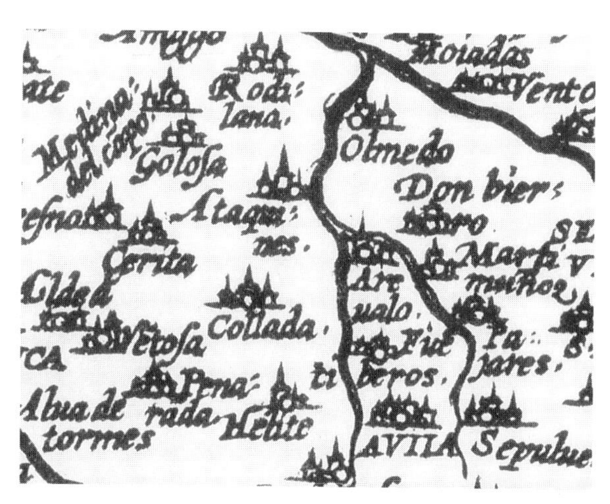

FIG. 5.1 Detail from map of Spain from Abraham Ortelius, *Theatrum Orbis Terrarum* (Antwerp, 1573). NL

Other foreign compilers of maps in the later sixteenth century were less informed about the peninsula. The celebrated cartographer Gerard Mercator was unable to find reliable maps of Spain to include in his original *Atlas* of 1585, and there were still none in the second edition of 1595. Only when Mercator's plates were acquired by the Amsterdam mapmaker Jodocus Hondius was this anomaly rectified: the 1606 edition of his *Atlas* included seven maps of the peninsula (admittedly, six of them plagiarized from inferior earlier works), while the 1611 edition had nine and the 1638 edition had fourteen.[17]

And yet none of Mercator's Spanish maps was particularly good—few could be favorably compared with those available at the time for France, Italy, and the Netherlands—and the historian is bound to wonder why. An initial explanation is that, from the work of Geminus onward, all the published maps of Spain were made by outsiders; they were neither commissioned by government nor compiled from official resources. Yet, even so, they could certainly have been better, for relatively accurate data on the topography of Spain were easily available. For example, two excellent repertories of roads were compiled and printed by Juan Villuga (1546) and Alonso de Meneses (1576), while itineraries, guides, and journals written by travelers to Spain were abundant.[18] But the great sixteenth-century mapmakers seem to have been invincibly opposed to using noncartographic materials, and most of them also showed little interest in visiting the area (let alone in measuring or surveying it) for themselves. Instead, they preferred to use secondhand cartographic data, often combining a number of small-scale maps to create a sort of mosaic to represent larger areas. Admittedly, they strove diligently to select only the best of the secondhand data; but that was still no substitute for personal reconnaissance.[19]

No doubt the explanation for this curious failure was in part intellectual: this was how maps had always been made, so that Ortelius, Mercator, and the others saw nothing wrong with it. But it was also partly due to government obstruction. For example, between 1517 and 1523 almost seven thousand villages in Castile were visited, and their size, economy, appearance, and relative position noted, by Fernando Colón (son of Christopher Columbus) and a team of associates. Their findings were recorded in eight volumes of description (which have survived) and three volumes of pictures or sketches (now, sadly, lost). Colón's "Description and Cosmography of Spain" proceeded systematically, unaffected even by the outbreak of the *comunero* revolt in Castile, until in 1523—perhaps precisely because it had found favor with the *comuneros*—the royal council peremptorily ordered the preregrinations to stop, and a public proclamation was made that anyone found aiding or informing Colón or his team would be fined. The survey was left incomplete and remained unpublished and forgotten in the Columbus Library in Seville for four centuries.[20]

The lack of support for Colón's project is surprising, for Charles V would have been aware of the spate of similar publications in Germany and the Netherlands, all of them seeking to "illustrate" the past and to "display" the land and its people. Admittedly, the pictures included in some early examples of these works were crude—merely summarizing the text in a visual image rather than seeking to create an accurate picture—but by the sixteenth century recognizable representations of individual cities were offered. Sometimes a "prospect" was provided, often as a bird's-eye view; but this gradually gave way to a "perspective" or "profile" taken from a slight elevation and showing not only the facades of the principal monuments but a view of the street plan as well. By the time Sebastian Münster's *Cosmography* was published in 1540, both genres were firmly established, and there was little further technical advance in the art of the cityscape until the days of Matthaeus Merian and Wenceslaus Hollar a century later.[21]

But Spain, as usual, was different. Eight years after Münster, Pedro de Medina dedicated to the young Philip II his *Grandezas y cosas notables de España;* it described each town in

Spain, with its history illustrated by a panoramic woodcut—often, however, it was the same panoramic woodcut used over and over again. Even the map of Spain that served as a frontispiece was merely a reproduction of the "Ptolemaic" matrix composed by the school of Ancona roughly a century before. A second edition, which appeared in 1595 (again with royal approval), also unashamedly reproduced the Ptolemaic map, although it did offer some slight improvement in the cityscapes. There was now, for example, a recognizable likeness of Seville, with the Arenal, the Giralda, and the Golden Tower all clearly shown; but unfortunately the Seville woodblock was also used for the entries on Gibraltar and Aragon, and the woodblock for Madrid, which also served for Lisbon and Cartagena, looked like none of them.[22]

It is possible that the more accurate cityscapes in the second edition of Medina were inspired by the views made of all major Spanish cities by the Dutch artist Anton van den Wyngaerde, who traveled around Spain in the 1560s at Philip II's express "command and instruction to paint the pictures of several of my principal cities."[23] Now this commission was part of a major effort by the Spanish government, the first since Colón's incomplete *Itinerario,* to collect data systematically on the territories over which it ruled. It seems likely that the project was a personal initiative by Philip II, who had just returned from five years in northern Europe. Certainly a short time before he left the Netherlands he commissioned a very similar operation there. The cartographer Jacob van Deventer, who had already produced a series of excellent, detailed maps of each province of the Habsburg Netherlands (color plate 6), was ordered by the king on 29 May 1559 "to visit, measure and draw all the towns of these provinces, with the rivers and villages adjoining, likewise the frontier crossings and passes. The whole work is to be made into a book containing a panorama of each province, followed by a representation of each individual town."[24] The execution of this commission occupied van Deventer until he fled to Cologne at the outbreak of war in the Netherlands in 1572, by which time he had completed a total of almost three hundred bird's-eye plans. One hundred seventy-nine have survived, all oriented on north and all done to a scale of 1:10,645, and perhaps a further one hundred have been lost.[25]

It was not long after commissioning van Deventer that Philip II asked Anton van den Wyngaerde to come to Spain and perform a similar function there. His technique was somewhat different, for he worked from a slight elevation and in panoramic format, rather than from bird's-eye perspective; but the record was just as impressive. We know today of sixty-two finished views of some fifty Spanish cities by Wyngaerde, and preparatory sketches survive for several more. After the artist's death in 1571, Philip II decided to send his oeuvre to the Netherlands to be engraved, but the outbreak of the Dutch Revolt the following year made it impossible.[26] The originals were therefore displayed in the royal palace in Madrid until, in 1734, many were destroyed by fire.[27]

Wyngaerde's series of topographic plans was one of three major geographic projects commissioned by Philip II. The second was the set of government questionnaires, later known as the *Relaciones topográficas,* sent out to various communities in the 1570s. The idea seems to have originated with the royal chronicler Juan Paez de Castro, who prepared a questionnaire to be sent to each village in Spain requesting information about its geography, history, economy, population, and "antiquities."[28] But it was never distributed, and Paez died in 1570. Then four years later, under the direction of one of Philip II's most energetic yet least known ministers—Juan de Ovando, president of the Councils of Finance and the Indies—a questionnaire was sent to settlements in the diocese of Coria, followed up the next year by a fifty-seven-point *interrogatorio* for all communities in Castile.[29] Returns for some six hundred villages of New Castile, and for the city of Toledo, have survived in eight vast manuscript volumes preserved in the library of the monastery of San Lorenzo de El Escorial. Originally there were at least thirteen or fourteen.[30] But once again, the government seems to have made little use of them.[31]

The same is true of the third, simultaneous, project: a complete map of the Iberian peninsula. Its principal cartographer, Pedro de Esquivel, professor of mathematics at the University of Alcalá de Henares and an expert in surveying, embarked on this ambitious project in the 1560s. According to Felipe de Guevara (whose son Diego was one of Esquivel's assistants), "it was without exaggeration the most careful and accurate description ever to be undertaken for any province since the creation of the world. . . . There is not an inch of ground in all of Spain that has not been seen, walked over or tramped upon by [Esquivel], checking the accuracy of everything (insofar as mathematical instruments make it possible) with his own hands and eyes."[32]

In 1575, just after Esquivel's death, his colleague and friend the chronicler Ambrosio de Morales added further details on the late professor's cartographic achievement:

> He located places in the map or picture he was making by means of the tables of Ptolemy, covering the paper with squares and calibrating the sides by longitude and latitude. But he also worked with elegant devices [*primores*] so that everything would be as accurate and fine as possible. His friends and I admired these devices in his room and I would say something about them here except for the fact that they can only be understood by looking with one's own eyes, not by writing.[33]

According to Morales, Esquivel left his work largely "completed when he died, and His Majesty has it in his Chamber."[34]

Morales's encomium implied that Esquivel was working on just one map, which had not been completed by the time of his death. It was still not finished two years later, for when Philip II heard of the death of another member of the team (Diego de Guevara) in 1577, he wrote to his secretary: "I am reminded that he possessed the instruments and other papers of Esquivel. If this is so, I would like them to be collected . . . so that they should not be lost and so that the map of Spain, which he was making, should be continued."[35] It seems most probable that the "Map of Spain" project eventually resulted in the remarkable atlas of twenty-one maps now in the Escorial library.[36] The first (and most complete) map in the collection covers the whole peninsula (figure 5.2); the rest form a series of sectional surveys, done to the same scale, among which Portugal is the best covered and Aragon and Catalonia the worst. It is worth lingering over this achievement because at 1:430,000, the Escorial atlas contains by far the largest European maps of their day to be based on a detailed ground survey. No other major western state of the sixteenth century possessed anything like it, for where Philip Apian's celebrated map of Bavaria, based on a survey carried out between 1554 and 1561, covered 44,000 square kilometers, and Seco's map of Portugal (printed in 1560) cov-

Fig. 5.2 Map of the Iberian Peninsula from the Escorial Atlas. Library of the Monastery of El Escorial (LME)

ered almost 89,000, the Escorial atlas covered an area of no less than 497,000.[37]

How was it made? One important clue comes from the existence in Stockholm of a late sixteenth-century codex containing the coordinates of some three thousand locations in Spain—roughly half of those covered in the Escorial atlas. Each set of data consists of bearings from two separate observation stations, with a rough estimate of distances. However, these initial entries do not stand alone. There are also a host of annotations in the hands of the royal cosmographers Juan Bautista de Lavanha (whose name is on the flyleaf of the volume) and Juan López de Velasco.[38] Lavanha's comments were largely in the form of correc-

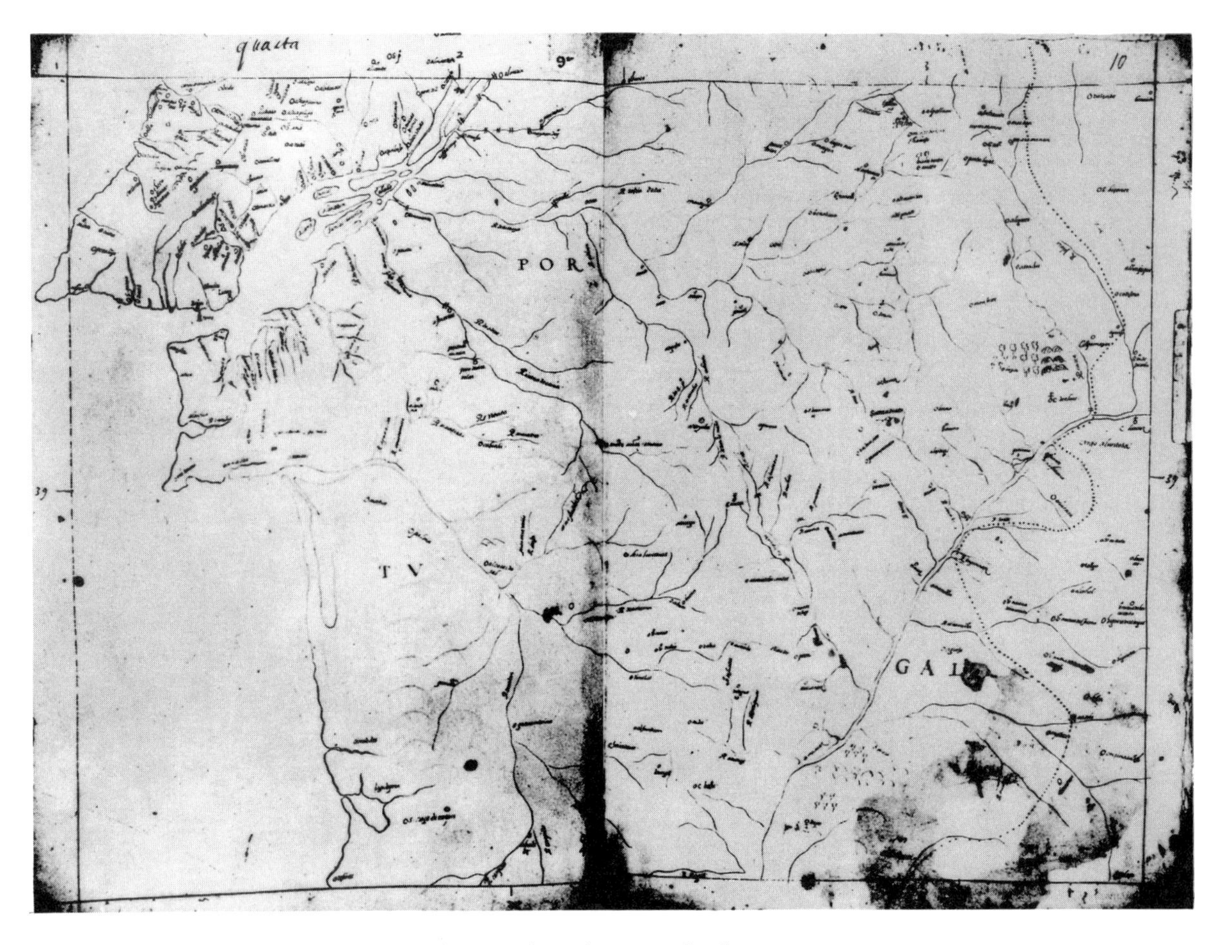

Fig. 5.3 Map of the Portuguese-Spanish frontier, from the Escorial Atlas. LME

tions to coordinates and geographical data—"this can't be right" (bk. 1, f. 28), "our informant was wrong" (bk. 1, f. 15), "this observation coincides with mine" (bk. 4, unfoliated), and so on. Velasco, by contrast, attempted to improve the accuracy of the whole work by adding several pages of stellar observations and declinations for various different latitudes, as well as the longitudes of a number of cities—Toledo, Madrid, Valladolid, Seville—calculated from the time recorded for the solar eclipse in 1577.[39] This is important, because the hand of López de Velasco has also been noted on the maps in the Escorial atlas—probably precisely to add the longitude of at least some places, without which detailed maps on this scale could scarcely have been completed.

There is also another reason to suppose that the maps were not finished for some years after 1577: Portugal, which is included as a fully integrated part of the Escorial atlas, was until 1580 an independent state into which Esquivel and his team were forbidden to go.[40] Even then it seems likely that the Portugal map

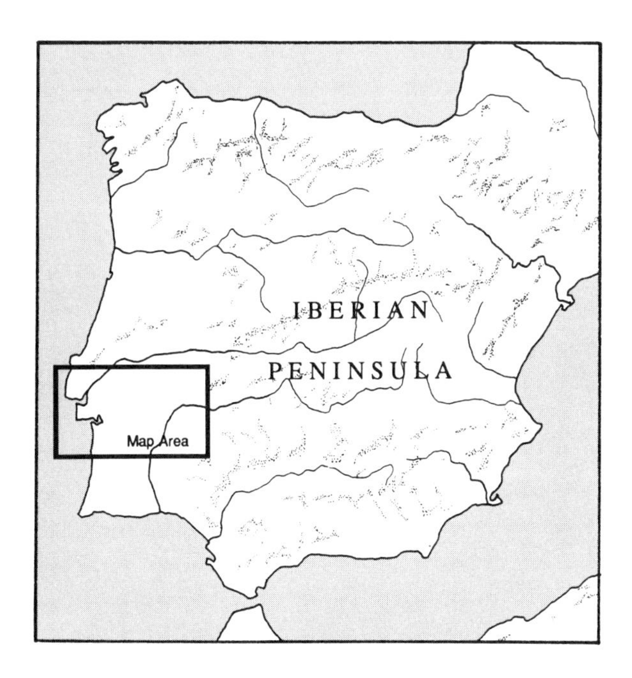

in the Escorial atlas derived from a separate and far superior survey, for there is a clear disparity in standard. Thus the rivers that cross the frontier are shown in far more detail on the Portuguese than on the Spanish side (see figure 5.3), which suggests that the Escorial cartographers incorporated the results of a prior survey for areas across the border. At first sight, the progenitor seems obvious: Pedro Alvares Seco's fine map of Portugal engraved and published in Italy in 1560 by Aquileo Estau, a Portuguese resident in Rome, with a dedication to a munificent prince of the church, Cardinal Sforza. But Seco's map is not as straightforward as it looks. In the first place, it is so accurate that it clearly stemmed from a detailed ground survey that would have taken several years to complete (if Apian took seven years to survey Bavaria, it must have taken Seco at least ten to cover Portugal). Such a protracted exercise could have taken place only under government license, almost certainly as part of a state enterprise. And yet, if Seco's work was official, why was it published in Rome instead of in Lisbon, with a dedication to an Italian cleric rather than to the Portuguese king? All the evidence suggests that the published map of 1560 was, in fact, a pirated edition of some master-map made by the Portuguese government.[41] Unfortunately, this presumed official map has not survived, but the observations on which it was based, probably dating from the 1530s, are to be found in a codex (now preserved in Hamburg) that lists the geographical coordinates of about one thousand places in Portugal (the Seco map has 1,154). It is clear that these coordinates were calculated by a geometrical survey from fixed points of observation, and that is undoubtedly why both the Escorial and the Seco maps of Portugal appear somewhat elongated and skewed: no correction was made for the curvature of the earth's surface.[42]

It seems probable that the Escorial cartographers, like Seco, had access to the official map that was compiled from these data, because the similarities between their map and Seco's are overwhelming (even the errors—mistaken configuration of rivers and so on—are the same). But they are not identical. The Escorial maps of Portugal contain several details that Seco's lacks, so the prototype must have been on a larger scale than Seco's, and it must have been completed before he began—certainly before 1560, when the pirated edition was made, and probably before 1550. The Portuguese were thus the first to produce a national map based on modern surveying techniques.

Given this chronology, and given the similarity in scale and technique employed for Castile and Portugal, it seems likely that the sectional maps of the Escorial atlas were made only after the Spanish conquest of Portugal in 1580, which gave Philip II's ministers full access to the rich cartographic materials in Lisbon. Only that would adequately explain why the Castilian spreads were drawn on the same scale as the earlier survey available for Portugal.[43] So it is suggested that the maps of the Escorial atlas were begun in 1580, or soon after, with a team of cartographers plotting the observations made by Esquivel into the framework laid down by López de Velasco, and Lavanha checking the information as it was mapped. The work no doubt continued for a few years, until it was abandoned (probably in the 1590s) with several of the maps still unfinished, to lie forgotten in the Escorial.[44] Nevertheless, by the end of Philip II's reign the Iberian peninsula was better represented in maps than any other European area of comparable size.

Naturally, cartography in Habsburg Spain did not cease with the Escorial atlas. The 1606 edition of Mercator's *Atlas,* as noted above, contained nine rather elementary maps of Spanish provinces (Valencia, Catalonia, Aragon, Navarre, Vizcaya, Galicia, Leon, Castile, and Granada); but little is known of their composition or their authorship. Quite different was the map of the kingdom of Aragon undertaken by Lavanha at the request of the representatives of the Cortes of Aragon. It was the fruit of some six months of peregrinations around the kingdom (4 November 1610–16 April 1611), all recorded in the surviving volume of observations, which shows that Lavanha not only visited in person each community (except for some in the Pyrenees) but also used some new surveying techniques. He brought with him special instruments for measurement (in order, he claimed, to avoid depending on other people's assessments, "which vary as often as the people who give them"), and he also made observations at 120 different stations, taking azimuth bearings on the other stations at each of them.[45] Thanks to these careful readings, Lavanha had a magnificently detailed map ready in parchment by the autumn of 1615. Engraving began the following year, and a limited edition, on a scale of 1:278,195, was published in six sheets in 1620. Unfortunately, not a single copy of this version seems to have survived, although the map was soon reproduced in most atlases (it appeared in the 1633 French edition of the Mercator-Hondius collection, for example) and was frequently reproduced from the original plates until 1777, when it was updated. Yet even then it was not replaced: roads were inserted, 326 new settlements were added, and 122 old ones were altered, but the basic cartography remained unchanged, despite the inevitable transformations that the passage of 170 years had caused.[46]

The initial publication of Lavanha's map of Aragon in 1620 seems to have evoked a positive response from the central government. Two years later, Philip IV ordered another Portuguese cartographer, Pedro Teixeira Albernaz, to undertake a survey of all the coasts and coastal settlements of the peninsula. He was to start at Fuentarrabía and work his way round, noting in each place

> the details, whether strengths or weaknesses, the gates, whether entries or exits, the quality and quantity of private and public estates, the services rendered to the crown; the antiquities and foundation of the town itself, and of the convents, the major houses, families and lineages; the climate of the land, its altitude, any flooding, the navigability of the rivers; and many other matters and details which are greatly important to the reputation, honour and existence of the said place.[47]

In other words, Teixeira was to produce a more detailed version of Pedro de Medina's *Grande-*

zas, and the text of his *Descripción de España*, completed in 1630 and preserved in manuscript in several European libraries, indicates that he performed this task faithfully. But there was more: Teixeira also prepared a series of coastal views and detailed maps. Unfortunately, all of these seem to have perished, although there is an indication of what might have been produced in his brother's *Atlas of the Coasts of Portugal*, published in 1648, and in his own two cartographic monuments, the *Descripcão e mappa geral do reino de Portugal* of 1662—which at last replaced Seco's map of 1560—and the amazing, gigantic map of Madrid, printed in 1656 at a size of 178 × 186 cm.[48]

At first, the Spanish Habsburgs' collection of maps covering the possessions of other rulers was less impressive. Indeed, many contemporary observers argued that Charles V (at least) placed insufficient reliance on maps. Thus after his abortive and costly invasion of Provence in 1536, according to Michelangelo, many Italians jested, "If only the emperor . . . had ordered a drawing to be made of the course of the river Rhône, he would not have met with losses so severe, nor retired with his army so disarrayed."[49] And yet this was manifestly unjust, because a contemporary French source records that, on this very campaign, the emperor "normally had in his hand or before his eyes a map of the Alps and Lower Provence, which the marquis of Saluzzo had given him. He studied it so often and so intently, using it to further his designs and his desires, that he began to think he had the country in his grasp instead of just the maps."[50] The problem, therefore, was apparently not the lack of a map, but the possession of too good a map—and certainly, if the campaign map of Champagne in 1544 (color plate 7) is anything to go by, it is easy to see why. Northeastern France is portrayed exactly as it would have appeared to Charles V's army as they entered from Lorraine on the rapid and victorious thrust that forced the humiliating peace of Crepy upon France. The emperor's ambitious plan is reflected in the inclusion of Fontainebleau, the favorite palace of France's king, at the upper right.[51]

Special Maps

This striking piece of imaginative cartography takes us into the realm of the specific: to the detailed maps prepared in connection with a deliberate policy or an individual campaign. And here we meet the problem of subsequent loss at its most acute, for maps of this type were undoubtedly the most common, and yet hardly any of them have survived. Some idea of the prevailing standards may be gathered, however, from a set of the charts made of the Azores archipelago by Luis Teixeira of Lisbon in 1587, apparently for the expedition led by the marquis of Santa Cruz to convoy the Indies treasure fleet safely back to Spain. They are all done to the same scale, replete with useful descriptive information, and were conveniently mounted on hinged wooden boards for use at sea.[52] Equally interesting are the surviving maps and plans prepared for Philip II's projected invasion of England in 1588. The earliest in date is the sketch of the theater of operations drawn by Bernardino de Escalante (figure 5.4).[53] Although a priest when he forwarded his project for the conquest of England, Escalante was the son of an army captain and had himself served as a soldier in Flanders in the 1550s; he had also composed an excellent manual of military practice.[54] Escalante's map showed three possible strategies for conquest. The first (shown on the left of the map) was a daring voyage into the North Atlantic directly to Scotland, where the invasion fleet would re-

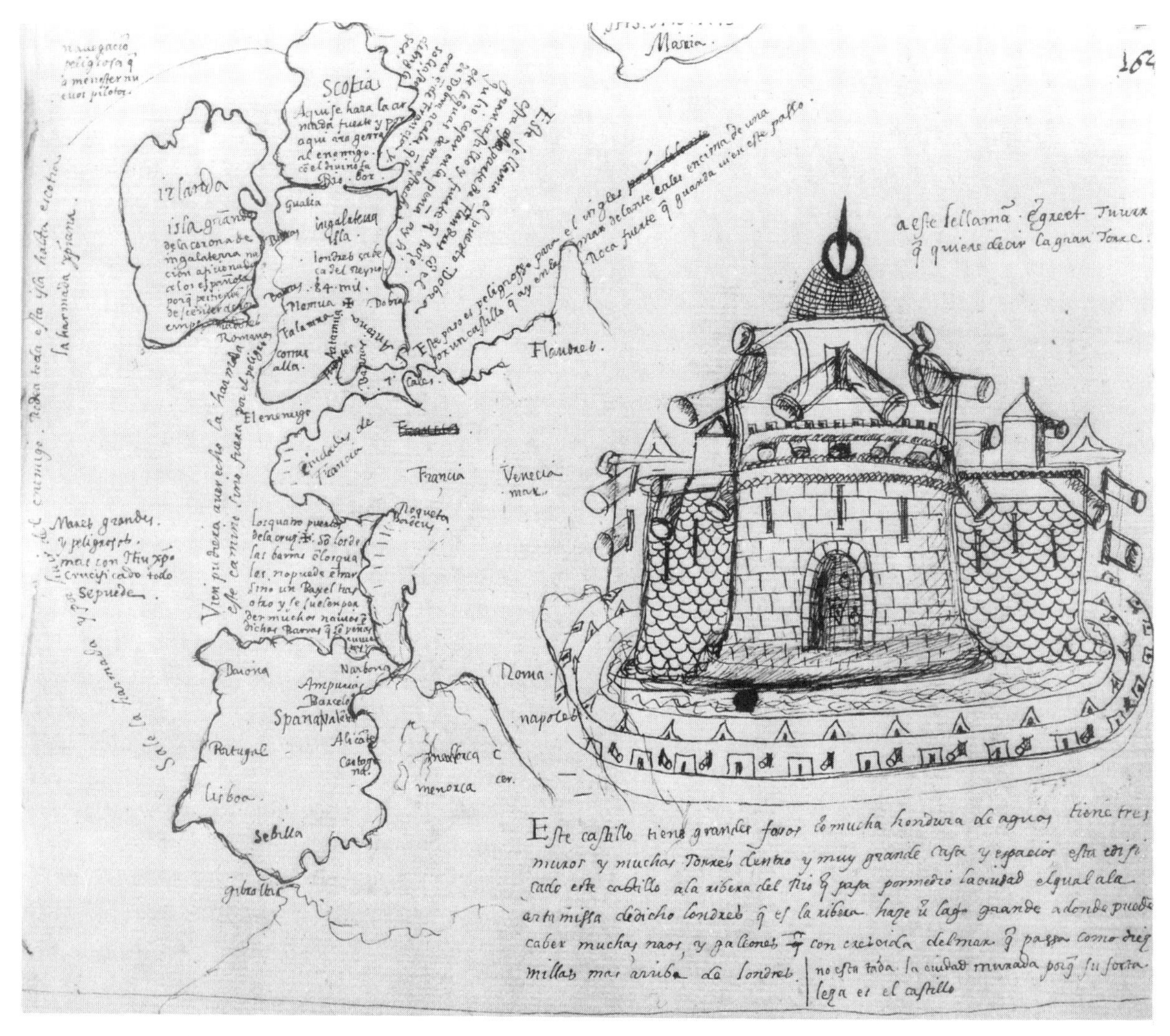

FIG. 5.4 Bernardino de Escalante, Sketch Map of possible invasion routes against England, June 1586. Biblioteca Nacional, Madrid

form before launching its main attack ("The seas are high and dangerous," Escalante observed, "but through Jesus Christ crucified everything is possible"). The second was a direct attack into the Irish Sea, a route "which the Armada could well have undertaken but for the danger" (presumably of "the enemy," whose forces are shown at the entrance to the English Channel). The third suggested route was a direct attack from Calais to Dover, and on to London. In the event, Philip II decided to adopt a combination of these suggestions, with the Armada leaving Lisbon for the Channel in order to link up with a second expeditionary force assembled in the Spanish Netherlands.[55]

This was easier said than done, however. To begin with, to ensure success, each of the 130 vessels of the Grand Fleet assembled in Lisbon required appropriate charts, sailing directions, and pilots to be sure of reaching the rendezvous. The last was the hardest, for pilots with experience of the Channel and the North Sea were scarce in Spain (and, even then, to draft them all for the Armada would have revealed too clearly the king's grand strategy). Recognizing this problem, the duke of Medina Sidonia, who arrived in Lisbon to take command of the fleet on 15 March 1588, took immediate steps to provide each ship with a set of navigational aids that could be used for guidance even without an experienced pilot. Almost at once he ordered his principal nautical advisers to put together some sailing instructions, and the result (published as a pamphlet of ten folios by a local printer on 30 March) was the *Derrotero de las costas de Bretaña, Normandía, Picardía hasta Flandes*.[56] A copy was sent to the king, who took a personal interest in its accuracy, checking its information against his own collection of maps. At one point he became concerned that he could not find Le Conquet, a place noted in the *Derrotero*, on one of his maps (and sent a secretary off to check its location on the map of France included in Ortelius's *Theatrum*), and that another place in Brittany was spelled differently in the *Derrotero* and on one of his maps ("but they must be the same place, which is very close to Ushant").[57] The king also fretted that one of his maps showed Dunkirk to be "25 to 40 leagues" from the Isle of Wight, whereas another showed the distance as fifty, "and I think this conforms more to the *Derrotero*."[58]

But for all that, the Armada's *Derrotero* was not what it seemed. It was not a guide specially made for the Grand Fleet (which, given that it was put together in less than a fortnight, was hardly to be expected), but a hybrid compilation of information from the rutters normally used by merchants traveling between the Iberian peninsula and the North Sea. Thus the Armada captains were sagely informed that, once they reached the anchorage bounded by the Goodwin Sands, "there you will wait for a pilot to take you either to London or Dover," while if they wished to sail from Dover to Flushing, they should cross the Channel to the Flemish coast, "and when you are off Blankenburgh you will fire a gun and a pilot will come out to take you to Flushing." At a time when Spain was at war with both England and the Dutch Republic, such counsel was clearly worse than useless.[59] The true value of the *Derrotero* lay in the detailed information it provided on the different depths and "bottoms" to be found in various parts of the Channel, and on how to recognize different landmarks ashore, so that pilots without direct experience of the area (or even ships without pilots at all) might have a chance of fixing their position as the fleet advanced.[60]

Together with the *Derrotero*, on 2 April 1588 Medina Sidonia sent the king a drawing ("un cartón").[61] Perhaps it was a map, for certainly the duke had already commissioned "80 mounted sea charts, with soundings around the coasts of Spain, England and Flanders" from Ciprián Sanchez, a Lisbon chartmaker. They were delivered just two weeks before the Armada sailed.[62] Given the unusual specification that "soundings" should be included, there is good reason to believe that two of these charts have survived (see color plate 8). Both are little more than Portuguese manuscript copies of the

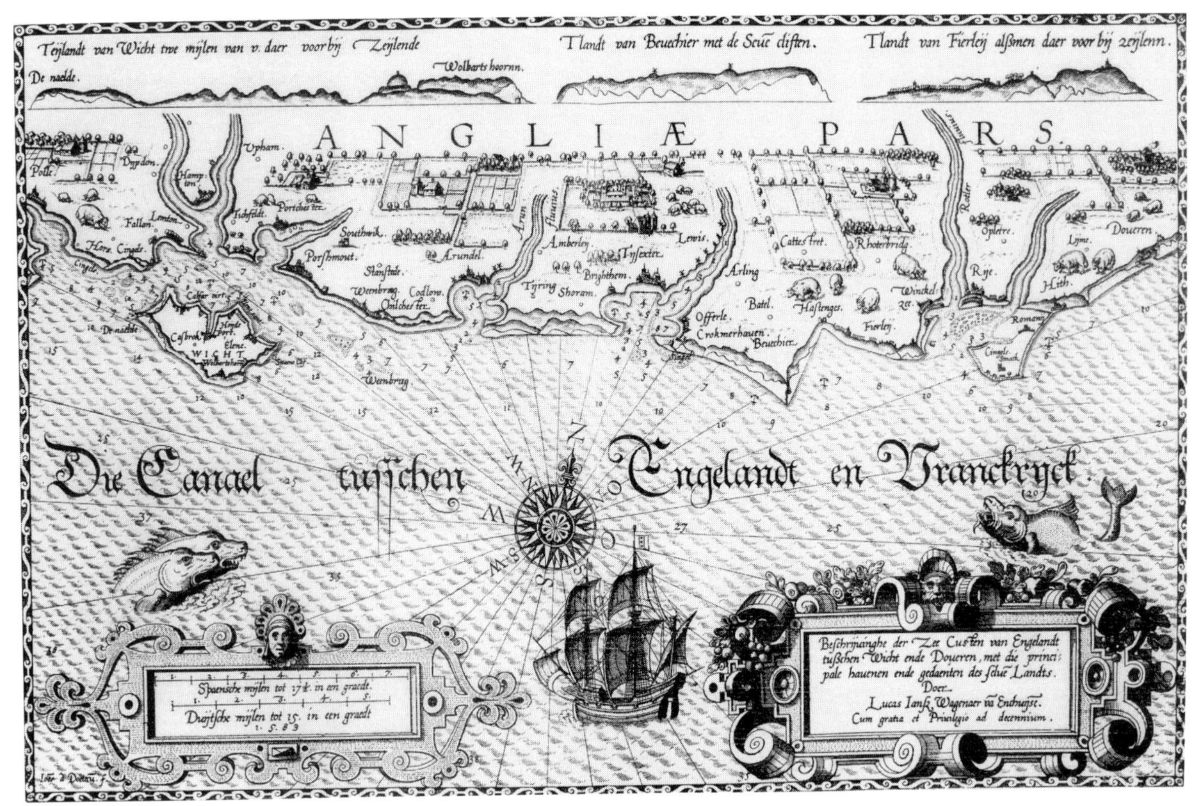

Fig. 5.5 Chart of part of the English Channel from Lucas Waghenaer, *Spieghel der Zeevaerdt* (Leiden, 1584–85).

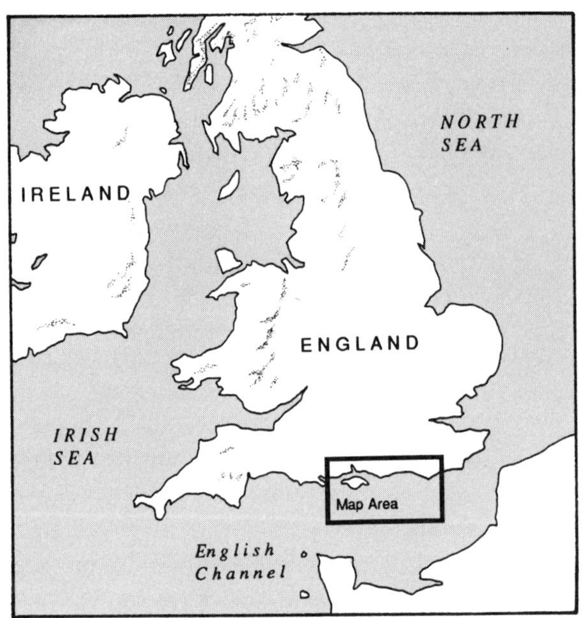

corresponding maps printed in Lucas Waghenaer's *Mariner's Mirror* (of which a Latin edition was published in October 1586), but as such they represented the most up-to-date aid to navigation available at the time (figure 5.5).[63]

The same cartographic care with which Philip II prepared for Spain's wars at sea was evident in preparations for its wars on land. A particularly clear picture emerges from the use of maps made by Spain's most famous military commander, Don Fernando Álvarez de Toledo, the third duke of Alba. A large number of maps arising from his campaigns in northern Europe between 1567 and 1573 have been preserved,

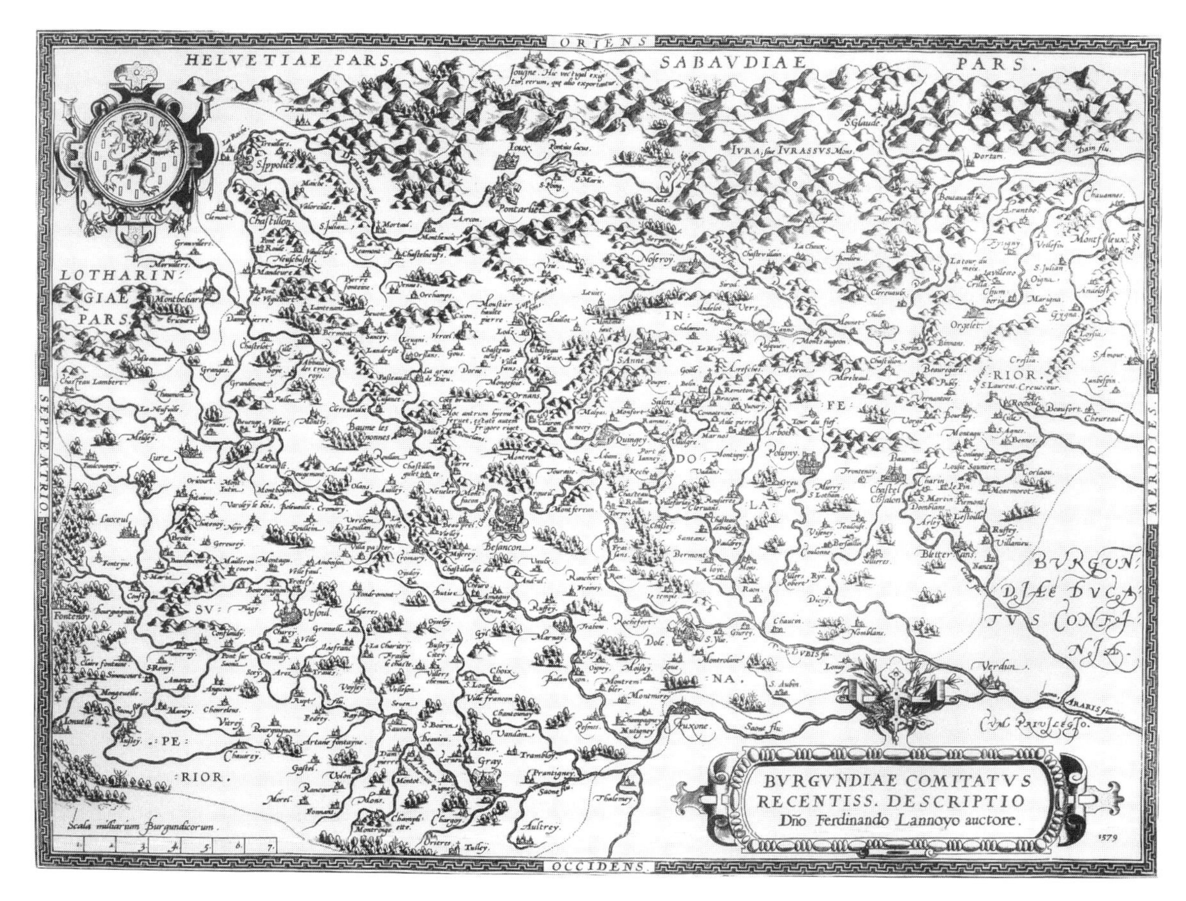

FIG. 5.6 Ferdinand Lannoy, map of Franche-Comté, from Abraham Ortelius, *Theatrum Orbis Terrarum* (Antwerp, 1579). The Newberry Library

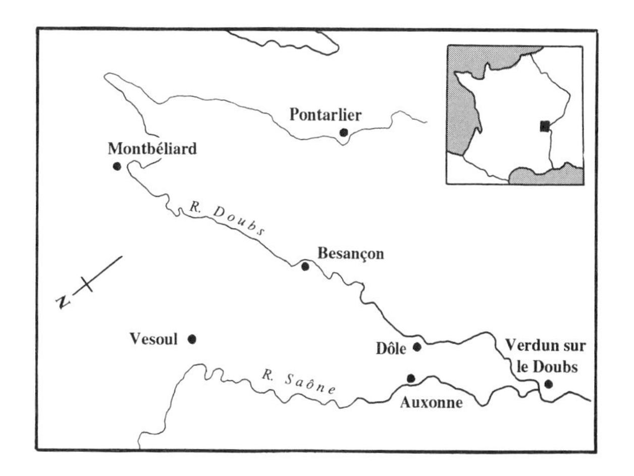

and, although Alba was reputed to be an unusually careful and methodical general, there is no reason to suppose that others would necessarily have used maps any less.[64] After all, as Marshal Vieilleville of France wrote slightly later, "A military commander must no more move without a map than a pilot or a galley captain, unless he wants to court disaster."[65]

Two of the most striking maps connected with Alba's military operations are today to be found among the records of the government of Franche-Comté at Besançon. The duke had

FIG. 5.7 Map of Franche-Comté prepared for Alba in 1573. Archives Départementales du Doubs, France

himself marched overland through the province from Lombardy to the Low Countries in 1567 at the head of ten thousand troops, and he made use of a copy of a newly completed map of the area prepared by Don Fernando de Lannoy. Although the map may seem somewhat vague to us, Alba considered it so accurate and useful that he forbade its publication, and it was only printed twelve years later, in the third edition of Ortelius's *Theatrum* (figure 5.6).[66] However, the duke evidently realized that something more specific would have been useful, and so for the next major military expedition up the "Spanish Road" linking Milan and Brussels, in 1573, he ordered two special maps

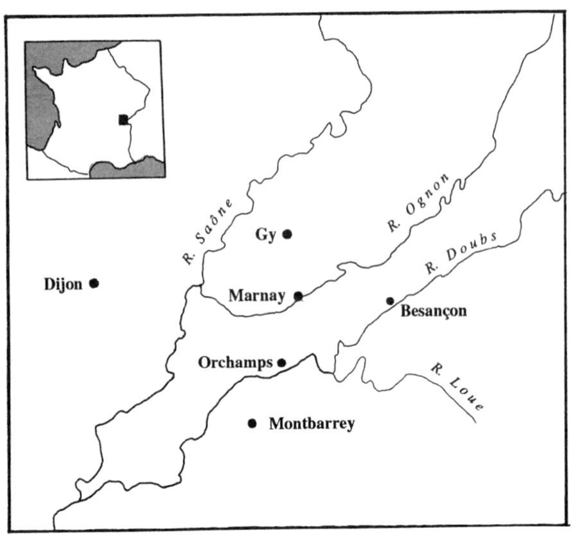

of Franche-Comté to be prepared, showing everything that an army on the march needed to know: the route it should follow, in a numbered sequence; the major rivers and forests, and the ways to traverse them; the alternative itineraries, in case of need; and the position of the nearest towns. It is difficult to see how, in the sixteenth century, better summary guides could have been produced at short notice (figures 5.7 and 5.8).[67] It is worth remarking that the expedition of 1573 was able to move up the Spanish Road with unusual speed, covering the 680 miles between Milan and Brussels in forty-two days.

During his governance of the Netherlands, Alba faced two military attempts to overthrow his regime, which necessitated major campaigns by his troops. The first, in 1568, he confronted in person, moving slowly but inexorably against his enemies—in Friesland in the spring, in Brabant and Limburg in the autumn. Once again, the duke commissioned

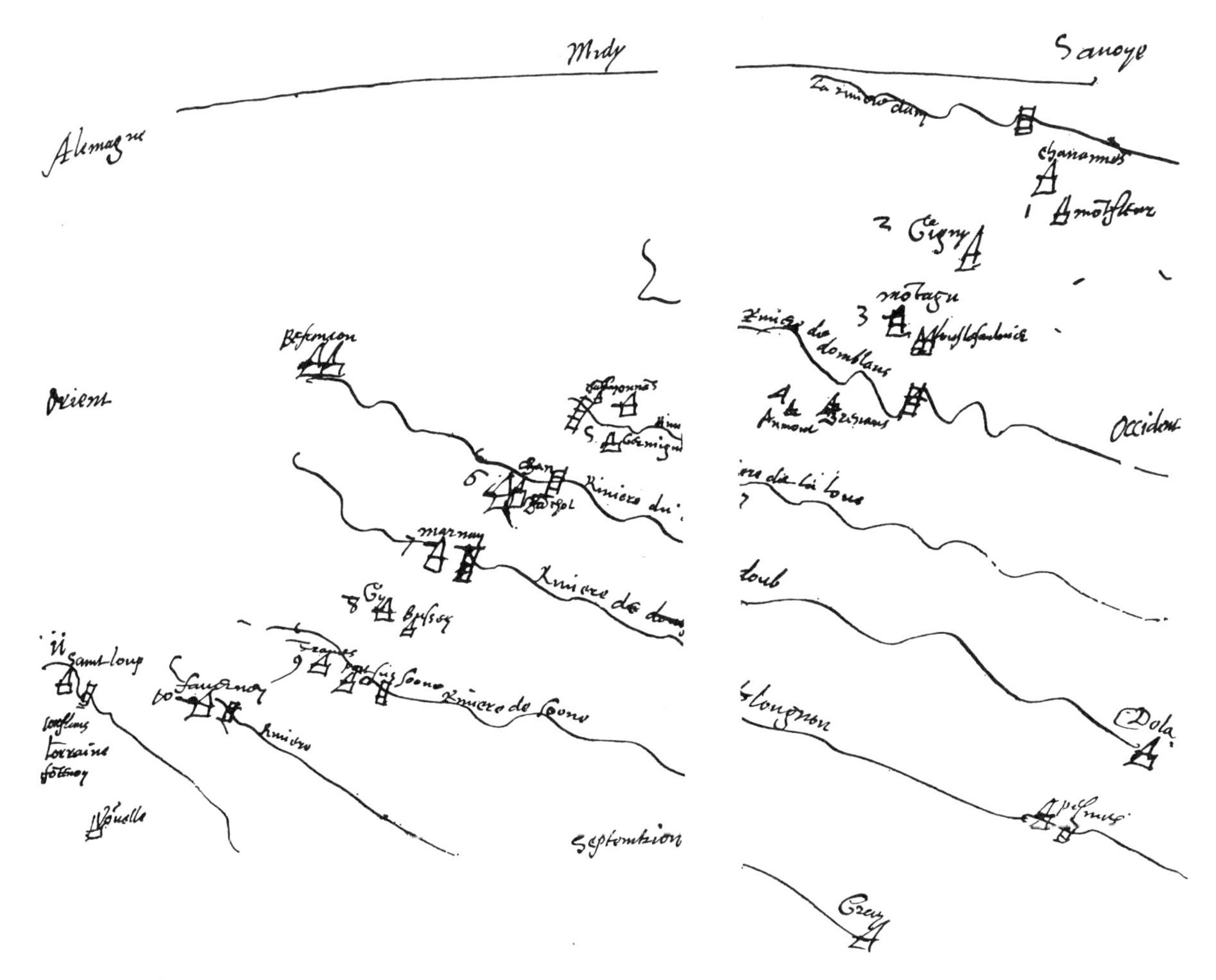

FIG. 5.8 Further map of Franche-Comté prepared for Alba in 1573. Archives Départementales du Doubs, France

maps in order to improve his understanding of the military situation. This was particularly important in the autumn campaign, against the prince of Orange in person, for the duke decided to adopt a Fabian strategy designed to avoid a battle, which might go against him, and instead sought to harass the invasion forces constantly and wear them down through attrition and hunger. In the words of one of his field commanders, Don Sancho de Londoño, "The duke has labored specifically to avoid fighting a battle, despite pressure from those who forget that victory is a gift of Fortune, which can favor the Bad as well as the Good. If Orange were a powerful monarch who could maintain a mighty army for longer, I would be in favor of fighting a battle; but since it is certain that shortage of money will cause his forces to crumble, and that he will not then be able to regroup, I am against it."

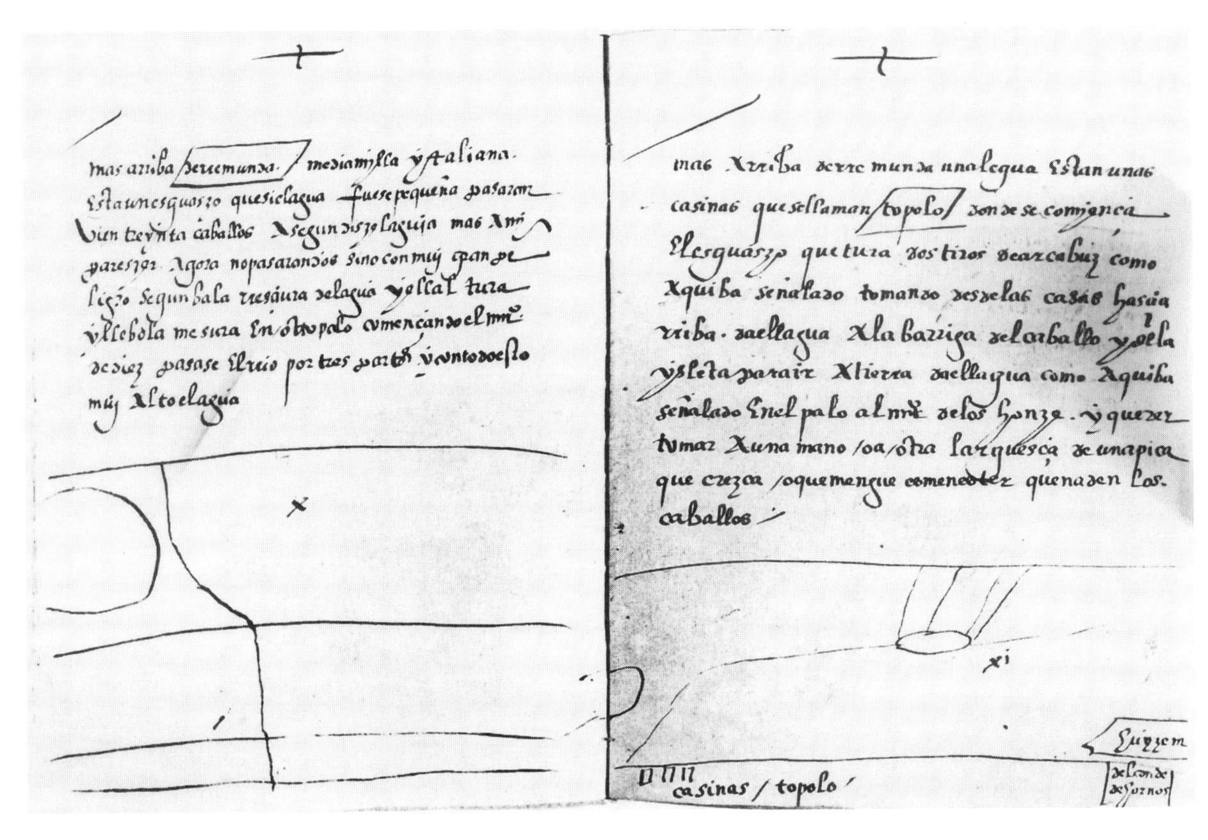

Fig. 5.9 Map of the Maas Valley, prepared for Alba, 1568. Alba Archives, Madrid

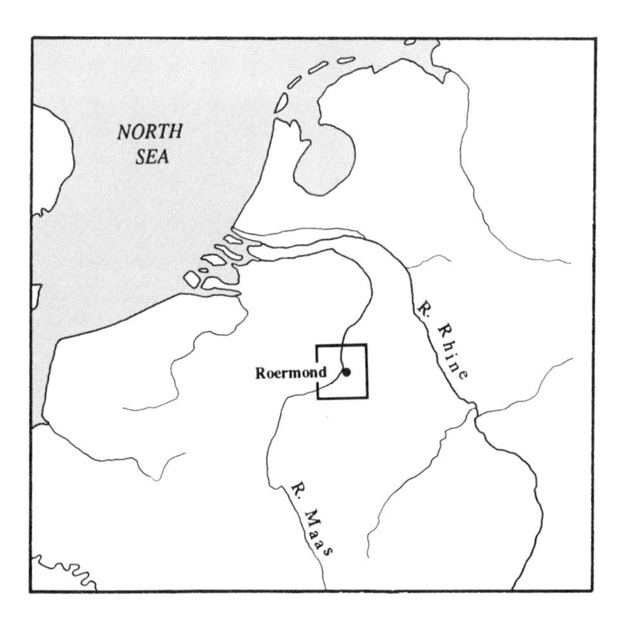

In the event, the duke's men skirmished for twenty-nine days continuously in order to keep Orange away from the heart of the Netherlands.[68] But such a strategy required a very detailed knowledge of local topography, so a series of maps and descriptions of the Maas valley, now in the Alba Archive in Madrid, were prepared for the duke by his scoutmasters Juan Despuche and Don Alonso de Vargas. The dossier is called "A Description of the River and Valley," and, like the later Besançon maps, everything a local commander would need to know is shown: maps and text describe how easy it was to cross the Maas at any given point

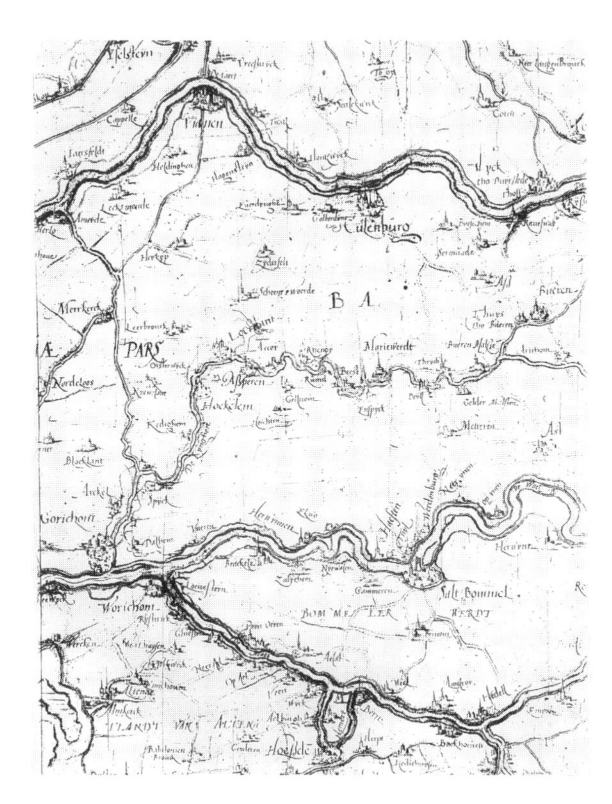

Fig. 5.10 Detail from Christopher Sgrooten's map of the Low Countries, 1573. Bibliothèque Royale, Brussels

(in case an opportunity suddenly arose to take the enemy in the flank by surprise) and which local features should be noted (figure 5.9).[69]

Once the invaders had been defeated, Alba set about reorganizing the government of the Netherlands in order to prevent such troubles from recurring.[70] One aspect of this process was the preparation of a series of maps and plans covering every province and all major towns. The work of van Deventer was continued and intensified, reaching completion by 1572, while in 1569 a special topographical survey of military installations in Holland and Zealand was carried out by the experienced engineers Gabriel Serbelloni, Chiappino Vitelli, and Bartolomeo Campi.[71] At the same time, the northwest coast of Flanders was surveyed by the Bruges painter Pieter Pourbus, who in 1571 completed a map of the Brugse Vrij of extraordinary accuracy and measuring a massive 335 × 620 cm.[72] Finally, Alba commissioned the noted cartographer Christopher 'Sgrooten to prepare a complete set of strategic maps of the provinces. Between 1568 and 1573 a marvelous collection of thirty-eight large maps of the Low Countries and adjacent areas of Germany was compiled, showing roads, river crossings, and sea routes (see figure 5.10). Spain itself would never have anything like this until the nineteenth century.[73]

Finally, in 1572, when the prince of Orange and his supporters launched another series of invasions of the Netherlands, a collection of twenty-one sketches of the campaign in Friesland was prepared by someone in the entourage of the royalist governor Gaspar de Robles, baron of Billy. Eleven of the pen-and-ink drawings are conventional "views" of the main strong points in the region; the other ten are "action pictures," showing the leading events in the campaign (figures 5.11 and 5.12).[74] Why they were made we do not know; but they were probably drawn at the time, for a pencil draft is clearly visible under the ink. If so, they are among the first known examples of pictorial war-reporting.[75] We are ignorant of the artist's identity, although he was clearly an Italian (since all the commentary is in Italian); presumably he was a draftsman or engineer attached to Robles's staff. Whether as a commemorative record of the campaign or as a visual reminder of the military geography of the province, however, these cartographic views would have served a useful purpose.

In retrospect, Alba's regime seems to have

FIG. 5.11 Drawing made for Gaspar de Robles in 1572. Humanities Research Center, Austin, Texas

been the high point of Spanish military cartography in the Netherlands. It is perhaps significant that neither of the duke's immediate successors as governors of the Netherlands possessed much in the way of cartographic aids before they reached Brussels. It is true that the duke of Medina Celi, appointed in 1571, owned provincial maps of Holland, Brabant, Flanders, and Zealand, one of France, an Ortelius "World Map," and a "map on which to make compass measurements" when he died four years later; but whether he acquired them before or after he went to the Netherlands is unknown. Certainly Don Luis de Requeséns, appointed in 1573, possessed only "a pair of compasses to make measurements on maps."[76] It was no way to win a war; and yet the situation did not improve with the passage of time. For the next thirty years, continuous hostilities in the Netherlands seem to have prevented the preparation of new military maps to complement and replace those of Pourbus and 'Sgrooten. It

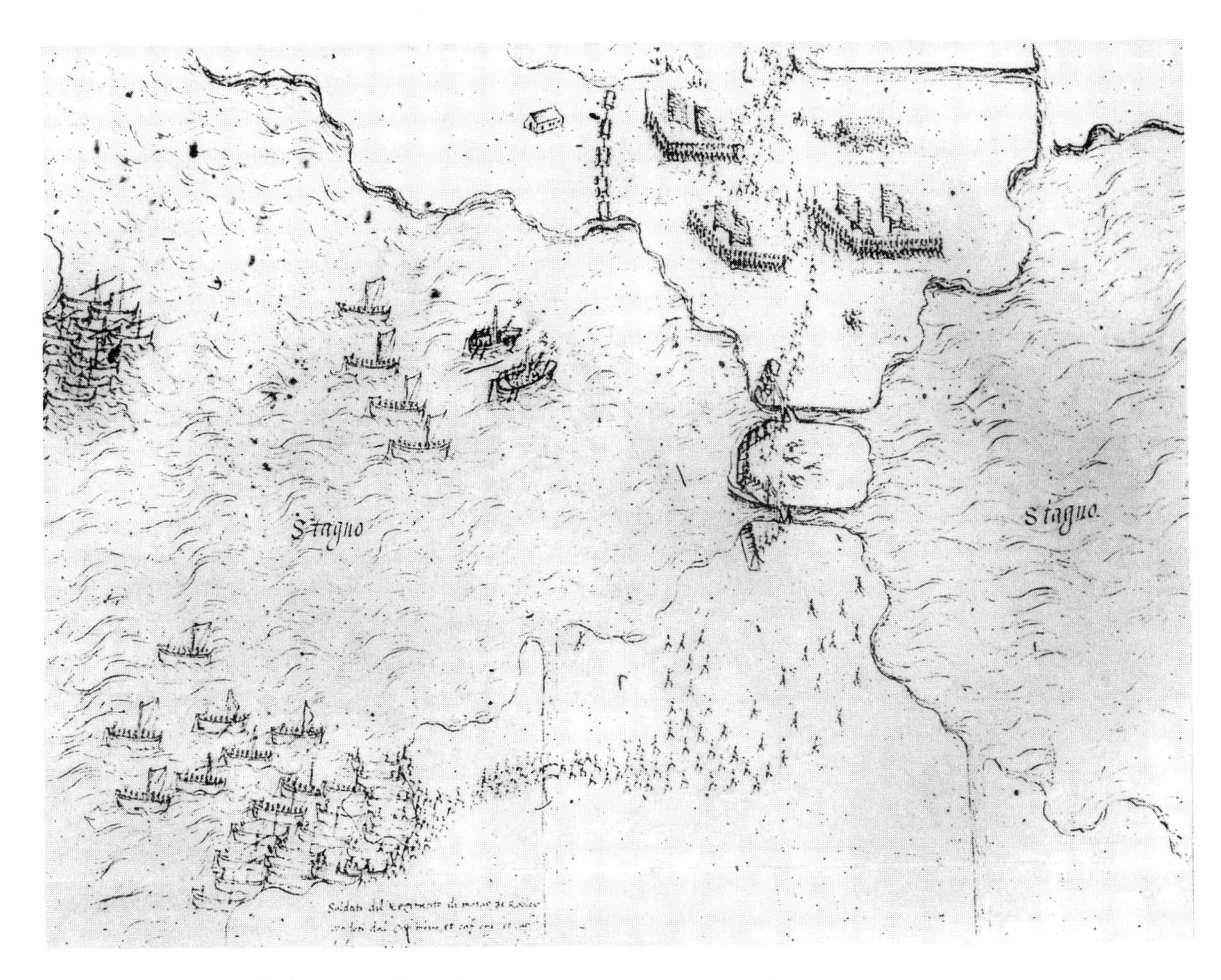

Fig. 5.12 Drawing made for Gaspar de Robles in 1572. Humanities Research Center, Austin, Texas

was much the same in Spain. No native school of terrestrial cartography seems to have been fostered or encouraged within the peninsula. After the deaths of Esquivel and Santa Cruz, mapmaking in Spain apparently became totally dependent upon the imported skills of other nations: at first the Portuguese, then increasingly the Italians, and finally the Dutch.[77]

Perhaps, indeed, the abundance of foreign talent helped to stifle the native tradition. Given the easy availability of foreign maps, the government may not have seen the point of investing in local alternatives, preferring instead to commission maps of those areas closed (at least in theory) to foreigners—the ocean sea and the overseas empire. Or did the Spanish Habsburgs, with the shortest land frontier of almost any state in Europe between 1580 and 1640, perhaps give up domestic mapmaking because they felt more secure than rulers of territories with long and vulnerable borders, such as France and the Dutch Republic? Either way, by 1642 (as Jean-Charles della Faille's letter indicated) there were no up-to-date maps available to ministers, and it is tempting to see in this failure a reflection of the "decline of Spain." Indeed, given the role of military factors in that process, it might be argued that the lack of maps with which to organize successful campaigns was not so much a reflection as a major cause of that decline. A government that attempted to formulate its strategies from the pages of a seventy-year-old general atlas was no longer a convincing imperial power.

Notes

1. Letter of 19 October 1642, from Cuidad Rodrigo near the Portuguese frontier, in O. van der Vijver, "Lettres de J.-Ch. della Faille, S.J., Cosmographe du Roi à Madrid, à M.-F. van Langren, Cosmographe du Roi à Bruxelles, 1634–45," *Archivum Historicum Societatis Iesu* 46 (1977): 73–183, at 172. The situation was, in fact, soon remedied, for a map of the Hispano-Portuguese frontier was in preparation as della Faille wrote: see M. C. Álvarez Terán, *Archivo General de Simancas, Catálogo XXIX: Mapas, planos, y dibujos* (Valladolid, 1980), 829. Such unpreparedness was not confined to Spain. The position in England at that moment, immediately after the unexpected outbreak of war in 1642, was not dissimilar: Christopher Saxton's wall map of 1583 was reprinted in 1644 as it stood, rechristened "The Quartermaster's Map," and used in action: see Sarah Tyacke, "Useful Maps: Themes in European Cartography," *The Map Collector* 28 (September 1984): 34–39, at 34.

2. Details from F. J. Sánchez Cantón, *Inventarios reales: Bienes muebles que pertenecieron a Felipe II*, Archivo Documental Español, 2 vols. (Madrid, 1956–59), 10–11, nos. 3206–9 (toothbrushes, etc.); 3385–86 ("xortijas de hueso que dicen ser buenos para las almorranas"); 1356–58 (Ortelius); and 4137–50 (maps and globes). Some further details may be found in the originals: Archivo del Palacio Real, Madrid, Sección Administrativa, 235–37: Inventarios 1–3. However, the inventories were compiled for the public auction of his goods specified in Philip II's will; "classified" items (like manuscript maps) would not have been included.

3. For examples of the king's acquisition of maps (both "printed and painted on canvas") from the Netherlands in the 1560s, see Richard L. Kagan, ed., *Spanish Cities of the Golden Age: The Views of Anton van den Wyngaerde* (Berkeley, 1989), 48f. Note also the cartographic forays to Portugal recorded by G. de Andrés, "Juan Bautista Gesio, cosmógrafo de Felipe II y portador de documentos geográficos desde Lisboa para la Biblioteca de El Escorial en 1573," *Boletín de la Real Sociedad Geográfica* 103 (1967): 365–74. For maps in the royal collection, see C. Fernández Duro, "Noticia breve de las cartas y planos existentes en la biblioteca particular de S.M. el Rey," *Boletín de la Sociedad Geográfica de Madrid* 26 (1889): 361–96, and 27 (1889): 102–65 (reprinted in *Acta Cartographica* 5 [1969]: 100–199).

4. Even the splendid atlas made by Leonardo Turriano of the Canaries, filled with views, plans, and maps compiled in the 1590s, has been excluded. These give a good idea of the highly sophisticated cartography of Spain's overseas possessions during the Habsburg period, but unfortunately there is as yet no satisfactory edition. See L. Torriano (sic), *Die Kanärischen Inseln*, ed. J. Wolfel (Leipzig, 1940) with the Italian original and a German translation, but by no means all of the maps and pictures.

5. For example, the great map collection of the count-duke of Olivares no doubt perished in the fires that consumed his archive in 1794 and 1795: see J. H. Elliott, *The Revolt of the Catalans: A Study in the Decline of Spain, 1598–1640* (Cambridge, 1963), 579. See also n. 77 below.

6. A Cortesão and A. Teixeira de Mota, *Portugaliae Monumenta Cartographica*, 6 vols. (Lisbon, 1960), 1:xlv; hereafter PMC. The five enormous and magnificent volumes of text and plates cover all the surviving works prepared by Portuguese cartographers between the late fifteenth and the eighteenth centuries; the sixth volume contains full indices. It took the two authors between five and six years to compile their masterpiece (they regretted that they had not had eight or ten); it was worth every minute.

7. A. Teixeira de Mota, "Some Notes on the Organization of the Hydrographical Services in Portugal before the Beginnings of the Nineteenth Century," *Imago Mundi* 28 (1976): 51–60, at 53.

8. Jean Nicot, *Thrésor de la Langue françoyse* (Paris, 1606), quoted in PMC 2:85 (my translation). The cartographer may have been André Homem.

9. Royal apostil to a *consulta* of 5 July 1566, in A. Heredía Herrera, *Catálogo de las consultas del consejo de Indias (1529–99)*, 2 vols. (Madrid, 1972), 1:121–22. The

unfortunate Legazpi had been dispatched entirely without maps, being instructed to get hold of Portuguese charts whenever he could, "even by buying them," in order to find out where he was going (see C. Quirino, *Philippine Cartography,* ed. R. A. Skelton, 2nd ed. [Amsterdam, 1964], x).

10. See, for example, the otherwise admirable survey of G. Marcel, "Les Origines de la carte d'Espagne," *Revue Hispanique* 6 (1899): 163–93.

11. See the classic article of R. Almagià, "The First 'Modern' Map of Spain," *Imago Mundi* 5 (1948): 27–31. The wide circulation of identical, fully illustrated manuscript copies of Ptolemy's *Geography* lends support to the view of T. K. Rabb that the initial impact of printing as an agent of change has been somewhat exaggerated. See T. K. Rabb and E. L. Eisenstein, "Debate: The Advent of Printing and the Problem of the Renaissance," *Past and Present* 51 (1971): 135–44.

12. The Gastaldi and Geminus maps are reproduced and discussed in D. Buisseret, "Spanish Terrestrial Cartography in Europe under the Habsburgs," in *The history of cartography,* ed. D. Woodward, vol. 3 (forthcoming).

13. Unfortunately, the fate of this item is not known; it does not appear in the inventories of Philip II and Santa Cruz (who died in 1572). See F. Picatoste y Rodríguez, *Apuntes para una biblioteca científica española* (Madrid, 1891), 291; and "Minuta del inventario de los papeles que quedaron por muerte de Alonso de Santa Cruz," in M. Jiménez de la Espada, *Relaciones geográficas de Indias: Peru* (Madrid, 1965), 1:284–87, listing only globes and maps of non-European areas.

14. Copies of all these maps may be found in the British Library. On the Palentino map, see R. Gallo, "Fra Vincenzo Palentino da Curzola e la sua carta della Spagna," *Atti della Accademia nazionale dei Lincei, Anno CCCXLIV (1947): Serie VIII Rendiconti: Classe di scienze morali, storiche, e filologiche,* 2:259–67.

15. Jerónimo de Roda to Ortelius, Brussels, 30 September 1571, communicating Espinosa's command, in *Ecclesiae Londino-Batavae Archivum,* ed. J. H. Hessels (Cambridge, 1887), 1:81f.

16. The correspondence of Ortelius, printed in the enormous (1xxv + 966 pages) volume edited by Hessels, is a fascinating source on the various revisions of the *Theatrum* (and many other matters). For something more compact, see J. Denucé, *Oud-Nederlandsche kaartmakers in betrekking met Plantijn,* 2 vols. (rpt. Amsterdam, 1964), 2:45–47.

17. See full details in J. Keuning, "The History of an Atlas: Mercator-Hondius," *Imago Mundi* 4 (1947): 37–62.

18. See J. Villuga, *Repertorio de todos los caminos de España* (Medina del Campo, 1546); and Alonso de Meneses, *Repertorio de caminos* (Alcalá de Henares, 1576)—although the latter is in fact little more than a slavish copy of the former. Both are discussed in G. Menéndez Pidal, *Los caminos en la historia de España* (Madrid, 1951), 86–87. Some 1730 editions of early modern travel-journals covering Spain are listed by R. Foulché-Delbosc, "Bibliographie des voyages en Espagne et au Portugal," *Revue Hispanique* 3 (1896):1–349. Extracts from some are printed by J. Garciá Mercadal, *Viajes de extranjeros por España y Portugal* (Madrid, 1952).

19. See the penetrating remarks on cartographic methods at this time by M. A. de Lavis-Trafford, *L'Evolution de la cartographie de la région du Mont-Cenis et de ses abords aux 15e et 16e siècles* (Chambéry, 1949), 121–25.

20. F. Colón, *Descripción y cosmografia de España,* ed. A. Blázquez y Delgado, 3 vols. (Madrid, 1908–15). See also the critical remarks of T. Marín Martinez, *Obras y libros de Hernando Colón* (Madrid, 1970), 159–314. P. Miquélez, *Catálogo de los códices españoles de la Biblioteca del Escorial* (Madrid, 1917), 251, noted that Pedro de Medina used the "Itinerario" (as Colón's survey was sometimes known) in his *Grandezas de España* of 1548.

21. The best account available is in G. L. Strauss, *Sixteenth-Century Germany: Its Topography and Topographers* (Madison, Wis., 1959); see especially 138–39.

22. Pedro de Medina, *Primera y segunda parte de las grandezas y cosas notables de España* (Alcalá de Henares, 1548; 2d ed., ed. Diego Pérez de Mesa, Alcalá, 1595).

23. Royal *cédula* of 8 August 1570, quoted in Kagan, *Spanish Cities,* 57.

24. Royal order printed by B. van 't Hoff, *Jacob van Deventer: Keizerlijk-koninklijk geograaf* (The Hague, 1953), 36. This amplified an earlier command issued in April 1558.

25. Three volumes of plans were recovered from Cologne, where van Deventer died in 1575, and sent to Spain; but only two have survived. Biblioteca Nacional Madrid (hereafter BNM), MS Res 200, contains 105 unfinished plans from the provinces of Holland, Zealand, and Gelderland; and MS Res 207 contains 74 plans from Hainaut, Cambrai, Artois, and Flanders. A third volume (no doubt containing town plans from the remaining provinces) has been lost. See van 't Hoff, *Jacob van De-*

venter, 50ff.; and J. M. Depluvrez, "Les Représentations cartographiques de villes de la province de Hainaut dans l'oeuvre de Jacques de Deventer," *Annales du Cercle Royal d'Histoire et d'Archéologie d'Ath et de la Région et Musées Athois* 49 (1983): 419–26. These plans have never been published. It used to be thought that another collection of 152 views by van Deventer, in a rather less finished state, were the "minutes" of the version sent to Spain; but it seems more likely that these were part of an earlier and quite separate commission. Topographical evidence makes it clear that most if not all of the "minutes" were painted before 1558, and it has been plausibly suggested that they were in fact prepared as "cartoons" or patterns for a vast series of tapestries portraying the towns of the Netherlands commissioned by the governor-general (Emmanuel Philibert of Savoy) in the mid-1550s. According to this interpretation, it was when he saw the cartoons that Philip II decided on the new series of cityscapes for the Netherlands. See C. Ruelens, "Plans topographiques des villes des Pays-Bas au 16e siècle," *Bibliophile belge* (1867): 280–90 (reprinted in *Acta Cartographica* 7 [1970]: 368–78). The "minutes" (now housed in the Koninklijke Bibliotheek, Brussels) have all been published in facsimile: see C. Ruelens, *Atlas des villes de la Belgique au 16e siècle* (Brussels, 1884); and R. Fruin, *Nederlandsche Steden in den 16e eeuw: Plattegronden van Jacob van Deventer,* 2 vols. (The Hague, 1916–23). Some of van Deventer's material was printed in 1572 by Georg Braun and Franz Hogenburg in volumes 3 and 4 of their *Civitates Orbis Terrarum* (see J. Keuning, "Sixteenth-Century Cartography in the Netherlands [Mainly in the Northern Provinces]," *Imago Mundi* 9 [1952]: 35–63). So accurate were these ichnographic views, which showed every house in each settlement, that they have even been used to calculate population levels. See J. C. Ramaer, "De middelpunten van bewoning in Nederland voorheen en thans," *Tijdschrift der Aardrijkskundige Genootschap,* 2d ser., 38 (1921): 1–38, 174–214.

26. Perhaps the decision to print Wyngaerde's views was also abandoned because of the publication of an alternative (but inferior) series of "views" done of Spanish cities by Joris Hoefnagel, also in the 1560s, in the *Civitates Orbis Terrarum* of 1572.

27. Thirty of Wyngaerde's lost paintings were described in an inventory of the Madrid palace made before the fire; see Y. Bottineau, "L'Alcázar de Madrid et l'inventaire de 1686," *Bulletin Hispanique* 60 (1958): 450–83, at 469 (items 1297–1326). On Wyngaerde's surviving corpus, see the definitive edition of Kagan, *Spanish Cities.*

28. Paez's *Interrogatorio* is printed in Miquélez, *Catálogo,* 252–54.

29. See N. Salomon, *La Campagne de la Nouvelle Castille à la fin du 16e siècle d'après les Relaciones Topográficas* (Paris, 1964), with full bibliographical references. At more or less the same time, Ovando supervised a similar operation in the New World. See the richly documented studies of H. F. Cline, "The *Relaciones Geográficas* of the Spanish Indies, 1577–84," *Hispanic-American Historical Review* 44 (1964): 341–74; C. R. Edwards, "Mapping by Geographical Positions," *Imago Mundi* 23 (1969): 17–28; D. Goodman, *Power and Penury: Government, Technology, and Science in Philip II's Spain* (Cambridge, 1988), 65–72; and J. M. López Piñero, *Ciencias y técnica en la sociedad española de los siglos XVI y XVII* (Madrid, 1979), 217–19.

30. See BNM, MS 5589 f. 64: "Ynterrogatorio," which mentions "13 or 14 volumes." A connection between the maps and the *relaciones* is suggested by Escorial MS K.I.1 f. 13v, which is endorsed "Itinerarios [de c]omo se a podido hazer las relaciones." Gonzalo de Reparaz Ruiz, the only scholar to have studied the Escorial maps in any depth, suggested that "relaciones" must refer to something else, because folio 13 is a map of León and Galicia, and the surviving *relaciones topográficas* never covered that area. But if there were once almost twice as many volumes (as the above-mentioned manuscript states), the picture changes. See G. de Reparaz Ruiz, "La Cartographie terrestre dans la péninsule ibérique au 16e et 17e siècle, et l'oeuvre des cartographes portugais en Espagne," *Revue Géographique des Pyrénées et du Sud-Ouest* 9 (1940): 167–202, at 188 n. 25.

31. However, the second edition of Pedro de Medina's *Grandezas de España* (1595) certainly did make use of them; see the evidence in Miquélez, *Catálogo,* 257.

32. See F. de Guevara, *Comentarios de la pintura* (c. 1564; Madrid, 1788), 219–21. According to Esquivel's friend Ambrosio de Morales, "he was commissioned by the king to travel throughout his kingdoms, seeking out with his own eyes all the villages, rivers and mountains—both large and small—so that a description of Spain could be made as certain and as complete, as detailed and as perfect, as His Majesty desired and Master Esquivel could execute." (A. de Morales, *Las antegüedades de las cuidades de España* [Alcalá de Henares, 1575], f. 4v.) See also Kagan, *Spanish Cities,* 44f.

33. Morales, *Antegüedades,* f. 4v-5. There is a useful biography of Esquivel—mostly drawn from Guevara and Morales—in Picatoste y Rodríguez, *Apuntes,* 86–89. See also the interesting remarks on Esquivel and his methods in F. Vázquez Maure, "Cartografía de la península: Siglos 16 a 18," in *Historia de la cartografía española* (Madrid, 1982), 59–74, at 61.

34. Morales (*Antegüedades,* f. 5) also wrote that Esquivel used special surveying equipment designed and built by himself, the two main pieces (made of wood) being by themselves heavy enough to constitute an entire muleload. But although the surveyor may indeed have made his own instruments, it seems more likely that they were copies of others used elsewhere for triangulation from fixed points. After all, the method of surveying by triangles was fully discussed in the pioneering work of Johann Regiomontanus, *De triangulis,* printed in 1533, and in Gemma Frisius, *Libellus,* and Apian, *Cosmographia,* both of which appeared in Spanish translation in 1548. All three were thus available before Esquivel began his labors.

35. Quoted in Picatoste, *Apuntes,* 88.

36. Biblioteca del Real Monasterio de San Lorenzo de El Escorial, MS K.I.1: each map measures approximately 30 × 45 cm. The relationship between Esquivel's survey and the Escorial atlas is inferred from the following data: (1) The calligraphy and cartography of the Escorial atlas clearly date from the reign of Philip II. (2) The maps in the atlas, except for one of the Portugal, appear to have no precursors and must therefore be the result of some special ground survey carried out by triangulation. (3) Such a survey would have taken several years and could have been undertaken only with a license from the Crown. (4) Only Esquivel is known to have undertaken such a survey during the reign of Philip II.

37. See the important articles of G. de Reparaz-Ruiz, "Cartographie terrestre" and "Le Plus Ancien Levé d'une nation européene: Une carte topographique du Portugal au 16e siècle," in *Mélanges d'études portugaises offerts à M. Georges le Gentil,* ed. L. de Matos and R. Ricard (Lisbon, 1949), 271–313; and "The Topographical Maps of Portugal and Spain in the Sixteenth Century," *Imago Mundi* 7 (1950): 75–82. The 1:430,000 calculation comes from Vázquez Maure, "Política cartográfica," 62.

38. Kungliga Bibliotheket, Stockholm, MS M.163 (formerly MS. Sparwenfeldt 17). According to the entry on its flyleaf, the volume was acquired in Madrid in the 1690s by Johan Gabriel Sparwenfeldt, along with twenty-four maps by Lavanha. No trace of these maps has been found, so conceivably they were another (perhaps finished?) copy of the twenty-one maps now in the Escorial, plus three more now lost. Certainly the Stockholm codex refers to the Escorial maps, because (1) much of the writing in both sources is the same; (2) both deal with the same places; and (3) some such set of coordinates would have been essential for the compilation of the Escorial atlas. To the best of my knowledge, the Stockholm codex has never been studied in detail. It is worth noting that Lavanha kept very similar records for his survey of Aragon in 1610–11 (see the source cited in no. 45), which suggests that Esquivel may have learned his technique from the same school of Portuguese cartography as Lavanha.

39. The lunar eclipse of 17 October 1584 was observed with the same meticulous coordination: sheets preprinted with disks of the same radius were sent out in advance by the Academia de Mathematica in Madrid to observers in Antwerp, Venice, Toledo, Seville, and Mexico City. (See U. Lamb, "The Spanish Cosmographic Juntas of the Sixteenth Century," *Terrae Incognitae* 6 [1974]: 51–62.). Goodman, *Power and Penury,* 67, records that observations were also coordinated with colleagues in China.

40. There were some exceptions, however; thus Juan Bautista Gesio secreted several Portuguese maps back to Spain in 1573, but they seem to have concerned only the demarcation line in the East Indies. See Andrés, "Juan Bautista Gesio."

41. See the articles of Reparaz Ruiz, particularly "Une Carte topographique," 310–14, where it is suggested that the Portuguese map was principally the work of Pedro Nunes, a noted cartographer, assisted by Seco and others.

42. A. Ferreira, C. de Morais, J. de Silveira, and A. Girao, "O mais antiga mapa de Portugal (1561)," *Boletim do Centro de Estudos Geograficos* (Coimbra) 2, nos. 12–13 (1956) and 14–15 (1957). See also the cautionary comments in PMC 2:79–86. The parallels between the Seco map and the Escorial "Portugal" maps were first noted by Reparaz Ruiz, "Une Carte topographique," 307f.

43. It is possible that Esquivel's "carta o pintura" with longitude and latitude calibration, described by Morales in 1575, in fact became the "key map" of the Escorial atlas; but if so, hitting upon exactly the same scale as the Portuguese master map would have been a remarkable coincidence.

44. It is of course possible that the Escorial atlas is a rough working copy of which the finished version was

purchased by Sparwenfeldt in the 1690s; see n. 38 above.

45. The copy, now in the Universiteitsbibliothek, Leiden, was printed by F. Sancho y Gil, *Itinerario del Reino de Aragón pr Juan Bautista Labaña*, Biblioteca de Escritores Aragoneses, Sección Histórica-Doctrinal 7 (Zaragoza, 1895). See also n. 38 above.

46. For a discussion (and reproduction) of the map, see PMC 4:69f. and pl. 423; for details on Lavanha, see A. Cortesão, *Cartografia y cartografos portugueses dos seculos XV e XVI*, 2 vols. (Lisbon, 1935), 2:294–361. The largest reproduction of the "Map of Aragon" seems to be in J. Blaeu, *Parte del Atlas Mayor, o Geographia Blaviana, que continene las cartas y descripciones de España* (Amsterdam, 1672), 306–28.

47. Taken from the royal order for San Sebastián to assist Teixeira, quoted in PMC 4:153f. (my translation).

48. The cartographic work of Teixeira is described and illustrated in PMC 4:141–43 and pl. 509–10 (Joao's coastal atlas); and 153–60 and pl. 518 (Pedro's Madrid map). A detailed map of Catalonia, with sixteen plans of fortified places and five views of cities, was compiled by Ambrosio Borsano and presented to King Charles II of Spain in 1687. It was the fruit of twelve years' surveying and, according to the author of a recent catalog, "deserves a monographic study": E. Santiago Paez, *La historia en los mapas manuscritos de la Biblioteca Nacional* (Madrid, 1984), 66–68. The main map, like Teixeira's vast map of Madrid, is reproduced in Buisseret, "Spanish Terrestrial Cartography."

49. Quoted in Francisco de Holanda, *Four Dialogues on Painting*, ed. A. F. G. Bell (London, 1928), 52f.

50. Martin and Guillaume du Bellay, *Mémoires*, ed. V. L. Bourrilly and F. Vindry (Paris, 1912), 3:118–19. My thanks to Richard Boulind for this splendid reference.

51. Santiago Paez, *La historia en los mapas*, 246, attributes this anonymous view entitled "Descrizion de la Francia por donde entro el Perador" to the year 1539, but, since the emperor did not "enter" Champagne until 1544, the latter seems the more likely date for the map. For the precocity of this panoramic technique of landscape mapping, see S. Alpers, "The Mapping Impulse in Dutch Art," in *Art and Cartography: Six Historical Essays*, ed. D. Woodward (Chicago, 1987), 51–96, at 72f.

52. See reproduction and description in PMC 3:51 and pls. 357–59.

53. There may have been earlier maps prepared in 1583, however, when Philip II requested and received position papers on how best to conquer England. For details see C. Martin and G. Parker, *The Spanish Armada* (London, 1988), 109, 282 n. 6. Philip II also possessed a copy of the 1579 Latin edition of Saxton's *Descriptio Angliae*: see Fernández Duro, "Noticia breve," *Sociedad geográfica de Madrid Boletín* 27 (1889): 164.

54. See BNM MS 5785 ff. 76–79, Escalante's "Discurso de Inglaterra," to which the map at f. 168 is doubtless related. I am most grateful to José Luis Casado Soto for bringing this reference to my attention, and for dating it to June 1586. See also ibid., ff. 90–92, Don Francisco de Bobadilla to Garcia de Loaysa, 22 July 1586, about Escalante's career; B. de Escalante, *Diálogos militares* (Madrid, 1583); and British Library Additional MS 28,456 ff. 123–27, Escalante to Don Cristobal de Moura (1591–92), stating that he had been in England (presumably in the 1550s, which would explain his defective memory of the Tower of London).

55. See details in Martin and Parker, *Spanish Armada*, 116ff.

56. Only the copy of the *Derrotero* sent to the king by Medina Sidonia on 2 April 1588 has survived. It is to be found at Archivo General de Simancas (hereafter AGS), Sección de Estado, 431 f. 17, and was printed by E. Herrera Oria, *La Armada invencible* Archivo Histórico Español 2 (Valladolid, 1929), 156–80.

57. Ibid., 180: undated note of Philip II and Don Juan de Idiáquez.

58. Ibid., 155: undated note of Philip II and Don Juan de Idiáquez.

59. Ibid., 168f.: the *Derrotero*.

60. See the pertinent remarks of M. J. Rodríguez Salgado, "Preparándose para zarpar: Pilotos, marineros, y navegación en la Armada española de 1588," in *La Gran Armada: Simposio Hispano-Británico*, Cuadernos Monográficos del Instituto de Historia y Cultura Naval (Madrid, 1989), 21–31.

61. C. Fernández Duro, *La Armada invencible*, 2 vols. (Madrid, 1888), 1:499, Medina Sidonia to the king, 2 April 1588.

62. AGS, Contaduría del Sueldo, 2d ser., unfol., "Asiento con Ceprián Sánchex maestre de hazer cartas de marear." Sánchez was paid forty escudos on 4 April "por 80 quarterones de cartas de marear con sus braxeajes de las costas de España, Ynglatierra y Flandes, quo por horden del duque de Medina Sidonia estava aziendo para repartir en los navios de la dicha armada." In the event, eighty-five charts were delivered in 12 May, and Sánchez was paid a further 297 escudos. It should be noted that the document does not specify whether the charts should

be all the same or not.

63. The maps are discussed in PMC 3:81, where it is noted that "both are folded down the middle, which suggests that they belonged to an atlas"; however, as in the case of the Terceira charts the centerfold may indicate instead that they too were mounted on a hinged board (*cuarteron*) for use on ships, as Medina Sidonia had intended. On stylistic grounds, PMC convincingly attributes the unsigned charts to Luis Teixeira rather than to Ciprián Sánchez. However, there may have been a separate contract with Teixeira—perhaps for the rest of the charts needed for the 130 ships in the fleet—and these survivors come from his quota; alternatively, Teixeira may have supplied the five extra charts delivered by Sánchez. The link between the Spanish Armada and the two charts (both of them sold at Sotheby's in London in 1937), may be inferred from the following: (1) given the offices the makers held (both Sánchez and Teixeira were "royal cosmographers"), these manuscript charts would doubtless have been official, not private, commissions; (2) these are the only known Portuguese charts of the period which include "soundings"—something specified in Medina Sidonia's contract; (3) the only occasion on which the king's fleet sailed to the Narrow Seas, and would therefore have needed such special charts, was in 1588. It is possible that these items came from the Andalusian flagship *Nuestra Señora del Rosario*, which was captured intact by Sir Francis Drake in August 1588.

64. An excellent life of the duke is available: W. S. Maltby, *Alba: A Biography of Fernando Álvarez de Toledo, Third Duke of Alba, 1507–82* (Berkeley, Calif., 1983).

65. Quoted by F. de Dainville, *Les Jésuites et l'éducation de la société française*, vol. 1, *La Géographie des humanistes* (Montpellier, 1940), 345.

66. Lannoy's map is discussed in L. Febvre, *Philippe II et la Franche Comté* (Paris, 1911), 114, n. 1; and G. Parker, *The Army of Flanders and the Spanish Road, 1567–1659: The Logistics of Spanish Victory and Defeat in the Low Countries' Wars* (Cambridge, 1972), 83. It is clear that Lannoy had completed the map *before* he knew the Spaniards were coming.

67. The maps are discussed in Parker, *Army of Flanders*, 102–5. The originals are in Archives Départementales du Doubs, Besançon, C264, unfol., endorsed: "Pour le passage de la gendarmerie et pour les estappes."

68. Archivio di Stato, Parma, *Carteggio Farnesiane* 109 (*Paesi Bassi* 4), unfol., Don Sancho de Londoño to the duke of Parma, 21 November 1568. There are two outstanding eyewitness accounts of the campaign: Bernardino de Mendoza, *Commentarios de lo sucedido en las guerras de los Países Bajos* (Madrid, 1592), bk. 4; and Josse de Courtewille, "Relation de l'expédition du prince d'Orange," in *Correspondance de Guillaume le Taciturne*, ed. L. P. Gachard (Brussels, 1851), 3:319–37.

69. Archivo de la Casa de los Duques de Alba, Madrid, Caja 166/2, unfol., "Relación de Juan Despuche y Don Alonso de Vargas sobre el país y rió" (the title is in Alba's own hand).

70. For details see G. Parker, *The Dutch Revolt*, 3d ed. (Harmondsworth, 1985), 105–17.

71. See S. Groenveld and J. Vermaera, "Zeeland en Holland in 1569: Een rapport voor de hertog van Alva," *Nederlandse Historische Bronnen* 2 (1980): 103–74. A similar survey had been commissioned by Charles V in 1552 to provide maps of all the coasts on which enemies might land.

72. Details in A. de Smet, "A Note on the Cartographic Work of Pierre Pourbus, Painter of Bruges," *Imago Mundi* 4 (1947): 33–36. Pourbus took soundings and sightings from a boat as well as questioning seafarers in order to get the most accurate information. Of course, individual ministers in Brussels also had their own map collections: for example, Viglius van Aytta, president of the Privy Council, possessed 189 maps (see E. H. Waterbolk, "Viglius of Aytta, 16th-Century Map Collector," *Imago Mundi* 29 (1977): 45–48.

73. 'Sgrooten's collection is now in the Bibliothèque Royale, Brussels, MS 21,596. It is described, along with 'Sgrooten's other work (such as another atlas for Philip II, completed in 1592), by L. Bagrow, *A. Ortelii catalogus cartographorum*, Petermann's Mitteilungen, Ergänzungsheft 199 (Gotha, 1928), 1:58–69; and A. Bayot, "Les Deux Atlas manuscrits de Chrétien Sgrooten," *Revue des Bibliothèques et Archives de Belgique* 5 (1907): 183–304.

74. The views are now in the Humanities Research Center at Austin, Texas, and are briefly described in H. P. Kraus, *Catalogue 124: Monumenta Cartographica* (New York, 1970), item 36 (pp. 72–75). My attention was generously drawn to this little-known collection by R. H. Boulind.

75. One must not infer from this that every soldier carried a sketch pad in his knapsack. It has recently been demonstrated that the apparently eyewitness illustrations in the "History" of Walter Morgan, an English captain fighting in Holland 1572–74, were in fact copied from contemporary prints. (See S. Groenveld, "Het Engelse

kroniekje van Walter Morgan en een onbekende reeks historieprenten (1572–4)," *Bijdragen en Mededelingen voor de Geschiedenis der Nederlanden* 98 (1983): 19–74.

76. Details from A. Paz y Melía, *Series de los más importantes documentos del archivo y biblioteca del Exmo: Señor Duque de Medina Celi* (Madrid, 1915), 162–65 ("Inventario de los bienes que quedaron por muerte del duque Don Juan de la Cerda," 1575); and J. M. March, *Don Luis de Requeséns en el gobierno de Milán, 1571–3*, 2d ed. (Madrid, 1946), 27–28 (inventory of papers left at Requeséns's death in 1576).

77. This impression, however, may be the result solely of the destruction of the map collection of the count-duke of Olivares, chief minister of Philip IV. Clearly, in its day, it was both extensive and impressive, because (to give a single example) in 1627, when Spain was again planning the invasion of England and Ireland, the Genoese ambassador in Madrid stated that he had seen in Olivares's office "six very clear and distinct maps" of the proposed invasion area, with the aid of which Olivares explained his grand strategy (J. H. Elliott, *The Count-Duke of Olivares: The Statesman in an Age of Decline* [New Haven, Conn., 1987], 282). It would be interesting to know who had drawn these maps. Another example of the riches of Olivares's map room, this time concerning the Netherlands, may be found in J. H. Elliott, *Richelieu and Olivares* (Cambridge, 1984), 28.

SIX

Mapping under the Austrian Habsburgs

James Vann

As a student of government, I am very much concerned with questions of political perception. What, I ask myself, did the early modern rulers and their ministers think of the lands over which they ruled and of the administrative organizations within which they worked? I raise these questions in the belief that if we can identify the governors' attitudes, we can better understand their political strategy and its effect upon the growth of the modern state. Particularly in the Austrian Habsburg case, identification of the political *mentalité* creates an intelligible context for the twists and turns in the building of an empire in which foreign policy so often appeared out of tune with the rest of Europe. To understand is not necessarily to approve, but it does help us to explain patterns in the Austrian monarchy that, from the perspective of the western European chancelleries, often contained more than a whiff of the irrational.

What I suggest is that a relation existed between the kinds of maps commissioned by the Austrian rulers and those princes' notions of their territories and government. The concept of the unified, centrally administered state developed late in Austrian Habsburg consciousness. Down to the eighteenth century they conceptualized their territories in dynastic or personal terms. For them the state meant the *casa habsburga* rather than any unified, centrally administered territorial configuration, and they developed their government accordingly. Their approach to the mapping of their territories documents this attitude.

Genetics rather than conquest determined traditional Habsburg political values. Ferdinand I, founder of the Austrian branch of the family's sixteenth-century empire, assembled his possessions through gift and marriage. His German and Austrian lands came from his brother, who had been their grandfather's heir; his wife brought claims to the crowns of Hungary and Bohemia. What resulted was a disparate assortment of lordships with nothing in common but a sovereign. The various lands did not share tradition, language, or law. In fact, each sovereignty had its own distinct historic existence. That the archduke of Upper Austria, for example, also claimed the throne of Bohemia was a matter of indifference to his Austrian subjects. Bohemia was his affair; their concern was with their own principality.

Curiously, the Habsburgs themselves took no particular steps to counter this feeling. In part, of course, they tolerated separatism because they lacked the money, personnel, and technology to overcome it. But inadequate resources is not really the point. The Austrian Habsburgs were a good deal better off than

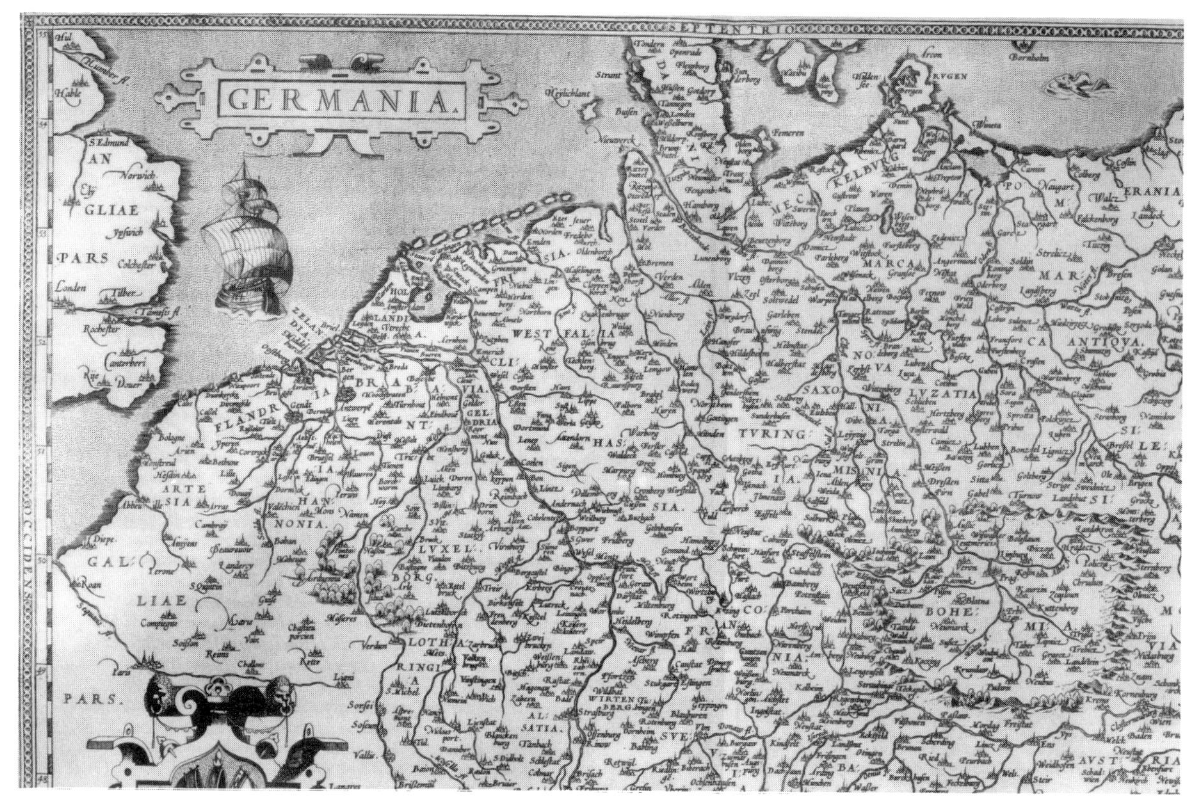

FIG. 6.1 Detail from "Germania" in Abraham Ortelius, *Theatrum Orbis Terrarum* (Antwerp, 1570). NL

many of their neighbors whose single-minded determination welded diverse lands into a coherent whole. Something quite different was operating in Vienna. There the stress lay not upon administrative unity but dynastic loyalty. And the obverse of that dynastic loyalty was the monarch's obligation to guarantee the historic political integrity of each of his territories.

Reinforcing an intensely personal philosophy of government was the Austrian dynasty's virtual monopoly of the imperial crown. As Holy Roman emperors, the Habsburgs were the first lords of Christendom, heirs to the Caesars and Charlemagne, guarantors of the international papal monarchy, and Europe's defenders against the Turks. True, these high-blown responsibilities were confined generally to ceremonial rhetoric. Yet even so, the imperial dream retained a surprising vividness for early modern Europeans, and for none more so than the Austrian Habsburgs. And in that political dream maps constitute statements of exclusion as well as inclusion. It is not too far-fetched, I think, to suggest that imperial pretensions militated against too precise a geographic description of Habsburg rule.

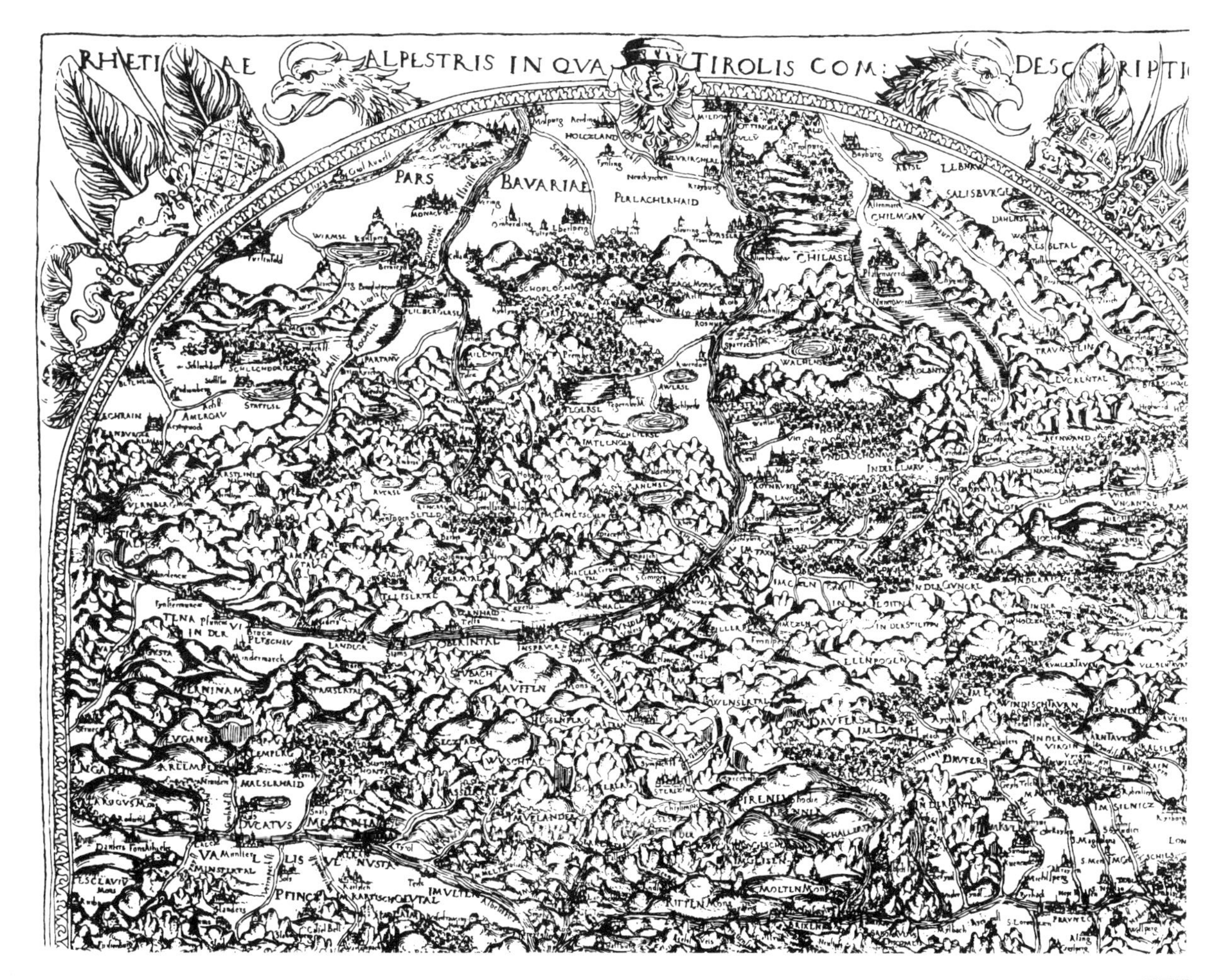

FIG. 6.2 Detail from Wolfgang Lazius, *Rhetiae Alpestris in qua Tirolis com: descriptio* (Vienna, 1561).

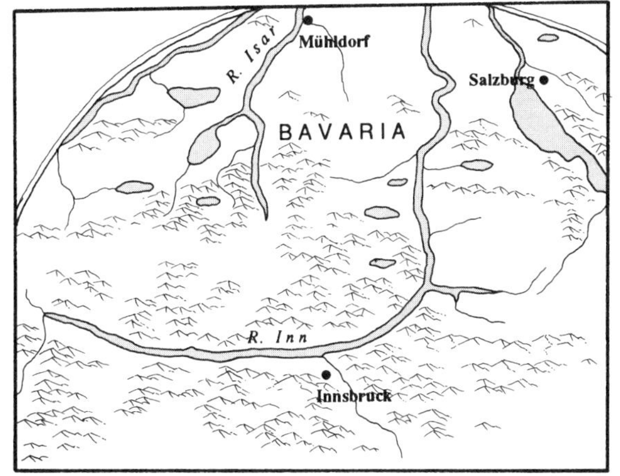

The Sixteenth and Seventeenth Centuries

Indeed, the few general maps produced in the sixteenth century with the Habsburgs' imprimatur focused, as does figure 6.1, upon Germany as an imperial concept. Maps of the family lands remained local studies, free from any suggestion of an alternative empire to that of the Reich. Thus we have Wolfgang Lazius's mid-sixteenth-century map of the County of

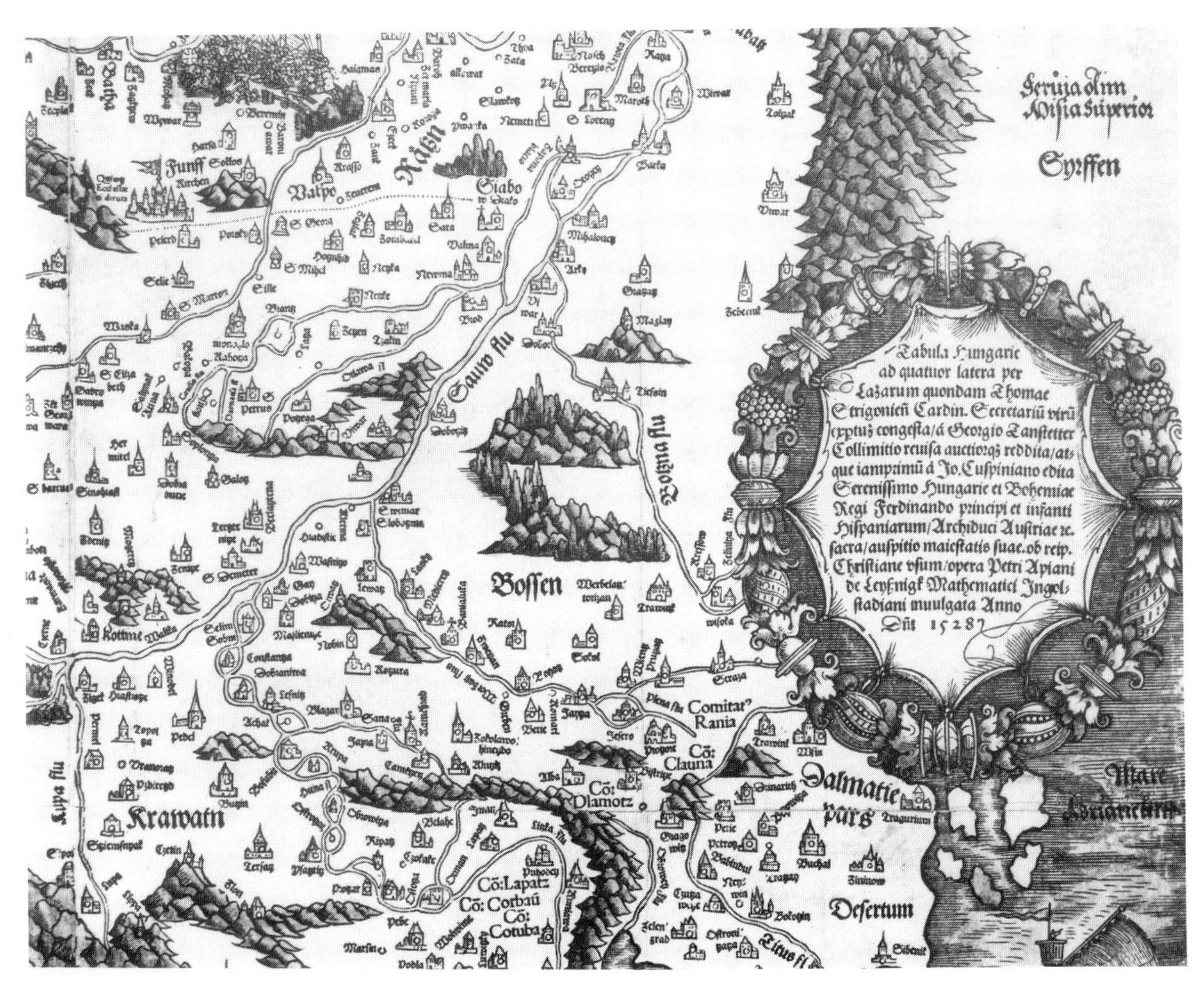

FIG. 6.3 Detail from the map of Hungary by Lazarus Secretarius (Vienna, 1528). Österreichischer Nationalbibliothek (ÖN), Vienna

Tyrol (figure 6.2). Note the heraldic renderings here: the arms of the estates and of the Habsburgs. But in the latter case it is the arms of the Habsburgs as counts of Tyrol. There is no incorporation of Tyrol as an element in a coat of arms embracing the totality of the family's lordships. A comprehensive arms had not yet been devised.

Given these highly personal yet universalist views of kingship, it is not surprising that the Habsburgs evidenced a keen appreciation for geography. Maximilian I was said to know the geography of his lands so well that "he could jot down an impromptu map of any region." His grandson Charles V and that emperor's chief Burgundian adviser both had extensive collections of virtually all the interesting cartographic documents of the time. But an in-

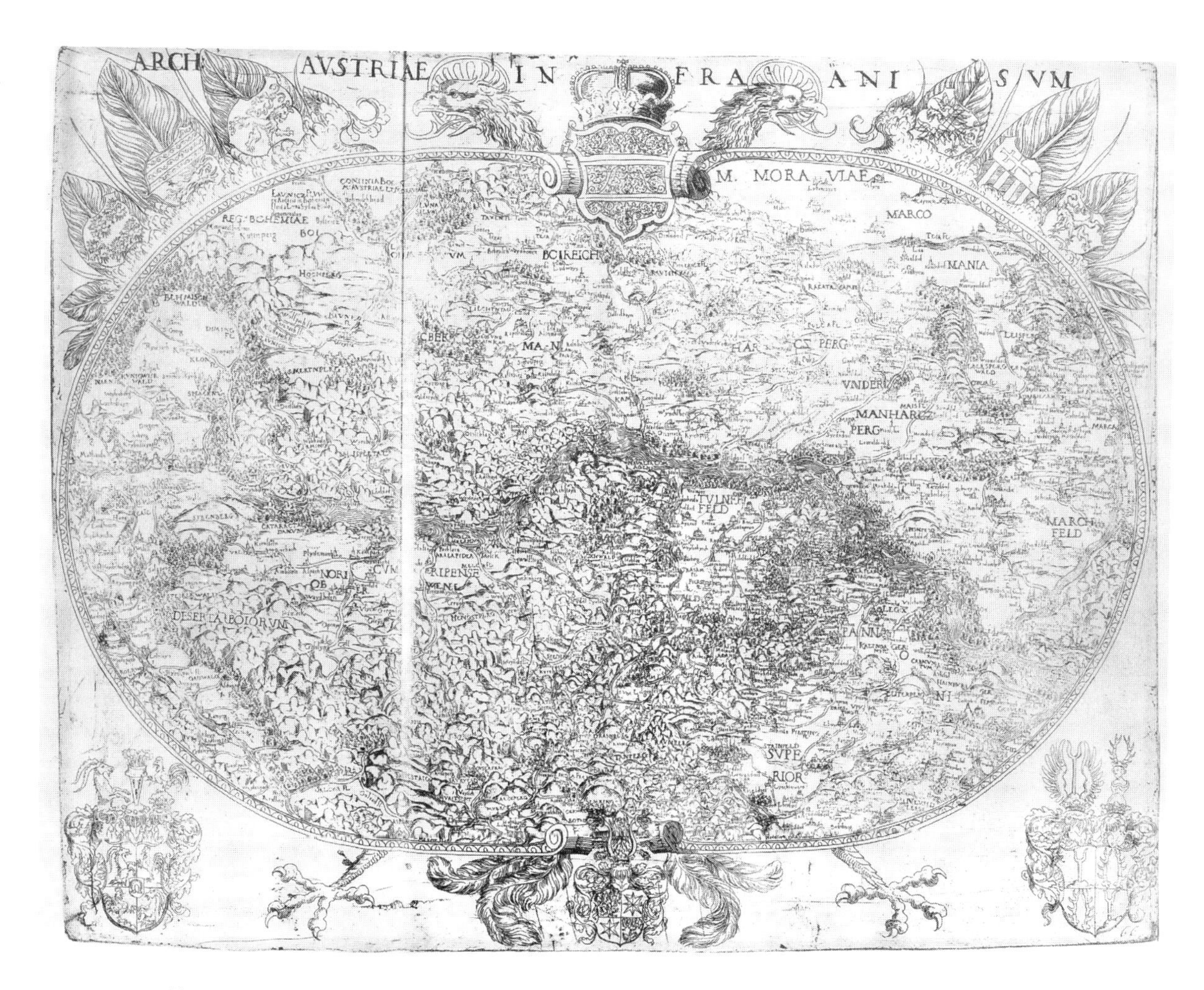

FIG. 6.4 Wolfgang Lazius, Lower Austria (Vienna, 1560). NL

terest in geography must be distinguished from a systematic use of or dependence upon maps as instruments of national statecraft.

In this latter regard the Habsburgs moved very slowly. The maps they commissioned for their lands stressed the particular territory and ignored the possibility of a larger unit. Thus Ferdinand commissioned a map of Hungary in 1628 that described that land in its historic

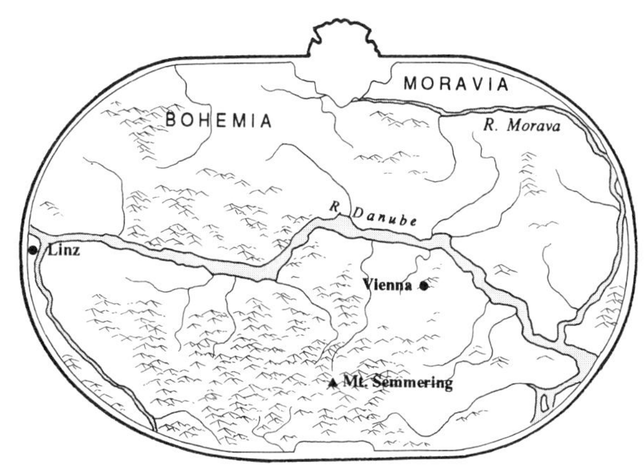

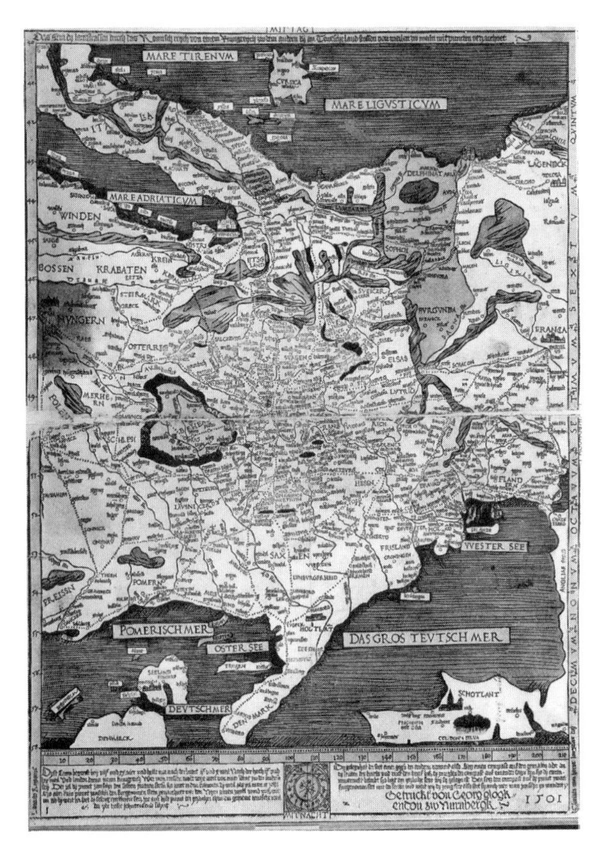

FIG. 6.5 Erhard Etzlaub, the "Romweg" map (Nürnberg, 1501). NL

uniqueness (figure 6.3). There was no effort, either in the map or the accompanying text, to conceptualize the kingdom as part of a larger political entity. Ferdinand's other commissions went for similarly specific enterprises, to maps of Vienna and of the individual archduchies. Figure 6.4, for example, was the result of a 1562 imperial commission to Wolfgang Lazius for a map of Lower Austria. Missing here, as in the other maps of this type, was the concept of the principality as an individual element in a centralized state.

I am not trying to suggest that there was no place in statecraft for the particular map, or that just because the Habsburgs did not commission national maps they saw no use for mapping. That is not my point at all. What I find interesting is that, just as their national politics were highly personalized and local, so too were their mapping commissions. Yet the Habsburgs and their ministers made good use of maps for local military strategy, for religious devotion, and for political propaganda, especially as regards the dynasty. That same Lazius who created the territorial maps also received imperial commissions for tactical maps depicting the great Habsburg victories in the Schmalkaldic War of the mid-sixteenth century.

The Crown also subsidized Georg Tanstetter to draw up a map celebrating Charles V's victory in 1522 against the Turks, and we know from contemporary accounts that Charles relied upon maps to secure such victories. The French memoirist Martin du Bellay recorded just such an occasion. He visited Charles in 1536 during the emperor's war against the king of France and wrote at that time of seeing Charles "studying the maps of the Alps and the lower regions of Provence so enthusiastically that the emperor had convinced himself that he already possessed the land in the same way that he owned the map."

Etzlaub's famous "Rom Weg," published at the beginning of the sixteenth century, makes the point from another angle (figure 6.5). This map went through countless editions and received sustained imperial patronage. It depicted the main travel routes (pilgrim roads) through Germany to Rome, routes mapped by means of points from town to town, mile by mile. Its thrust, of course, was Rome, but implicit in the map for anyone concerned with state-building was the notion of a highway system. The Habsburgs picked up immediately on this idea for the Holy Roman Empire but never translated the insight for their own territorial lands.

In fact, the only ruler to think at all in this direction was the early seventeenth-century emperor Rudolph II. His fascination, however, was with technology, not administration. Impressed by the new measuring instruments being developed at his court by distinguished scientists like Tycho Brahe and Johannes Kepler, he actually traveled from Nuremberg to Vienna taking exact mileage measurements, which he subsequently published as a road map. The link between intellectual curiosity and administrative strategy was simply never drawn—not, I would suggest, because of limited intelligence, but rather because other concerns took precedence.

Aside from travel and military strategy, the Habsburgs saw maps as tools for Humanist scholarship, a scholarship that could be harnessed to celebrate the glories of the dynasty. Maximilian I had sponsored a college of poetry and mathematics at Vienna which produced a steady stream of sixteenth-century Humanist cartographers in the service of the Crown. The group was a distinguished one, founded by Conrad Celtis, Germany's first poet laureate. Subsequent figures like Johann Stabius made substantial cartographic contributions, but here again the concern lay elsewhere than with political administration.

Apart from their theological interests, the German-speaking Humanists were concerned to prove that their ancestry was no less glorious than that of the haughty Italian scholars. To that end they appealed to the notion of geographic environment, a sentiment the Habsburgs quickly latched onto as a means of extolling the dynasty's mission. Listen to Johann Cuspinian, whose famous *Austria Regionis Descriptio* deplored the fact that Italian writers "have deprecated Austria, partly from malice and partly out of jealousy." Cuspinian retaliated by saying, "The fame of the Holy Roman Empire and the attractiveness of Austria demand better treatment." With the enthusiastic support of the Habsburgs, he insisted that "Austrian history, scenery, beauty, and glory must be broadcast." But "Austria" in this case meant only the two archduchies—not the collective possessions of the monarchy.

Although seventeenth-century commissions suggest a redistribution of political emphasis—and I shall return to this point in a moment—the monarchy itself continued to regard mapping in sixteenth-century terms. That is, the Habsburgs and their ministers commissioned maps to provide records of important battles, to furnish propaganda for the dynasty, and to define particular territorial units. Imperial publications surrounding the 1683 siege of Vienna by the Turks illustrate the first two of these points.

Recall the drama of that great event: a seemingly invincible Turkish army had swept across Hungary into eastern Austria and was pounding away at the very gates of Vienna. Western Europe held its breath as the imperial army struggled to hold the city and thus to secure the defenses of Latin Christianity. The heroism of the Viennese, their dramatic, last-minute rescue by the Polish forces of John Sobieski, the crusading fervor of the French and German troops hastening toward Vienna to relieve the city—all combined to make the siege perhaps the most sensational single event of the seventeenth century. Once victory had been secured, the Habsburgs commissioned a rash of publicity to celebrate the triumph, including several cartographic contributions. Figure 6.6 is a detail from Leander Anguissola's

Fig. 6.6 Detail from the plan of Vienna by Leander Anguissola (Vienna, 1706). ÖN

scholarly reconstruction of the plan of the siege, with ballistic ranges figured from the tower of St. Stephen's Cathedral.

Our only hint of political innovation in the seventeenth-century maps comes indirectly—and is all the more intriguing for that. Government commissions reveal a shift in patronage patterns. Throughout the sixteenth century such mapping commissions as occurred came from the Crown. Now initiative passed from the monarchy to the territorial estates. The change is an interesting one. It suggests that heightened concern by the estates with boundaries and with the necessity of rendering political rights as a spatial abstraction might well have been triggered by a perception in the localities of outside forces' eroding territorial exemptions, breaking down the political integrity of the principality.

It may well be that, long before the Viennese government articulated for itself a coherent program of administrative centralization, the Crown had begun pushing in piecemeal fashion against territorial particularism. Or phrased another way, it might be useful to conceptualize seventeenth-century Habsburg domestic policy in terms of a gap between rhetoric and reality. For while the Crown's cartographic commissions suggest that the dynasty was continuing to promote conventional representations of its relations to the Crown lands, the dramatic upsurge in local commissions implies a quite different perception in the lands of the central government's policies. The irony, of course, is that the end product, the map, remained the same, a local study. It is only when analysis shifts from the map to the patron that the other insight suggests itself.

More research on the territorial estates must be done before we can speak authoritatively about their perceptions. I simply advance the point now as a suggestion. What I can say with certainty is that throughout the seven-

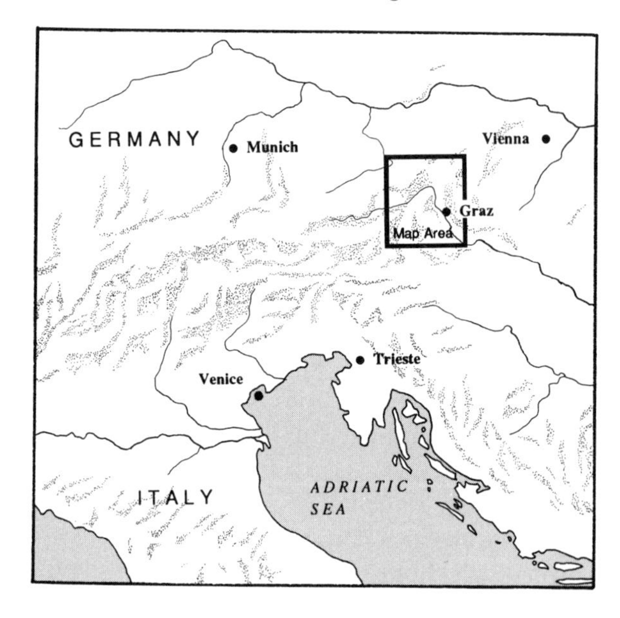

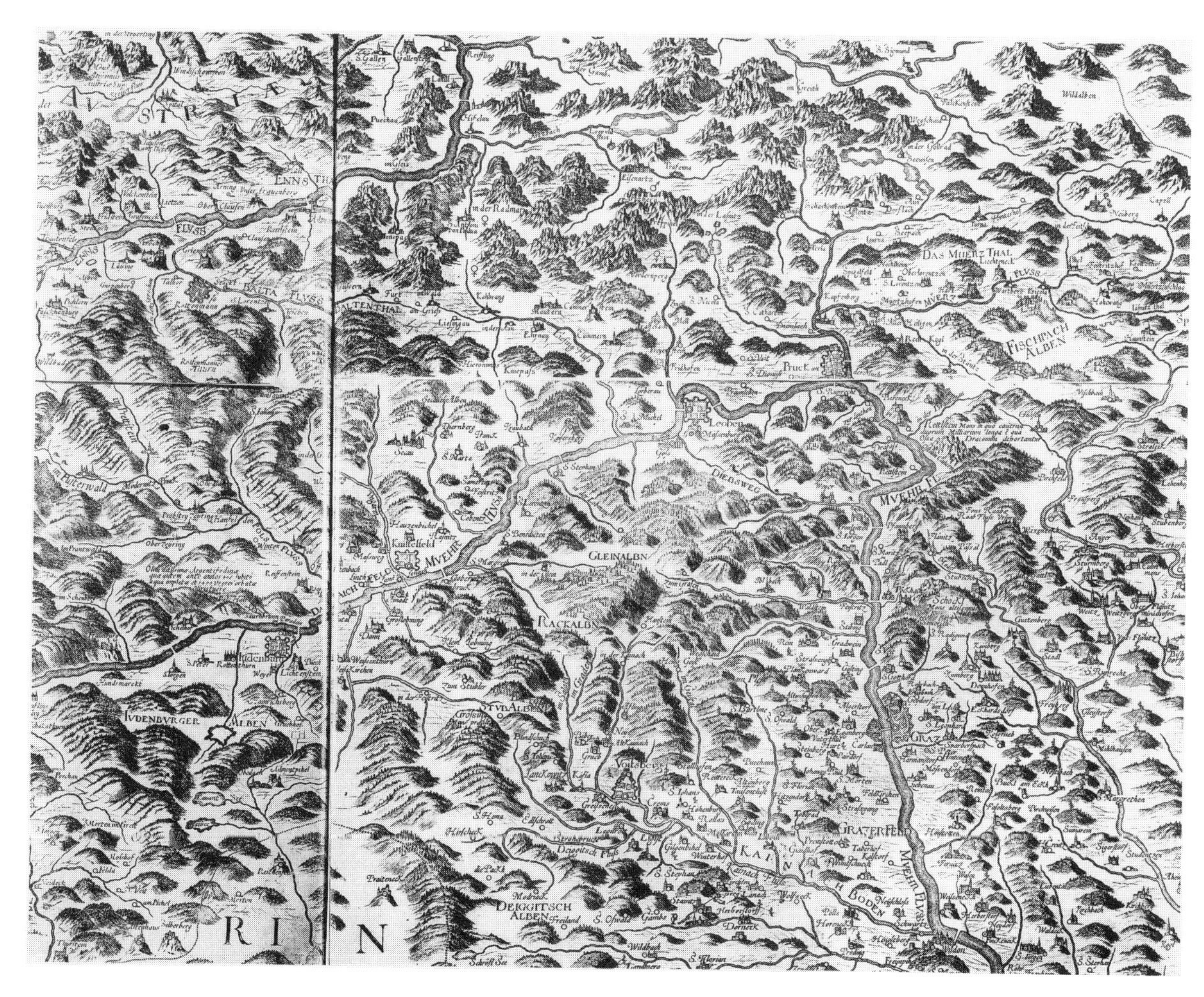

Fig. 6.7 Detail from Georg M. Vischer, *Styria Ducatus* (Graz, 1678). ÖN

teenth century the territorial estates expressed a concern for mapping that they had not evidenced earlier and that they dropped in the eighteenth century. In 1619, just at the beginning of the Thirty Years' War, the estates of Upper Austria had already commissioned Abraham Holzworm to produce a map. Just a few years earlier the Styrian estates had ordered a similar map of Inner Austria. The trend continued throughout the century. By 1636 the estates of Carinthia were printing and distributing a map of that principality; in 1664 the estates of Hungary followed suit. The 1670s saw a second round of expensive commissions by the estates of Upper Austria, Lower Austria, and Styria, and the 1680s were marked by elaborate mapping projects paid for by the estates of Siebenbergen, Moravia, Carinthia, and Carniola.

As the 1678 map of Styria illustrates (figure 6.7),the territorial maps were substantial undertakings. They defined internal administrative boundaries; they clarified external frontiers; they illustrated the hydrographic network and marked principal topographic features of the commonwealth. That such maps were being ordered in large and frequent editions by the estates, those very bodies in charge of preserving traditional political distinctions, bespeaks at least two motives: first, a desire to govern more effectively; second, a perceived need for a spatial abstraction of their own authority. These were not the kinds of maps useful for reforming the tax code, developing industry, or studying military tactics. Rather, they were concrete statements of political rights, statements now deemed important to those who felt compelled to define themselves by a marked separation from their neighbors. In this sense we may well have a case of perception and response from below to a strategy of centralized state-building not yet fully articulated—or even consciously identified as a systematic policy—at Vienna.

The Eighteenth Century

By the mid-eighteenth century all this had changed. The combined impact of three major European events had transformed the political consciousness of the Habsburgs, and the Viennese government was openly embarked upon a methodical program of state-building. The first stimulus came with the death in 1700 of Charles II, the last of the Spanish Habsburg kings. The major powers scrambled to divide the spoils, but in the end it was a French prince who acquired the Spanish throne with its vast overseas empire. The Austrian Habsburgs were fobbed off with the governorship of the old Spanish Netherlands (modern-day Belgium and Luxembourg) and extensive rights and properties in Italy. Meanwhile—and this is the second major event—Habsburg armies, led by those famous generals Eugene of Savoy and Wilhelm Ludwig of Baden, were chalking up impressive territorial gains in the east. They not only liberated Hungary from the Turks but also made substantial conquests in the area we now call the Balkans. Finally, this redistribution of territory and restructuring of internal alignments within the *case habsburga* coincided with a sharp reversal of the balance of power in the Holy Roman Empire. I refer here to the emergence of Brandenburg-Prussia as an aggressive counter to Austrian influence in the Germanies.

Cut off from the internationalism that the Spanish connection had provided, and checked in the Reich by a strong, hostile Prussian state, the Habsburgs turned increasingly to a consolidation of their own expanding possessions. Charles VI spent most of his reign securing the famous Pragmatic Sanction, a covenant that not only guaranteed the succession of his daughter Maria Theresa but also obtained European recognition of the fact that his family's possessions now constituted an indivisible political unit.

Frederick the Great's conquest of Silesia convinced Maria Theresa and her ministers of the absolute imperative of developing a strong central government for this unified inheritance. They quite correctly realized that a set of common political institutions would have to be created for the Austrian monarchy, such as had enabled Frederick to mobilize the resources necessary to sustain his Silesian conquest. "Austria," in other words, must no longer mean simply a collection of dynastic possessions. These possessions must be merged into a unified, centrally administered state. The task was difficult, if not impossible; and in the end the reformers fell short of their goal. But that is another story. The success or failure of the Austrian state is not our concern. What interests us is the fact that this new conception of the Habsburg state found expression in a bold, innovative use of maps.

It is surely more than mere coincidence that at the very moment in which they began articulating the need for a powerful central government, the Habsburgs and their ministers turned to the concept of a comprehensive mapping scheme for the empire. The initial step was the demand for and the publication of a series of maps, such as J. C. Homann's 1730 rendering of the Habsburgs' holdings in Europe. Here is a comprehensive statement of the monarchy's possessions, the bulk of which are shown to be a coherent geographic bloc. There is no reference as such to the Holy Roman Empire. Rather, the Austrian Habsburgs stood, as they had never represented themselves earlier, atop a clearly defined empire with fixed boundaries and a distinct geographic shape. The territorial arms were now arranged as a federal unity.

Significant though it may have been to promote a spatial concept of rule in Europe, the Habsburg decision in 1764 to commission a comprehensive mapping of the entire monarchy communicated even more of the government's new orientation. The project was a vast one and little known outside ministerial circles. It took twenty-three years to complete and, in the end, yielded some fifty-four hundred maps. In its scale and scope, it can only be compared to the much-celebrated map of France by Jacques-Dominique de Cassini. The 180 sheets of the Cassini map were engraved in 1789, bound, and published immediately throughout Europe. The maps produced something of an international sensation and became at once a cartographic landmark. Their comparatively greater fame is justified in that Cassini's work was more sophisticated than the Austrian survey. Whereas he employed a standard triangular measure for all of France, the Austrian measurements varied, so that the maps were neither uniform nor continuous in design and symbol.

The discrepancy in quality was reflected in the comparative cost of the two undertakings. While the French crown paid some 700,000 livres for the Cassini survey, the Austrian bill was less than half that amount, about 300,000 florins. The Austrian survey—known as the *Josephinische Aufnahme*—met a fate quite different from that of the French, and this more than anything else explains the fact that it is generally unknown. Only three hand-drawn copies were prepared of each map. They went, respectively, to the emperor, to the government archives, and to the president of the *Hofkriegsrat*, the military board that now constituted the apex of the Austrian bureaucratic pyramid. Fortunately for us, an exception occurred in the Austrian Netherlands, where in the late 1760s the imperial artillery officer Count Josef Ferraris (1726–1814) undertook the survey. As a result of political upheavals, his work found its way ultimately to the engraver, so we have evidence of both the detail and the quality of the undertaking. It has all been recently published in twelve volumes as the *Cartes de Cabinet des Pays-Bas autrichiens* (Brussels, 1965–1974), and contains a very large number of maps at a scale of roughly 1:25,000. Figure 6.8 shows us a detail from the sheet including Brussels. The city is on the left, behind its fortifications; note the prominent gardens of the royal palace. A very straight road leads out to the northeast through relatively open country with a few large farms. Figure 6.9 shows the quite differ-

Fig. 6.8 "Bruxelles" sheet from the *Cartes de Cabinet* composed by Count Josef Ferrarris. Bibliothèque Royale (BR), Brussels

ent country that lay round the town of Lier. Here the fields were very small and irregular, and the little town has some difficulty filling up the space behind the fortifications. Note the careful rendering of settlement patterns, fields, and roadways as well as of water, footpaths, and forests.

These engravings constituted a striking exception to Habsburg policy and resulted from political circumstances beyond the Austrian government's control. Where the Crown could dictate procedure, no engraving was permitted and access was strictly limited. Down to 1864, in fact, the maps were classified as top secret documents, and before a person could study any one of them he had to obtain signed permission from the sovereign. A handwritten text accompanied each sheet. One of them reads, "The town is a solidly built affair, with a large military barracks and stables on the edge. Principal buildings are the town hall, a convent, a church, and a large parish house. Outlying buildings are well constructed, especially those near the mill on the banks of the Crems river. That river joins the Danube just below the town, and at that point the Danube makes the area something of an island. The terrain is generally flat but dominated by the hill rising behind the town."

Tactical concerns obviously determined

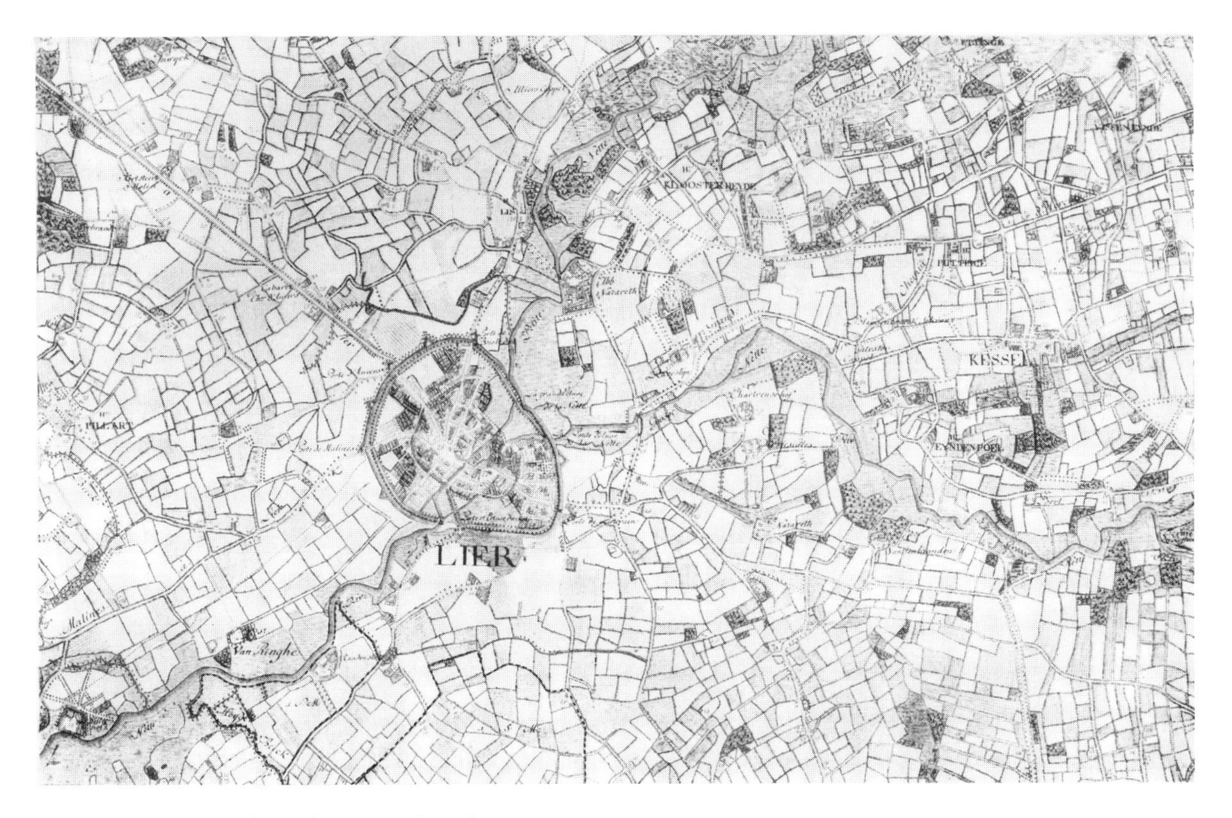

Fig. 6.9 "Lier" sheet from the *Cartes de Cabinet*. BR

this description. Indeed, the initial impulse behind the entire survey came from the Austrian General Staff. Horrified by the tactical and logistical blunders their officers had committed while still fighting on Habsburg territory during the Seven Years' War, the *Hofkriegsrat* had approached the Crown for more accurate maps. Otherwise, as Field Marshall Lacy wrote in an oft-quoted memorandum to the government, troops would be marched into areas where there was no adequate billeting, days would be wasted in abortive efforts to cross difficult terrain and battles would be lost for want of knowing where best to position the troops. The Crown's decision to initiate the survey must therefore be thought of as a military as well as a political statement—and in the military sense, less an effort at state-building than an innovative response to a conventional problem of defense.

But the monarchy clearly had other expectations as well. From the beginning, the emperor and his ministers looked to these maps as indispensable tools of government. They were particularly concerned to use them to improve state finances and to stimulate a national economy. To this end they insisted that the maps contain all information necessary to plan and execute a coherent economic strategy for the state. Thus, on 26 September 1766, in commissioning the mapping of Marmoros, a recently added province in Hungary, the Crown stipulated that in addition to the standard survey map done on a scale of 1:28,000, the engineers were to prepare a second set of maps for the imperial treasury, this time on a scale of 17,200. These latter maps were to contain all information necessary to determine a more accurate tax base and to implement a more equitable distribution of the tax burden.

Three years later, once again in connection with a Hungarian mapping enterprise, the Crown was even more specific: "Ownership must be indicated for every village and estate, together with an accompanying text stating by what right the property is held. Information must also be provided as to the amount of tax currently paid, the seigneurial rights attached to the property and the principal crops produced. Finally, indication must be made of industry and trade in the area, the languages spoken, the religions, the local customs, the population (divided by sex), and an estimate on the potential for demographic and industrial growth as well as for additional settlement."

Administrative emphasis of this sort came gradually to dominate the entire mapping enterprise and in some instances to supersede military concerns altogether. Take the case of the mapping of Bukovina. The government's directive of 4 February 1782 stated explicitly that the goal was not military but "the creation of a useful political administration for the province, one that would provide the basis for a new constitutional structure that would ensure a well-run government." Of course, well-run government for these eighteenth-century Austrian statesmen meant more than sound justice and strong defense. They placed the state at the center of society and invested it with sweeping social and economic responsibilities. Ambitious programs of educational reform, highway building, health care, and industrial growth flowed from the central administration. Contemporary maps both reflected and stimulated this expansion by the state.

Figure 6.10, for example, is the kind of postal map that the government began printing and circulating throughout the nation. This one is for Carinthia and identifies not only the mail route but also the travel time between the government-subsidized postal stations. Note the lines leading out of the province to principal clearing stations for the national mail. A Styrian industrial map makes the point from another direction. The Crown relied heavily upon such maps for determining its policies of industrial subsidy and investment. On a map such as this one a Viennese administrator could locate everything from salt and ore mines to breweries, mineral deposits, glass factories, tanneries, dairies, and fisheries. Market towns are designated with special symbols distinguishing those whose trade was national from those with only local or regional trade. The Habsburg government commissioned and published a series of such statements for the empire.

The map, in a word, had come into its own as an instrument of government. It now served the Austrian state and assisted the Habsburgs and their ministers in carrying out the social and political responsibilities prescribed by the eighteenth century for enlightened government. On the darker side, it furthered an unprecedented meddling by the government in society. In this sense it nurtured the despot. Astute contemporaries caught this paradox and in their political writings placed the map squarely in the service of the state and its ruler. Sixteenth-century Humanists had looked to

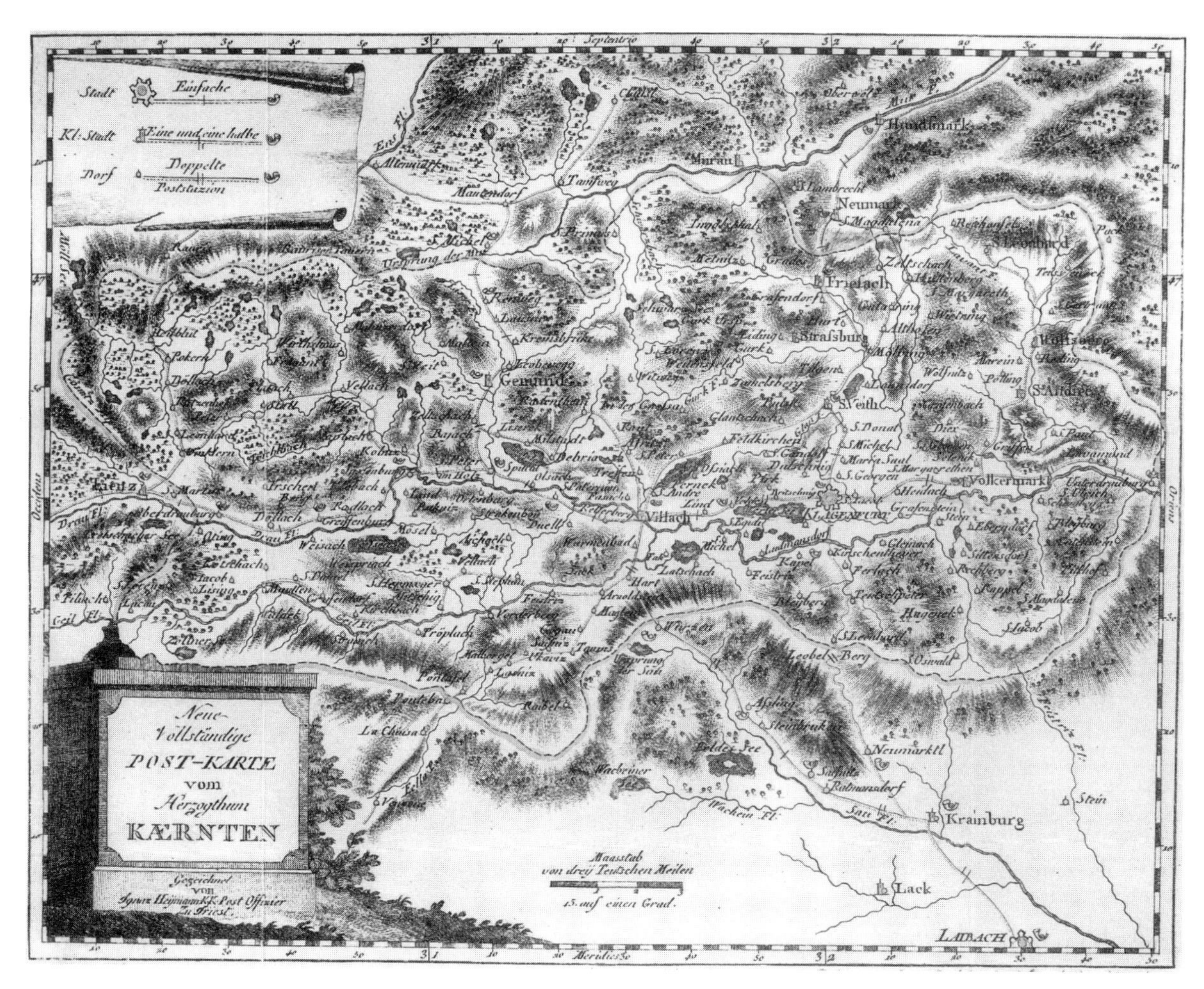

Fig. 6.10 Postal map of Carinthia, 1798. ÖN

the map as a means to learn about and thus better appreciate God's creation. Others had seen it as a means for establishing the geographic integrity of their cultural tradition. Such concerns had little place in the eighteenth-century geographic consciousness.

Johan Michael Franz, the first university professor hired explicitly for geography, spoke for the new order. He tied the map and its maker directly to the service of the state. "Our goal," as he defined it in his inaugural lecture of 1753, "is to prepare a description of the state, to illustrate it with good maps, and then to see to its publication." The Austrian Habsburgs and their ministers may have been ambivalent about publication, but they most certainly concurred completely with the belief that the state lay at the center of the cartographer's concern.

SEVEN

Monarchs and Magnates: Maps of Poland in the Sixteenth and Eighteenth Centuries

Michael J. Mikoś

In the sixteenth century, the Republic of Poland and Lithuania, one of the largest countries in Europe, enjoyed a period of economic prosperity and cultural splendor. It extended to the sources of the Oka and Donets rivers in the east, Brandenburg in the west, the Baltic shore with Courland in the north, and the Carpathians and the mouth of the Danube on the Black Sea. Its institutions, laws, and practices were inspired by deeply rooted beliefs in individual freedom and civil liberty, in sharp contrast to those of many of its neighbors, whose policies began to be dominated by the growth of the central, authoritarian power exercised by the ruler or state.

After the demise of the powerful Jagiellonian dynasty, the Polish throne became elective, and power shifted into the hands of the nobility,[1] which played the leading role in the political, social, and cultural life of the country, fiercely defending its "golden freedom." This numerous class, which constituted 6.6 percent of the population in the sixteenth century and 9 percent in the seventeenth century, obtained very quickly various immunities and numerous monopolies in government, administration, and church appointments.

In 1573, the nobility set the conditions for the coronation of Henry Valois, the king-elect. They insisted, for example, on the right to elect the king freely in the future and to approve all declarations of war and imposition of taxes, and on the right of resistance if the king contravened his oath. They restricted the rights of the clergy, the burghers, and the peasants to own land, while they themselves controlled about 60 percent of this major source of wealth in the country. They controlled the population by tying the peasants to the estate land. They abolished municipal taxes for their own property and duty taxes on commercial goods for their own use, and obtained the exclusive right to exploit a number of minerals from their land. They became a powerful closed estate, a ruling class in full control of the republic.

Although the noblemen succeeded in winning independence from the monarchy, they soon became dependent, politically and economically, on the oligarchy of great nobles, the magnates. A number of rich and influential families, for example the Tarnowskis and the Zamoyskis in Poland and the Radziwiłłs and the Sapiehas in Lithuania, tended to control the key offices of state and church and owned a disproportionate amount of land, organized as latifundia, and insured against dispersal by the Law of Entail (1589). Entire regions were run as single economic units, administered from palaces and linked with key properties from which the rest were managed. The magnates displayed their power and influence by erecting splendid palatial residences and commanding armies and retinues commensurate with their ambitions and status. They employed masters-of-robes, masters-of-horse, military captains, chaplains, secretaries, artists, ballet masters, engineers, architects, historians, carpenters, butlers, jesters, foreigners, dwarfs, and Tartar or African slaves.

The greatest of these families ruled entire regions in virtual sovereignty, built palaces, libraries, private chapels, and theaters, displayed their wealth in furniture, jewelry, and costume, and entertained lavishly at various ceremonies, festivals, and parties. In pursuit of their own ambitions, the Polish magnates weakened royal control as well as the powers of the legislative and judiciary bodies and, in their frequent disregard of national interests, precipitated a state of near anarchy.

But at the same time Polish magnates and monarchs excelled in giving support for education and for cultural achievements, leaving behind lasting monuments in literature, art, architecture, and cartography. The most successful in advancing the mapping of Polish territories were King Stefan Batory and Prince Krzysztof Radziwiłł in the sixteenth century and King Stanislas Augustus Poniatowski and Prince Józef Jabłonowski in the eighteenth century.

The Sixteenth Century: Batory and Radziwiłł

Stefan Batory, who reigned as an elected king of Poland from 1576 to 1586, was born in 1533, the youngest son of the palatine of Transylvania. He received an excellent education, traveled in western Europe, and served at the imperial court in Vienna. Having fought for fifteen years as commanding general in the struggle of Transylvania for independence, in 1571 he was elected prince of Transylvania and in 1576 king of Poland. He was knowledgeable about contemporary political affairs, well read in history, and eloquent in many languages, especially Latin.

Batory's greatest achievements were on the battlefield, however. A gifted strategist and an accomplished military leader, he reformed the royal army and fought for years against the forces of Ivan the Terrible. After defeating him at Duneberg and Połock, Batory gained for Poland Livonia and the region of Połock. He was aware of the role of public relations in carrying out his plans and obtained the support of Jan Kochanowski, a great Polish poet. During his campaigns, he was followed by a field press, called the "flying press," which publicized his successive victories.

The reign of Stefan Batory opened a new chapter in the development of Polish cartography. Batory laid the foundations of military mapping, linking closely the practical drawing and use of maps with military strategy.[2] Józef Wereszczyński, the bishop of Kiev, described in 1592 King Stefan Batory's preparations for the war with the Turks:

> King Stefan would not think about anything else day and night, nor would he talk more often about anything but the Turk, learning with great diligence both about the places and easier passages to his lands, which although in general he was aware of, if somebody began to tell him about it, for better knowledge, he willingly listened. And then he would always go to the map on which were described all *situs locorum*, Turkish cities and castles, seas, rivers, mountains, hills, forests and fields, and would spend a lot of time over it, looking for places of easier passage, through which Christian armies could proceed better, and in time he arrived at such knowl-

edge of such places that there is no mountain, no river, finally no place so well camouflaged by the Turk, that he would not have accurate knowledge of.[3]

In preparation for his war against Muscovy, Batory studied a plan of attacking the enemy from the direction of the White Sea, using the map of Europe by Mercator, most likely the second edition of 1572.[4] He also consulted Ortelius's *Theatrum Orbis Terrarum*; "I understand that the King of Poland Stefan nowadays has the *Theatrum* before his eyes," wrote Piotr Edling, superintendent of Kamień diocese churches, to Ortelius, from Kołobrzeg, 15 August 1580.[5]

King Stefan Batory's principal cartographer was Maciej Strubicz. He prepared a description of Livonia, consisting of two parts, historical and geographical. Completed and dedicated to the king in 1577, it was not published until 1727, probably because the king was not satisfied with its unoriginal content and requested a new cartographic picture of Livonia. After 1577, Strubicz served as a royal cartographer, and although he did not participate in military campaigns, he was in close contact with the king and with Jan Zamoyski, chancellor of the crown. In 1579 the king owed Strubicz his pay for two years. In a letter to Zamoyski, dated 25 October 1579, Strubicz wrote about his maps of the whole country: "According to the order of His Royal Majesty, our gracious Lord, I have meanwhile written out, with great effort and my own outlays, with the greatest diligence as for the location, possession, and dimension of the lands and domains belonging to His Royal Majesty, and returned them to His Royal Majesty, illuminated separately on different tables."[6] In November 1579, he wrote to Chancellor Zamoyski that he had sent one map of Lithuania to the king for his military campaign.

Strubicz's map *Magni Ducatus Lithuaniae, Livoniae, et Moscoviae descriptio* (figure 7.1), drawn about 1581 and published in 1589, is the only one of his maps that has survived. A later, improved version of the map advanced the development of cartographic representation of the Lithuanian territory, especially of its network of rivers and localities, and was used by Mercator (1595) and Mikołaj Radziwiłł (1603).

The Połock campaign of King Stefan Batory in 1579 gave rise to the first specimen of Polish military cartography, a map of the Połock region by Stanisław Pachołowiecki. It was fairly detailed, with a scale of 1:700,000 and sixty-nine place-names, and was published in Rome in 1580 by Giovanni de Cavalleriis (figure 7.2). The map is known from the 1837 and 1840 editions produced from a contemporaneous copy of the original map, which was lost.

Batory praised the cartographic skills of Pachołowiecki; in the royal privilege of 1581, awarding Pachołowiecki a new coat of arms, the king wrote, "We perceived certain divine seeds of that genius in his elegant painting of figures and also in his drawing of enemies' castles."[7] The map itself, however, was poor; its orientation indicates that the author did not use a compass, and the network of rivers, forests, and settlements was so inadequate that the map can be construed only as a first draft. Attached to the map were plans of seven castles in the Połock region which the king besieged and captured in 1579.

Another cartographer in the service of King Stefan Batory was Peter Francus. In the privilege of 19 September 1579, Francus was commissioned "to draw a picture, a very real one, of the site of Połock castle and of that siege and attack as well as the attack on the Sokol castle and other drawings pertaining to our

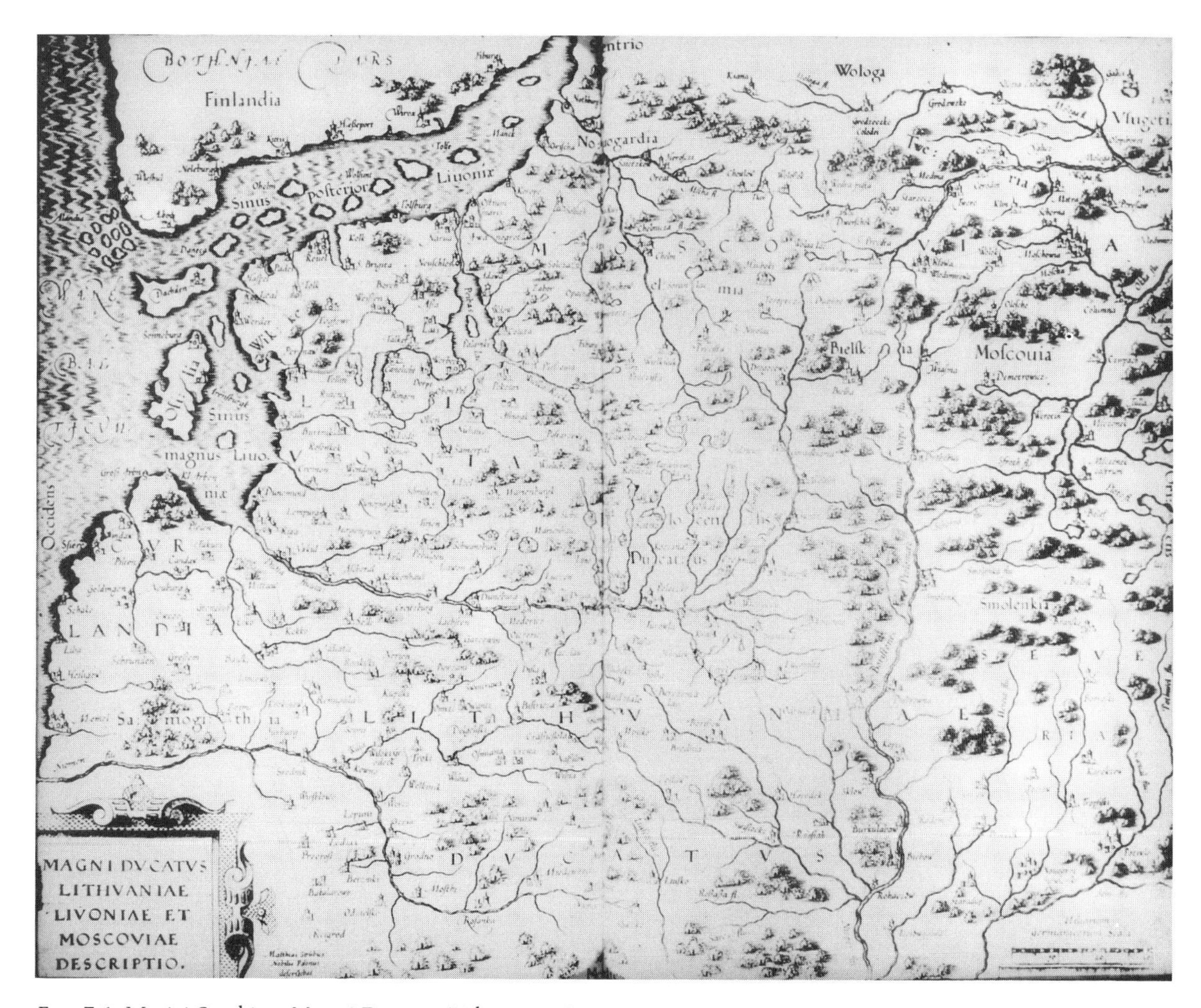

Fig. 7.1 Maciej Strubicz, *Magni Ducatus Lithuaniae, Livoniae, et Moscoviae Descriptio,* 1589

campaign, both in part and in its entirety."[8] For this he received a stipend of fifty florens and was knighted after the campaign.

The Połock campaign was not the end of cartographic work during Batory's reign. On 11 December 1580, the king wrote to Jan Zamoyski about a map of Livonia (1:700,000) by Sulimowski, "a description of those provinces which we took from the enemy,"[9] and sent the map to the Chancellor for correction and evaluation. Another map of Livonia by Stanisław Sarnicki (1585) originated from the same campaign. In 1578, Marcin Broniewski traveled as a royal emissary to the Chan and authored a description of Tartary and a map of Crimea, which he sent to the king in 1579. From the Pskov campaign of 1581 comes an anonymous *Civitatis Plescouiensis delineatio*. It is not a

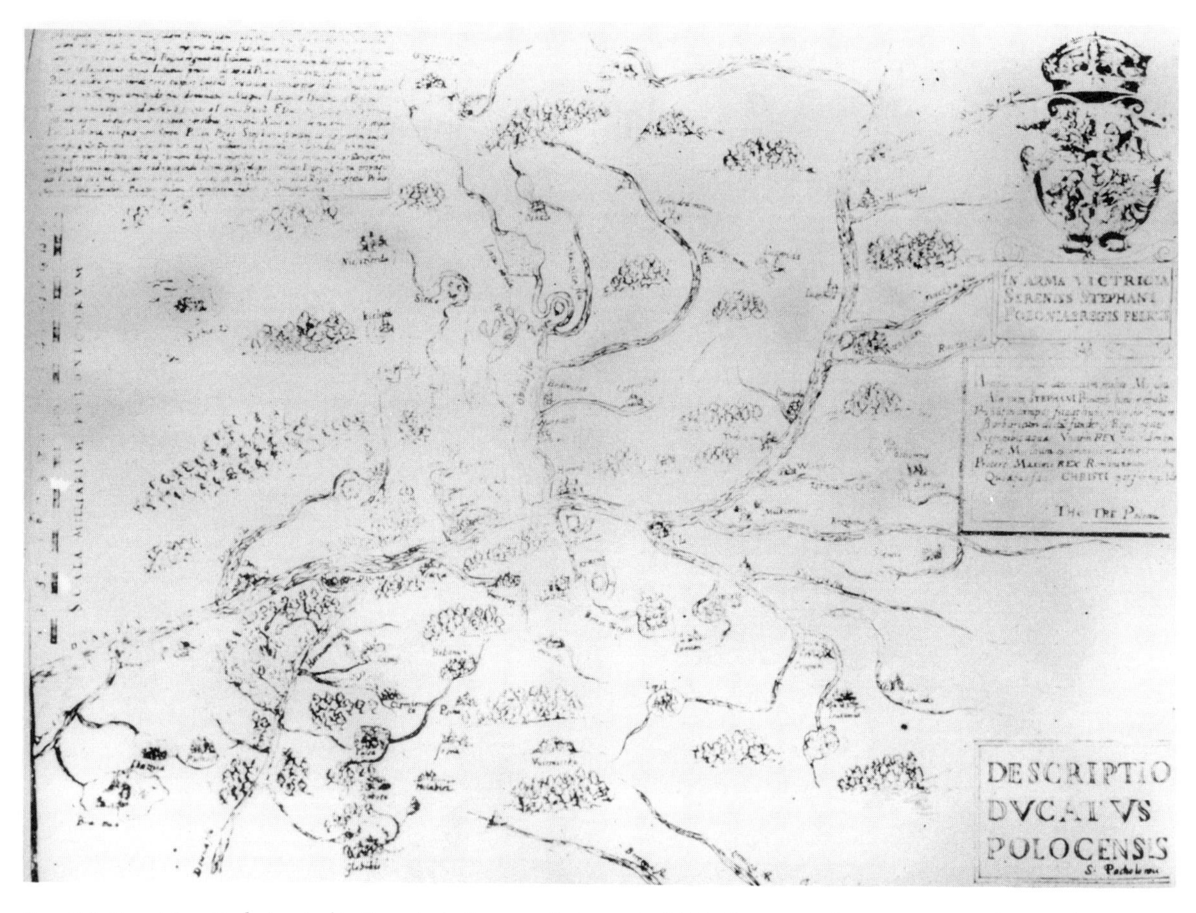

FIG. 7.2 Stanisław Pachołowiecki, *Descriptio Ducatus Polocensis*, 1580. Courtesy of Tomasz Niewodniczański

map but a pictorial itinerary oriented to the west, with an iconographic representation of small towns and the city of Pskov, showing their locations and distances.

During the last few years of Stefan Batory's reign, a new project of depicting Lithuania was undertaken by Prince Mikołaj Krzysztof Radziwiłł. Some of the cartographers who worked on the map also practiced their skills under the king, but the mapping of the Grand Duchy of Lithuania was conceived, financed, and directed by the prince himself.

Mikołaj Krzysztof Radziwiłł (1549–1616), nicknamed "Little Orphan," was born in Czmielów, a son of Mikołaj Radziwiłł the Black, the chancellor of Lithuania and palatine of Wilno. Up to the age of fourteen he was tutored at home; he then went abroad, studied in Strasbourg and Leipzig, and visited many royal courts. He participated in the war against Muscovy in 1568 and, after the election of Stefan Batory in 1576, took part in his campaigns. Radziwiłł was seriously wounded during the siege of Połock in 1579, and in 1582 he under-

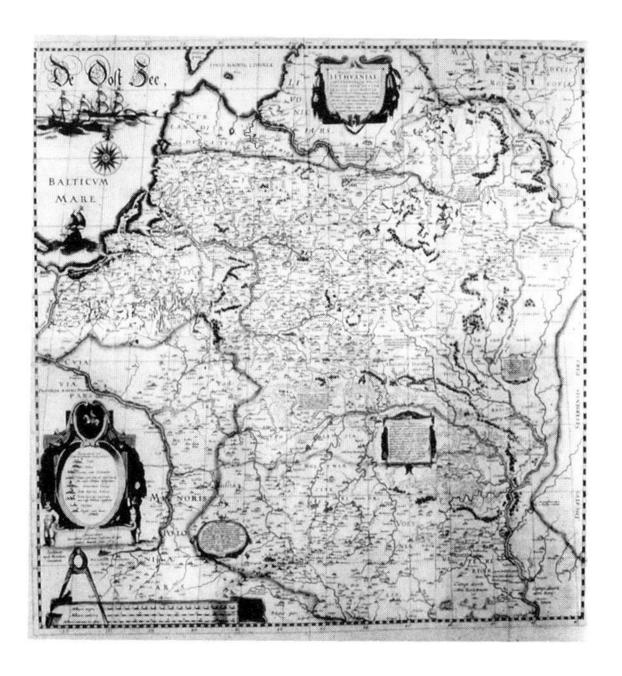

FIG. 7.3 Mikołaj Krzysztof Radziwiłł, *Magni Ducatus Lithuaniae*, 1613 American Geographical Society Collection (AGSC), University of Wisconsin–Milwaukee Library

took a pilgrimage to the Holy Land from which he returned in 1584. He described his peregrinations in a vivid narrative published in Polish, Latin, and German, in which he discussed the architectural monuments he had seen, as well as social and political conditions, traditions, and customs, especially the folklore of Syria and Palestine.

Radziwiłł was elevated successively to the positions of court marshall (1568), grand marshall of the Grand Duchy (1579), palatine of Troki (1590), and palatine of Wilno (1604). He donated large sums of money to the Wilno Academy, which had been founded in 1579 by Stefan Batory, and established a college for Jesuits as well as many Catholic churches, hospitals, and schools. In this way, he secured for himself a network of educational and administrative posts, strengthening the role of the Catholic church in Lithuania and, consequently, the union between Lithuania and Poland. He was also known for his modern methods of administering the economy of his large properties.

The first, lost edition of the Radziwiłł map *Magni Ducatus Lithuaniae . . .* was probably printed in 1603.[10] The map is known from its Amsterdam edition of 1613, of which only one copy is preserved in the University Library at Uppsala, and from numerous editions of the Blaeu atlas, beginning in 1631. It was engraved in four plates by Hessel Gerrits. The scale of the map is 1:1,293,000 and its size 109 x 84 cm (figures 7.3 and 7.4).

Preparatory works for the map of Lithuania may have begun on a large scale as early as 1584, and were carried out by a team of collaborators, including Maciej Strubicz, Tomasz Makowski (who prepared the manuscript engraving), and an unknown magnate from the Ukraine. They were supported by Jesuit schol-

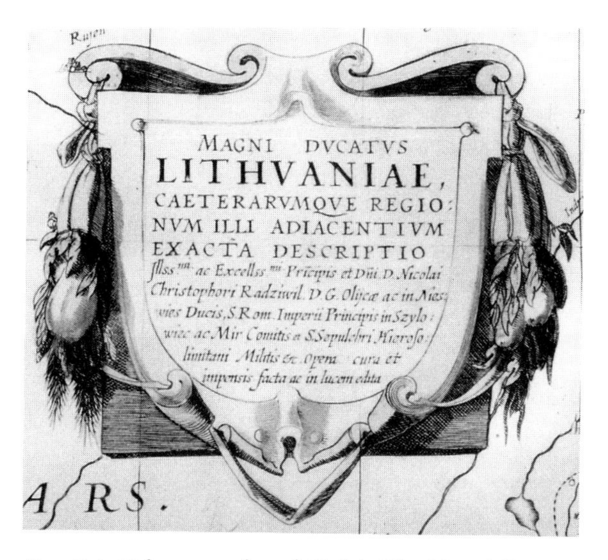

FIG. 7.4 Title cartouche of Radziwiłł's *Magni Ducatus Lithuaniae*. AGSC

ars and by administrative officials from the territories of Prince Radziwiłł. The map was based on original geographical measurements of the main towns in Lithuania, resulting in the first fairly faithful representation of the whole country; the error of latitude was limited to about 22 km and of longitude to only 11 km.

The map is an important source of historical information. It contains data on hydrography, forests, place-names, administrative and judicial divisions, seats of Catholic and Orthodox bishoprics, and residences of important families. The configuration of the terrain is shown by means of small mounds, and some small bridges and historical sites are marked as well.

Working from property acts, inventories, conscription registers, and population figures provided by local informants, the authors depict 1,020 localities on the map, about 543 of them in the territory of Lithuania, of which about 357 appeared for the first time. The map distinguishes, by size, large towns (*urbs*), towns (*civitas*), small towns (*oppidium*), and villages with a manor (*pagus cum domo nobilis*; figure 7.5). All localities that had more than fifty households (that is, more than 350 inhabitants) were listed on the map. Different symbols for various types of localities and miniature outlines of some castles provide rare iconographic information pertaining to the architecture of the area (figure 7.6).

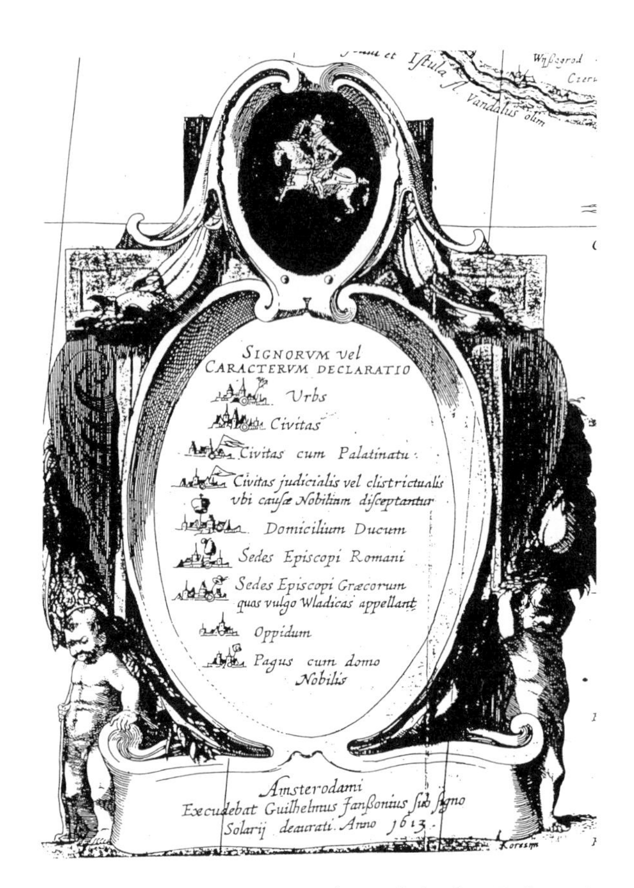

FIG. 7.5 Key to conventional symbols for Radziwiłł's *Magni Ducatus Lithuaniae*. AGSC

THE EIGHTEENTH CENTURY: PONIATOWSKI AND JABŁONSKI

Stanislas Augustus Poniatowski (1732–98), the last king of Poland (1764–95), was born into a wealthy, powerful, and cultured family. Having received a thorough education, at the age of sixteen he began his European travels, spending most time in France and England. After his return, he embarked on a political career, obtained the position of steward of Lithuania, and, in 1757, became ambassador to Russia. He returned to Poland in 1758, won a seat in the Chamber of Deputies, and in 1764 was elected king of Poland.

Although his tragic reign concluded with the partition of Poland, Poniatowski distinguished himself as a great patron of learning and the arts, and his efforts to develop Polish cartography produced important results. After his election, he undertook an ambitious program of creating appropriate conditions for the

FIG. 7.6 Detail from Radziwiłł's *Magni Ducatus Lithuaniae.* AGSC

development of cartography, surrounded himself with astronomers and cartographers, organized an office of cartography at the Royal Castle in Warsaw, and assembled there a sizable collection of 325 maps, globes, and astronomical instruments.

Poniatowski's keen interest in maps is well documented. In 1765, he did not hesitate to "steal" from Prince Jabłonowski an eminent cartographer, Francis Czaki, who joined the king's project together with 201 maps, and received three hundred ducats and the title "Captain of Crown Artillery" as a reward. When Tadeusz Kościuszko approached the king with a request to borrow his maps during the siege of Warsaw in 1794, the king replied: "If I still had diamonds left, I would prefer to give them, rather than these maps, which are the fruit of twenty years of my efforts."[11] He loaned the maps to Kościuszko, nevertheless. When he departed for his exile in Petersburg, Poniatowski carried with him some maps from his cherished collection.

To secure a sound mathematical base for his "geographical enterprise" of mapping the country, King Stanislas Augustus established the School of Cadets. The program of the school was designed to prepare the cadres of military cartographers. In 1774, the king instructed the cadets: "Meanwhile, not to lose time, I demand as a test of your skill that you send a map, not larger than one sheet, depicting those places in which you are located and half a mile of the country around them."[12] He also set up, in 1774, an astronomical observatory at the Royal Castle, directed by Father Jowin Bystrzycki, and relied on astronomical and geographical data provided by the Jesuit academies and schools. In 1775 he established the Engineering Corps of the Crown to build fortifications and, later, to organize cartographic surveys.

The principal cartographers who were engaged at the Royal Castle in drawing the atlas of Poland were Charles Perthées, Jan Bakałowicz, and Francis Czaki. The most prominent of them, Perthées, was named Geographer Royal and Lieutenant of the Crown Artillery (a colonel in 1783) and remained in the king's service until the end of his reign.

In 1766, Perthées had authored the first general map of Poland. In 1768 he made another map, *Mappa geographica generalis Regni Poloniae et Magni Ducatus Lithuaniae . . . ,* and in 1777 he produced a much larger general map of Poland, but all were unfortunately lost. On the basis of additional surveys of the country and some supportive work by Francis Czaki, who traveled in Lithuania assembling materials for a map of the Grand Duchy, Perthées continued his work and prepared for the king new versions of the maps of the whole country. The result was a large (1:934,000) map of Poland, in forty-eight sections, which was lost during the Second World War (only full-sized negatives have survived). Because of his reliance on

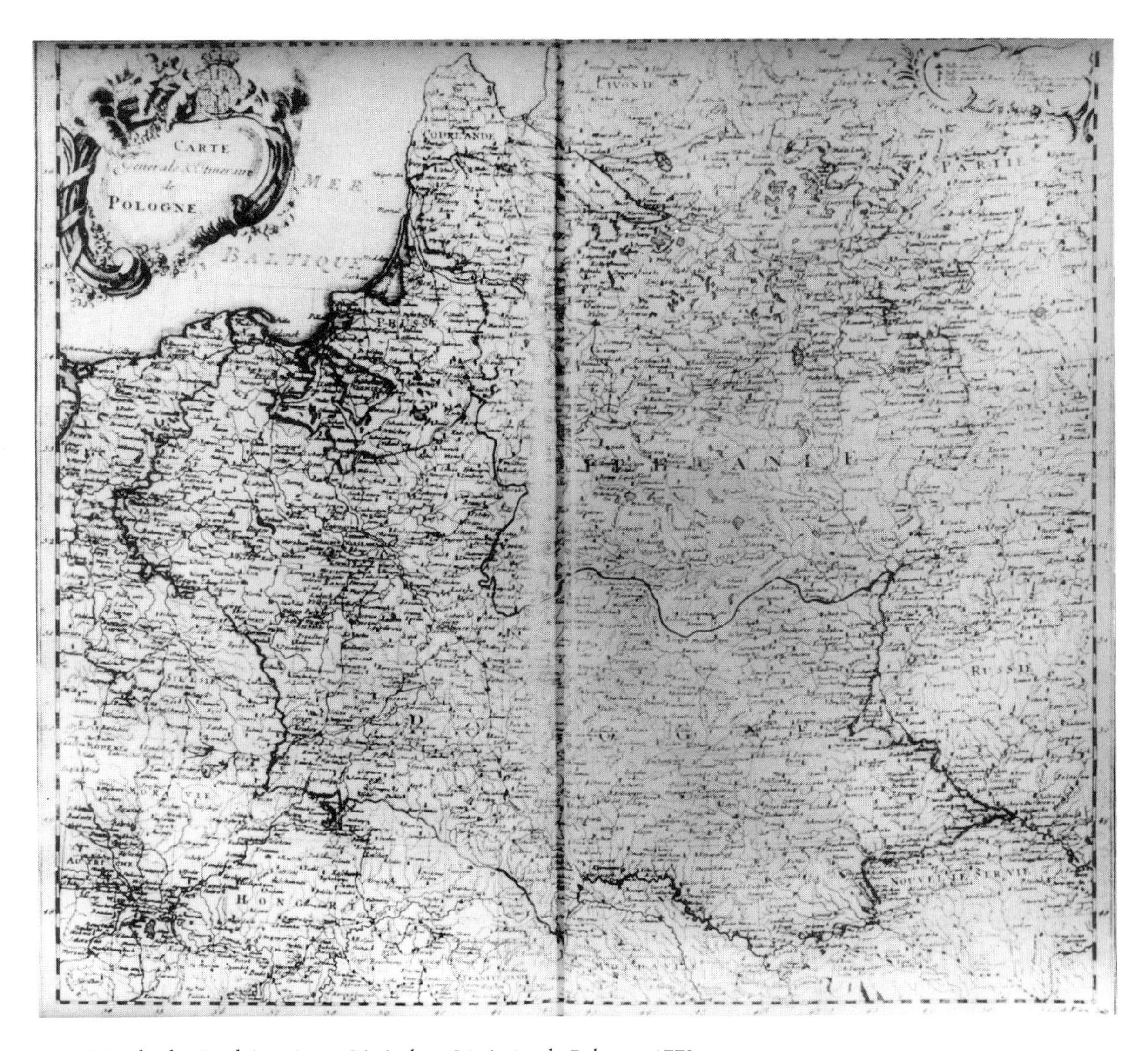

FIG. 7.7 Charles Perthées, *Carte Générale et Itinéraire de Pologne*, 1773

faulty determinations of longitude and overcrowding of the map with some 28,375 cities, towns, and villages, Perthées' map was not satisfactory. In 1773, he produced a small road map (1:4,700,000), entitled *Carte générale et itinéraire de Pologne* (figure 7.7).

In 1778, Michał Poniatowski, bishop of Płock and brother to the king, began sending geographical questionnaires to all parishes. On the basis of these data, Perthées drew a series of special maps of individual palatinates, on a scale of 1:255,000 (figure 7.8). Even though these maps were drawn by one man, without field surveys and without uniform and reliable

FIG. 7.8 Details from Charles Perthées's manuscript map of the Palatinate of Masovia, 1783. Institute of History of the Polish Academy of Sciences, Warsaw

sources, they provided a good representation of the hydrographic system, the network of roads, and the distribution of cities, towns, and villages.

In 1796, Perthées finished a new version of his map of Poland in twenty-four sections, but it was taken away to Russia after the fall of Poland and never returned. Similarly, Perthées' manuscript of *The Geographical and Statistical Description of the Parishes of the Polish Kingdom,* in twelve volumes, with 2,278 sketch-maps, a priceless historical and cartographic monument, was transferred to Kiev in the nineteenth century and has not been returned.

The story of lost maps and supporting documents of the royal geographer reflects fairly accurately the fate of Polish cartography under King Stanislas Augustus. In spite of all efforts, the king was aware of the odds and failures of his cartographic plans, when he wrote in 1777 to Poczobut Odlanicki, "Toiling for many years over my geographical projects, I have encountered constant obstacles in this enterprise; it seemed that the earth itself rose up in opposition so as not to be measured."[13]

Józef Aleksander Jabłonowski (1711–77), a son of the grand hetman of the Crown, was related to the Sobieskis, Radziwiłłs, and, through his wife, the royal house of the Bourbons. He was educated in his parents' home and as a young man traveled in France, Italy, Germany, Holland, and England. He was deputy to the Diet, steward of the Grand Duchy of Lithuania, and palatine of Nowogrodek from 1755 to 1772, when he resigned his office in protest at the first partition of Poland. Unlike his ancestors and many Polish noblemen, he did not seek glory on the battlefield, but devoted his life to scientific and learned activities, publishing at home and abroad his Latin, Polish, and French treatises on subjects of Polish and Slavic history, heraldry, geography, and astronomy. He assembled a large library and corresponded with scholars, scientists, and booksellers in Poland and abroad. He induced others to carry on research, subsidized various publications, and was a member of the academies of science and learning in Bologna, Padua, Rome, and Paris, having been recommended in 1761 for membership in the Académie Royale des Sciences in Paris by Joseph Nicolas Delisle for his works "pour perfectionner la géographie."

Jabłonowski's most enduring contribution was in the domain of Polish cartography. He undertook the task of mapping the entire country, resulting first in a map of Poland, probably never published, drawn in 1755 by Saint-Hilaire in Paris. In 1760, Jabłonowski engaged Francis Czaki, who had been traveling through Poland, visiting all the provinces and making detailed maps, since 1740. Some of the maps drawn by Czaki were sent in 1761 to Jean d'Anville, who made corrections, used them for his own map of the course of the Vistula, and returned them to Czaki and Jabłonowski. They in turn thanked the French cartographer for his assistance, as they were planning to make use of his observations in the general map, which was to be finished in four months and then taken by Czaki to be engraved in France.

Work on the map was greatly delayed, however, due to the defection of Czaki to the service of King Stanislas Augustus. Jabłonowski protested to the king, writing in 1765:

> Czaki, captain, several years ago having accepted my service, receiving promptly every year a high salary, holding my possession for his use and comfort, holding under his command my people, traveling at my substantial cost all over the country, drawing a geographical map of Poland, deserted me, having not accounted for the costs and taking with him up to two hundred of the most rare geographical maps, amassed with great effort, work, and cost, and took refuge under the protection of His Royal Majesty, which he should not have expected having committed such a shameful deed.[14]

The maps were returned, and the prince engaged new cartographers to complete his project. The prospectus of the map was in existence in 1770, when subscriptions were solicited. Father Teodor Waga participated at this stage of the project, supplying materials from the region of Mazowsze, and the task of compiling an atlas of Poland was entrusted to the

Fig. 7.9 Title page of the *Carte de la Pologne,* 1772 by Josef Aleksander Jabłonowski and Giovanni Rizzi Zannoni. AGSC

Italian cartographer Giovanni Rizzi Zannoni.

The *Carte de la Pologne divisée par provinces et palatinats et subdivisée par districts . . .* was published by Zannoni in 1772 in Paris (figure 7.9).Zannoni wrote a flattering dedication to Prince Jabłonowski but claimed credit for authorship of the atlas. The map was engraved by French artists in twenty-four sections, each section 52.5 x 33.5 cm, at a scale of 1:692,000. It was more accurate in the sections depicting Poland, with the mean error of latitude for all sections ranging from 7 to 24 km and the mean error of longitude from 15 to 28 km. Relief was represented by the use of a "bird's-eye view," and inscriptions were in Polish and French, and in Moldavia and along the Black Sea in Turkish as well (figure 7.10).

King Stefan Batory, King Stanislas Augustus, Prince Mikołaj Radziwiłł, and Prince Józef Jabłonowski all contributed significantly to the development of Polish cartography. Stefan Batory was not only a patron of cartography and founder of Polish military cartography but also a leader who used maps before his campaigns, carried on mapmaking projects during wartime, and tried to preserve records of his achievements for posterity by having them depicted on maps. Most of the cartographic projects he initiated were for immediate, practical purposes. Batory did not assemble enough

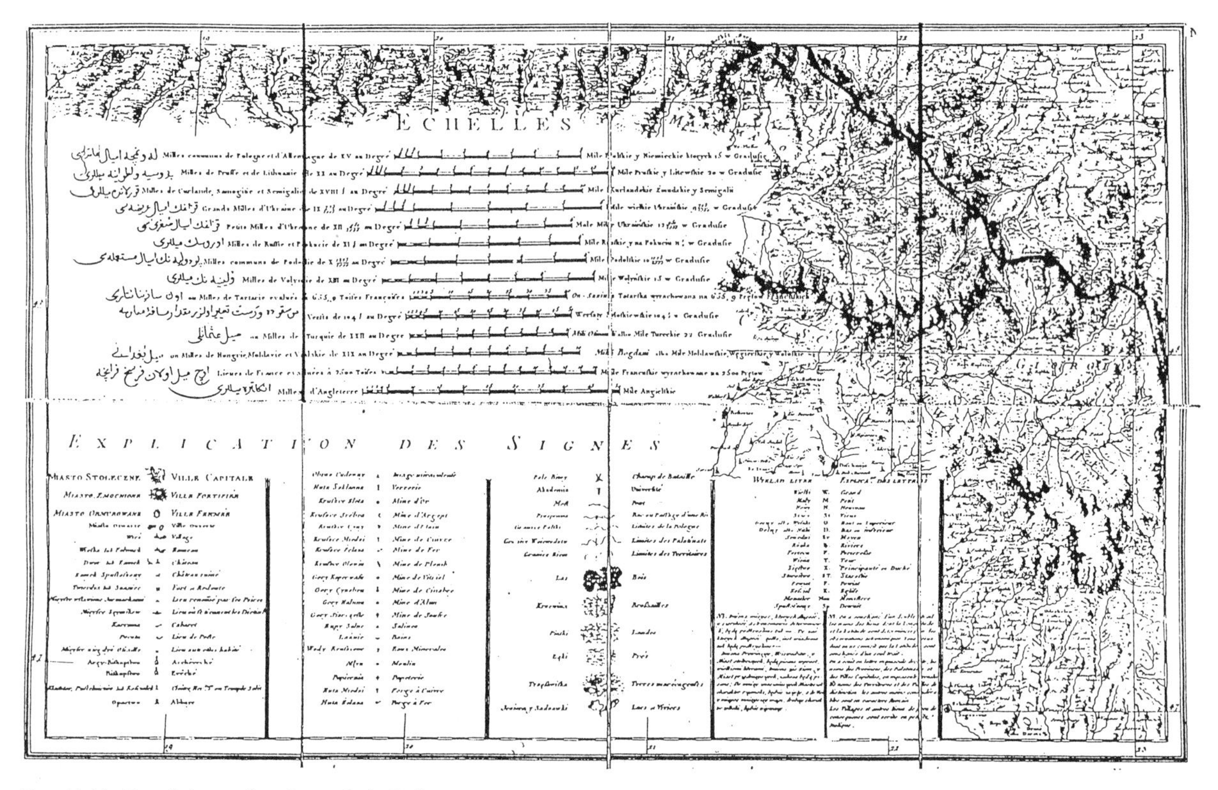

Fig. 7.10 Detail from the *Carte de la Pologne.* AGSC

qualified mapmakers, engravers, and publishers to draw a sufficient number of smaller maps to be compiled into a large map of the entire country.

It was Prince Mikołaj Radziwiłł who succeeded in designing and carrying through a national project by mapping the whole territory of Lithuania. Undeterred by the large size of the country, parts of which were inaccessible, he had the vision, will, and means to produce a map that served, with some corrections and additions, until the end of the eighteenth century as a source of basic geographical information about the territory of Lithuania. The map is a major achievement of Polish and European cartography of the sixteenth century, and an essential contribution to the development of the mapping of the continent.

Similarly, in the eighteenth century, King Stanislas Augustus' efforts stimulated activity in mapmaking and laid the foundation of the development of modern Polish cartography. It fell short, however, of bringing Polish cartography to the level then present in western Europe. The king's projects were limited in scope and were not supported by scientific and administrative institutions capable of producing large topographic maps of the whole country.

Although Prince Jabłonowski did not succeed in his plan to reform Polish cartography in the eighteenth century, his long-range project of producing an atlas of Poland resulted in a set of general maps of the country, based solely on Polish materials. Thanks to its mathematical basis and richness of detail, it is considered the largest and the best of all printed maps of Poland until the partitions.

Stefan Batory and Stanislas Augustus were limited in their use of maps as tools of governing the country. The power of the king and central government in Poland was so circumscribed that the production of large maps of Polish and Lithuanian provinces was beyond the ability of any head of state, and there was not sufficient political and scientific support for a national project of mapping the whole republic. This state of affairs become particularly visible in the last quarter of the eighteenth century, when Polish leaders were woefully short of adequate maps for the defense of the country against the military campaigns and diplomatic maneuvers that resulted in the partition of Poland among its three neighbors.

It is therefore a remarkable achievement of Mikołaj Radziwiłł and Józef Jabłonowski that they could sustain the long-range projects that resulted in the mapping of Lithuania and Poland. Thanks to their efforts, the European scientific community could become acquainted with Polish and Lithuanian geography and elements of its history, religion, and economy. The maps served also as monuments of Polish culture and of the class that produced them, attesting the power of the great Polish and Lithuanian magnates—the very power that limited the kings in their political and cartographic ambitions.

Notes

1. For a recent thorough description of the role of nobility in Polish life, see Norman Davies's chapter *"The Nobleman's Paradise"* in his *God's Playground: A History of Poland* (New York: Columbia University Press, 1982), 1:201–55, from which the following section is adapted.

2. In Polish literature, some theoretical references to the subject of the importance of mapmaking during the war can be found in Stanisław Łaski, *Księgi gotowości wojennej* (1550) and Jan Tarnowski, *Consilium rationis bellicae* (1558).

3. *Excitarz Xiędza Iosepha Wereszcyńskiego . . . do podniesienia Woyny S. przeciwko Turkom y Tatarom . . .* (Kraków, 1592), 101–2.

4. W. Zakrzewski, "Rodzina Łaskich w XV w.," *Ateneum* 4 (1883): 458.

5. J. H. Hessels, *Abrahamii Ortelii (Geographi antverpiensis) . . . Epistulae . . . (1542–1628)*, Ecclesiae Londino Batavae Archivum 1, no. 97, (Cambridge, 1887), 232–35.

6. *Archiwum Jana Zamoyskiego* (Warsaw, 1904), 1:371–74, no. 354.

7. B. Paprocki, *Herby rycerstwa polskiego* (Kraków, 1858), 276.

8. A. Pawinski, *Źródła dziejowe* (Warsaw, 1882), 11:66–67.

9. *Archiwum Jana Zamoyskiego* (Warsaw, 1909), 2:34.

10. For the history and thorough description of the map, see two works by Stanisław Alexandrowicz: "Mapa Wielkiego Księstwa Litewskiego Tomasza Makowskiego z 1613 r. tzw. 'radziwiłłowska,' jako źródło do dziejów Litwy i Bialorusi," *Studia Źródłoznawcze* 10 (1965):33–67; and *Rozwój kartografii Wielkiego Księstwa Litewskiego od XV do połowy XVIII wieku,* Uniwersytet im. Adama Mickiewicza, Wydział Filozoficzno-historyczny, Seria Historia 50 (Poznań, 1971).

11. J. Riabinin, "Rozmowy Króla z Naczelnikiem," in *Przegląd Historyczny* (Warsaw, 1911) 12:245.

12. K. Buczek, "Prace kartografów pruskich w Polsce za czasów króla Stanisława Augusta, na tle współczesnej kartografii polskiej," in *Prace Komisji Atlasu Historycznego Polski,* Zeszyt III, PAU (Kraków, 1935), 226.

13. L. Uziębło, "Pamięci Poczobuta," *Tygodnik Ilustrowany,* 1910, p. 1070.

14. Buczek, "Prace Kartografów," 258.

Index